Gecko Fauna of the USSR
and Contiguous Regions

Nikolai N. Szczerbak
and Michael L. Golubev

Translated from the Russian by
Michael L. Golubev and Sasha A. Malinsky

Editorial supervision by
Alan E. Leviton and George R. Zug

Gecko Fauna of the USSR
and Contiguous Regions

Published with the aid of a grant from the Atherton Seidell Endowment Fund by
Society for the Study of Amphibians and Reptiles

Series Editor's Note

The Society is grateful to Nikolai N. Szczerbak and Michael L. Golubev—authors of the original Russian-language edition of this book, published in Kiev in 1986—for their cooperation in producing this English version. Dr. Golubev has also graciously updated the book by correcting errors in the original text and by adding references to the recent literature (1986–1996). We thank Alan E. Leviton and George R. Zug for their careful editing of the English manuscript. The Atherton Seidell Endowment Fund, administered by the Smithsonian Institution, has generously provided financial support for translation services and to publish the color plates.

The illustration of the Persian spider gecko (*Agamura persica*) on the titlepage was drawn by George Henry Ford. It was taken from William T. Blanford's book, "Eastern Persia . . . volume II. The Zoology and Geology." Macmillan and Co., London, 1876.

Contributions to Herpetology, Volume 13

Kraig Adler, *Editor*
Timothy D. Perry, *Associate Editor*

Volumes in the *Contributions to Herpetology* series can be purchased from the Publications Secretary, Robert D. Aldridge, Department of Biology, Saint Louis University, 3507 Laclede, Saint Louis, Missouri 63103, USA (*telephone*: area code 314, 977–3916 or 977–1710; *fax*: area code 314, 977–3658; *e-mail*: ssar@sluvca.slu.edu). A list of all Society publications, including those of The Ohio Herpetological Society and the *Catalogue of American Amphibians and Reptiles*, is printed at the end of this book; additional copies of this list are available from Dr. Aldridge. Volumes in the *Contributions* series are published irregularly and ordered by separate subscription, although Society members receive a substantial pre-publication discount. Authors who wish to have manuscripts considered for publication in the *Contributions* series should contact the Editor: Kraig Adler, Cornell University, Section of Neurobiology and Behavior, Seeley G. Mudd Hall, Ithaca, New York 14853–2702, USA.

Members of the Society receive a quarterly technical journal (*Journal of Herpetology*) and a quarterly news-journal (*Herpetological Review*). Currently, dues are US$30.00 for students, $40.00 for all others, world-wide; institutional subscriptions are $70.00. Additional $35.00 for airmail delivery outside the USA. Society members receive substantial discounts on *Herpetological Circulars, Facsimile Reprints*, and on books in the *Contributions* series. The *Catalogue* is available by separate subscription. Apply to the Society's Treasurer, Robert D. Aldridge (address above). Overseas customers can make payments in USA funds or by International Money Order. All persons may charge to MasterCard or Visa (include account number and expiration date).

© 1996 Society for the Study of Amphibians and Reptiles
Library of Congress Catalog Number: 96–69974. ISBN: 0–916984–39–7.
Production specifications are given on the last page of this book.

EDITORS' PREFACE

In 1993, with the arrival of Michael L. Golubev and his family in the United States, the editors thought it opportune to arrange for an English edition of the major work on the geckos of Palearctic Asia, originally published in Russian and not readily accessible to herpetologists in Western Europe or the Americas. Reference to the book by Western herpetologists has been uncommon, even though a cursory examination of its contents suggested that it would be more widely consulted were it not for the language barrier.

Dr. Golubev, on being approached with the proposal to prepare an English translation, readily agreed. The editors then drafted a proposal to the Atherton Seidell Foundation for funds to defray the costs of the project. In short order, the Foundation agreed to provide a substantial portion of the funding necessary, the remainder being provided by the Society for the Study of Amphibians and Reptiles through arrangements concluded by Dr. Kraig Adler, editor of the Society's *Contributions to Herpetology* series.

In due course, Dr. Golubev, ably assisted by his Seattle-based nephew, Sasha Malinsky, his son, Sasha Golubev, and with the constant encouragement of his wife, Mila, completed the project within a year. Of course, Russian and English languages are not structurally interchangeable, and there are significant differences in the use of even such seemingly trivial elements as the definite and indefinite articles (i.e., "the" and "a"). Thus, it became the intention of the editors, working in concert with Dr. Golubev, to produce a *readable and grammatically acceptable* English translation while keeping as close to the structure of the original Russian text as possible. We believe we have done so, but because some changes were called for that could be interpreted as substantive deviations from the original text, such changes are highlighted by inserting comments in brackets [ ] or braces { }.

Four major changes are introduced in the English edition that require explanation, one of omission, one of substitution, and two of addition. The first is the omission of the explanation of abbreviations that appear at the bottom of page 4; the abbreviations are not used in the English edition and are, therefore, unnecessary. The second is more draconian and affects the 8 color plates (24 color photos of geckos), which are included in the original Russian publication. Many of the original color photos could not be located by the authors. But, with the help of Prof. Szczerbak and Drs. N. L. Orlov, Steven C. Anderson, and P. Kodym, and J. Robert Macey, a carefully chosen selection of alternative photographs, many of the same specimen but in slightly different poses, have been substituted. They are ordered in precisely the same order as those in the original publication. Each photo has been examined by Dr. Golubev, who approved their inclusion in the English edition. Third, Dr. Golubev assembled a supplemental bibliography of important papers on the geckos of Palearctic Asia published since 1986 that are germane to the contents of the monograph. Lastly, these new references are cited in supplemental footnotes that appear in this edition and are identified by the use of an asterisk (*) because numbered footnotes (i.e., 1, 2 . . .), were used in the original Russian edition.

Nikolai N. Szczerbak

Michael L. Golubev

The editors want to take special note of the contribution to this project of Dr. Kraig Adler (Cornell University), editor of the *Contributions to Herpetology* series published by the Society for the Study of Amphibians and Reptiles. Not only did Kraig arrange for the supplemental funding for publication, but he also painstakingly re-copyedited the entire translation, this after the near-countless passes by both Zug and Leviton, and came up with an appalling number of corrections that had been missed by the five pair of eyes that had so carefully scanned the manuscript before him. He also reworked the bibliographies, a non-trivial contribution in its own right, to insure accuracy and standardized format. And, quite apart from his editorial expertise, Kraig took complete charge of assembling the set of color photographs that are incorporated into this English edition and saw them through the separation and proof production process. Lastly, Kraig spent much time with the details of production to insure the highest quality possible, and this for little reward save that of our deepest gratitude and thanks.

In addition to Dr. Adler, Dr. Michele L. Aldrich read and critiqued selected sections of the final document and offered suggestions for improvements for which we are most appreciative.

Finally, to the authors of *Geckos of the USSR and Contiguous Regions*, Professor Nikolai N. Szczerbak and Dr. Michael L. Golubev, our sincere thanks for all you did to bring about the reality of an English edition of your important contribution in herpetology.

Alan E. Leviton
California Academy of Sciences
San Francisco, CA

George R. Zug
National Museum of Natural History
Washington, DC

14 September 1996

TRANSLATOR'S INTRODUCTION TO THE ENGLISH EDITION

The translation of the Russian text of *Gecko Fauna of the USSR and Contiguous Regions: A Field Guide*, was done by two translators, Michael Golubev and Sasha Malinsky. Dr. Golubev worked on the systematic portion of every section and Mr. Malinsky on the ecological parts. In addition, Mr. Malinsky also handled the "Preface," "Materials and Methods," and "Phylogenetic Relationships and Zoogeographic Analysis" sections. The two exchanged sections of text to check for accuracy and intelligibility.

In the original text of the section on "Materials and Methods," certain subjects were not explained. For instance, it was not stated that in compiling the map legends, when a locality was encountered by the authors not only in the literature but also in the examined material, preference was given to the museum collection references, because they could confirm that the material was identified in accordance with our viewpoint. Also not explained is the difference in meaning of the section headings "Life Color and Pattern" versus "Color and Pattern" under the species accounts. The former refers to the colors in life based on observations of living individuals; the latter, the color as observed in preserved specimens.

When some significant characters were observed in all the representatives of a given genus, they were cited in the generic diagnosis and not repeated subsequently in the species diagnoses (with the exception of the ones that had some variation within the species being described). These characters were grouped into "additional characters" in every generic and subgeneric section. They are not "legally" diagnostic but, in the opinion of the authors, should assist in the correct initial identification of the lizard. This leads to an important practical consideration. If you want to compare your specimen with the descriptions contained in this text, begin your comparison with the genus descriptions for greater reliability.

*****

Possibly, the readers may be interested in learning a little about the making of this book. The idea belongs to my former supervisor, Professor N. N. Szczerbak. In the fall of 1973, when I returned from compulsory service in the Soviet Armed Forces (people with a university education are still drafted for one year) to the Ukrainian S.S.R. Academy of Sciences' Zoological Museum, which he headed, Dr. Szczerbak called me into his office and proposed that we work with the Middle Asian geckos. As is known, their systematics were poorly developed, and the proposed conditions seemed quite acceptable: my Ph.D. thesis, his supervision, and a joint monograph. After we filed numerous pieces of required paperwork, at the beginning of 1974, the Presidium of the Ukrainian S.S.R. Academy of Sciences (a massive bureaucratic structure that commands science in that country to this day) decreed that the study of the U.S.S.R. gecko fauna was to be completed within the following five years. The result of this work would be a monograph in a volume of 20 "print sheets" (equivalent to signature; one print sheet equals 22-25 typewritten pages).

Once the project was begun, it became quite clear that it would be impossible to work successfully on the systematics of some part of the group without giving attention to the rest of its members. Then, we decided to include within the scope of our work all the species from neighboring countries that belong to the genera found in the deserts of the U.S.S.R. This also turned out to be insufficient. The case of the gecko "*Alsophylax pipiens*," from Afghanistan, underscored this limitation. It turned out to be a new species from another genus. So, we had to widen the bounds of research to include almost all the "unwiden-toed" geckos of the Palearctic. This led to a significant increase in the time required for the work, because the vast majority of the collections were outside the U.S.S.R., in American and European museums. At that time, neither Dr. Szczerbak nor I could travel abroad. Dr. Szczerbak, who had serious differences with the Stalinist regime, was not a party member. And, my lack of party membership was exacerbated by the fact that I am also an ethnic Jew. Such sins were not forgiven by the Soviet authorities. So, we were both "unreliable" and could only hope for help from our foreign colleagues. With few exceptions, they did not let us down and we were sent all the specimens we were interested in studying. Thanks only to Dr. Szczerbak's administrative skill and energy, the Presidium of the Academy of Sciences extended the book's publication deadline by five years, a rare exception.

The work was divided by the authors in the following manner. I studied the collections, compiled the diagnoses and descriptions (with the exception of the eublepharids' generic diagnosis and the *Tenuidactylus kotschyi* species section, which were done by Dr. Szczerbak), performed the statistical treatment of data, and photographed fragments of the lizards in the laboratory. Because only three genera were officially included in my doctoral thesis ("Palearctic Gecko Lizards Genera: *Alsophylax, Bunopus,* and *Tropiocolotes,*" Ph. D. thesis, Kiev, 1982, 151 p., in Russian), the intraspecific systematics sections in all the other articles were written by Dr. Szczerbak on the basis the data we collected in common. The first three chapters of the book, found in the "General Section," he also wrote himself (the sections of text concerning the genera *Alsophylax, Bunopus,* and *Tropiocolotes* were transferred from my Ph.D. thesis almost without changes). He collected the data for the ecological portions of the articles (down to "Y-a-zzz," "Tsok-tsok-tsok," and other sonograms). The maps and legends were put together by Mr. A. Tokar.

A separate problem arose for us with the genus *Stenodactylus*. On one hand, the "unwiden-toed" geckos of this group clearly fit into our research; on the other, not a single representative of the group is found in the U.S.S.R. Stormy debates on whether or not to include this genus in our study were interrupted in the nick of time by Dr. E. N. Arnold's (British Museum [Nat. Hist.]) 1980 publication reviewing this genus. However, even this timely present from London did not help us fit it into the boundaries, programmed by the Presidium. When our work was finally written, it exceeded the planned volume by a factor of one-and-a-half.

At this point I must digress from issues of herpetology and shift to another topic. In the land of my birth everyone is familiar with the eternal debate between East and West: who was the innovator, who is better? Popov or Marconi? Yablochkov or Edison? Rulier (a Russian biologist of German-French descent) or Darwin himself? Alyokhin or Casablanca? "Vostok" or the "Shuttle"? "Dynamo" Kiev or "Bavaria" Munich ("Dynamo," of course!)? There are even people who are absolutely convinced that Russia is the birthplace of elephants (well, at least mammoths). Have any of the readers ever considered whose bureaucracy is better? Without a doubt, the American bureaucracy, equipped with a formidable computer network, e-mail and faxes, microphones and calculators, appears quite confident and can allow itself a toothy smile towards a supplicant from the unreachable height of its unshakable governing well-being. The orthodox Soviet bureaucracy, devoid of many technical innovations and material excesses of its Western counterparts, never smiles. So, when it fanatically repeats the first commandment of all the world's bureaucracies, "It is not allowed!", with teeth set and a stony expression on its impersonal face, it is in no way inferior to the American model, and in some respects (in ideological conviction, for example) even surpasses it. Its logic is simple. Were 20 signatures planned? Yes! The appropriate amount of paper and ink, type-setting time, editing and printing time, resources for mailing have been allotted, man-hours have been computed (man-minutes, man-seconds?). And not an extra page, day, or gram are allowed. At that time, this gibberish had a smart name of "Planned Economy." Even the tremendous energy of Dr. Szczerbak could not re-plan the planned economy.

The first author began the task of cutting the text. The section "Differences from Related Species," as well as the specific unique data on geographical variation of the wide-ranging species together with tables of comparisons between our samples, were cut from every article. Other sections were condensed. Of course, the second author did not see this as a successful solution, but no one asked me anyway. Shortly after this "vivisection," the book was accepted for publishing. But a sigh of relief was still premature. In the spring of 1986, when our field equipment had been carefully packed, when the airplane tickets had been purchased and the faces of colleagues seemed to be from another world, when we could no longer sit still at the microscope and we could almost see the crests and peaks of Kopetdagh and Tjan-Shan, literally days before flying out, we received a draft of the first third of the book from the publisher for proofreading. Of course, it could not wait until our return. We were gripped by a feeling that a skydiver must experience when he discovers, halfway down, that a new parachute is still on the plane and there is no sense in going back for it. That part of the text was not proofread very thoroughly and that is why the majority of the typos is concentrated there.

Finally, at the very end of 1986, the book saw the light of day.

Incidentally, this monograph turned out to be symbolic, in a sense. It closes the final chapter in the story "Herpetology of the U.S.S.R." All subsequent researchers will be adding the adjective "former" to the name of this country in the titles of their works.

Even before this publication, our nomenclature errors, involved in the descriptions of the genus *Tenuidactylus*, were noted in the literature (Kluge, 1985; Böhme, 1985). Due to a number of circumstances, I have never before addressed this issue. Having such an opportunity now, I wish to note that both authors do not consider the law, applied in this case, to be good. However, the "law may be bad, but it is the law." Besides, I am fully satisfied by the fact that our idea on separating the Indo-Malay and Palearctic geckos has been accepted by our contemporaries, if not in form, at least in its essence.

Two negative reviews were written in response to the book (Bauer, 1987; Semenov, 1989, in Russian). Of course, it would be pointless to take on the role of a mother that defends her offspring with fanatical devotion. It is obvious that the book is not without its flaws. I would only like to quote a saying from ancient Rome, "Feci quod potui, faciant meliora potentes" (I did all that I could, if someone can, may they do better). I hope that the English language reader will find enough useful information here and draw his or her own conclusions about the book.

*****

Acknowledgments. First, I would like to note that the idea to translate the book belongs to Drs. A. Leviton (San Francisco, CA) and George Zug (Washington, DC) and the decision to publish the English edition was made by Dr. K. Adler (Ithaca, NY). The study of Asian herpetology has had unique and invaluable support from these people in the recent years. A grant from the Atherton Seidell Foundation (Smithsonian Institution) made the project a reality. During the translation, it was necessary to check the correct spellings of geographic names in the original sources, many of which were not available to the translators. In this area, enormous assistance was provided by Dr. G. Zug and Mr. R. Crombie (Washington, DC) and Dr. A. Leviton and Mr. J. Vindum (San Francisco, CA). Dr. V. Fet (New Orleans, LA) and Dr. A. Baranov (Moscow, Russia) helped with the Latin names of some plants and insects. Dr. S. Anderson (Stockton, CA) kindly provided a manuscript on the lizards of Iran and copies of a number of new publications. Dr. M. Khan (Rabwah, Pakistan) sent along some comments on the Russian text of the monograph. My son, Alex, typed the literature list, down to the last comma. In addition to her constant moral support, my wife, Mila, almost never went against my request not to clean up my desk. I would say that such restraint was not at all easy for her. I am deeply grateful to all these people for their help.

Michael L. Golubev
April 2, 1996

AKADEMIA NAUK UKRAINSKOJ SSR
ZOOLOGICHESKIJ INSTITUT im. I. I. SCHMALBGAUZENA

N. N. SZCZERBAK, M. L. GOLUBEV

GECKO FAUNA OF THE USSR AND CONTIGUOUS REGIONS

KIEV NAUKOVA DUMKA 1986

UDK 598.112.1 (4−013)

Gecko Fauna of the USSR and Contiguous Regions / Szczerbak N. N., Golubev M. L. — Kiev: Naukova Dumka, 1986 — 232 p.

В определителе обобщены сведения об одной из наиболее слабо изученных групп ящериц, ведущих сумеречно-ночной и скрытый образ жизни, среди которых много редких и исчезающих видов. Приведены детальные карты распространения с кадастром мест находок, материалы о филогенетических отношениях гекконов, а также оригинальные определительные таблицы. Изучена географическая изменчивость гекконов.

Для зоологов, герпетологов, специалистов в области охраны природы, работников музеев и заповедников, преподавателей и студентов вузов.

Табл. 7. Ил. 122. Библиогр.: 217—226 с.

Ответственный редактор
М. А. ВОИНСТВЕНСКИЙ

Рецензенты
И. С. ДАРЕВСКИЙ, В. А. КОТЛЯРЕВСКАЯ

Редакция общей биологии

НИКОЛАЙ НИКОЛАЕВИЧ ЩЕРБАК
МИХАИЛ ЛЕОНИДОВИЧ ГОЛУБЕВ

ГЕККОНЫ ФАУНЫ СССР И СОПРЕДЕЛЬНЫХ СТРАН

Определитель

Утверждено к печати ученым советом Института зоологии им. И. И. Шмальгаузена АН УССР

Редактор *С. В. Вечерская*
Оформление художника *В. М. Флакса*
Художественный редактор *Р. И. Калыш*
Технический редактор *А. М. Капустина*
Корректоры *П. С. Бородянская, И. В. Кривошеина*

ИБ. 7465

Сдано в набор 17.06.85. Подп. в печ. 29.01.86. БФ 04513. Формат 70×108/16. Бум. тип. № 1. Лит. гарн Выс. печ. Усл. печ. л. 21,0. Усл. кр.-отт. 23,1. Уч.-изд. л. 22,94. Тираж 1200 экз. Заказ 5—1760. Цена 2 р.

Издательство «Наукова думка», 252601, Киев 4, ул. Репина, 3.
Головное предприятие республиканского производственного объединения «Полиграфкнига». 252057 Киев, ул. Довженко, 3.

Щ $\frac{2005000000-045}{М221(04)-86}$ 368-86

© Издательство «Наукова думка», 1986

TABLE OF CONTENTS*

	Russian	English
Foreword .... page 3**		3**

GENERAL SECTION

	Russian	English
Brief review of investigations of geckos of the fauna of the USSR and adjoining countries	5	5
Material and methods	13	13
Phylogenetic relationships and zoogeographical analysis	16	16

SYSTEMATIC SECTION

	Russian	English
Family Eublepharids — Eublepharidae	25	25
Genus Fat-tailed Geckos — *Eublepharis*	25	25
Spotted Fat-tailed Gecko — *E. macularius*	26	26
Iranian Fat-tailed Gecko — *E. angramainyu*	29	29
Turkmenian Fat-tailed Gecko — *E. turcmenicus*	30	29
Family Geckos — Gekkonidae	33	32
Genus Skink Geckos — *Teratoscincus*	33	33
Common Skink Gecko — *T. scincus*	35	33
Common Skink Gecko — *T. s. scincus*	37	37
Keyserling's Skink Gecko — *T. s. keyserlingii*	38	37
Rustamov's Skink Gecko — *T. s. rustamowi*	38	38
Przewalski's Skink Gecko — *T. przewalskii*	42	41
Small-scaled Skink Gecko — *T. microlepis*	43	44
Bedriaga's Skink Gecko — *T. bedriagai*	45	45
Genus Fringe-toed Geckos — *Crossobamon*	46	45
Eversmann's Fringe-toed Gecko — *C. eversmanni*	47	46
Eversmann's Fringe-toed Gecko — *C. e. eversmanni*	49	48
Lumsden's Fringe-toed Gecko — *C. e. lumsdenii*	50	49
Eastern Fringe-toed Gecko — *C. orientalis*	54	54
Genus Pygmy Geckos — *Alsophylax*	55	55
Subgenus Lowland Pygmy Geckos — *Alsophylax*	56	56
Squeaky Pygmy Gecko — *A. pipiens*	56	56
Przewalski's Pygmy Gecko — *A. przewalskii*	62	62
Armored Pygmy Gecko — *A. loricatus*	65	64
Armored Pygmy Gecko — *A. l. loricatus*	67	67
Szczerbak's Armored Pygmy Gecko — *A. l. szczerbaki*	68	67
Smooth Pygmy Gecko — *A. laevis*	72	72
Tadjik Pygmy Gecko — *A. tadjikiensis*	78	78
Subgenus Mountain Pygmy Geckos — *Altiphylax*	81	80
Tjan-Shan Pygmy Gecko — *A. tokobajevi*	81	80
Genus Tuberculated Geckos — *Bunopus*	88	86
Southern Tuberculated Gecko — *B. tuberculatus*	89	87
Arabian Tuberculated Gecko — *B. spatalurus*	94	92
Arabian Tuberculated Gecko — *B. s. spatalurus*	97	95
Hajar Tuberculated Gecko — *B. s. hajarensis*	97	95
Thick-tailed Tuberculated Gecko — *B. crassicauda*	98	96
Genus Dwarf Geckos — *Tropiocolotes*	100	98
Subgenus African Dwarf Geckos — *Tropiocolotes*	101	99
Tripoli Dwarf Gecko — *T. tripolitanus*	102	99
Steudner's Dwarf Gecko — *T. steudneri*	106	103
Scortecci's Dwarf Gecko — *T. scortecci*	108	106
Subgenus Microgeckos — *Microgecko*	110	108
Khuzestan Dwarf Gecko — *T. helenae*	111	109

* For ease of use, this table of contents has been expanded from the original Russian-language version to include an entry for each taxon.

** The first column of numbers refers to pages in the original Russian publication; these are included intratext in the English translation in braces in bold typeface, {**p. 00**}. The second column of numbers refers to the pagination of this English edition; these page numbers are located in the lower, outer corner of each page.

	Russian	English
Persian Dwarf Gecko — *T. persicus*	page 115	111
Persian Dwarf Gecko — *T. p. persicus*	117	114
Bakhtiari Persian Dwarf Gecko — *T. p. bakhtiari*	117	115
Spurge Persian Dwarf Gecko — *T. p. euphorbiacola*	118	115
Latifi's Dwarf Gecko — *T. latifi*	119	116
Subgenus Asian Dwarf Geckos — *Asiocolotes*	120	120
Leviton's Dwarf Gecko — *T. levitoni*	121	118
Flat Dwarf Gecko — *T. depressus*	124	121
Genus Keel-scaled Geckos — *Carinatogecko*	126	123
Iranian Keel-scaled Gecko — *C. aspratilis*	127	124
Iraqi Keel-scaled Gecko — *C. heteropholis*	130	124
Genus Thin-toed Geckos — *Tenuidactylus*	130	126
Subgenus Thin-toed Geckos — *Tenuidactylus*	131	128
Caspian Thin-toed Gecko — *T. caspius*	132	128
Caspian Thin-toed Gecko — *T. c. caspius*	137	133
Island Thin-toed Gecko — *T. c. insularis*	137	134
Turkmenian Thin-toed Gecko — *T. turcmenicus*	141	138
Turkestan Thin-toed Gecko — *T. fedtschenkoi*	146	143
Long-legged Thin-toed Gecko — *T. longipes*	153	149
Long-legged Thin-toed Gecko — *T. l. longipes*	156	152
Small-scaled Thin-toed Gecko — *T. l. microlepis*	156	152
Southwest Thin-toed Gecko — *T. l. voraginosus*	157	153
Subgenus Mediterranean Thin-toed Geckos — *Mediodactylus*	159	155
Mediterranean Thin-toed Gecko — *T. kotschyi*	160	156
Danilewski's Mediterranean Thin-toed Gecko — *T. k. danilewskii*	163	159
Gray Thin-toed Gecko — *T. russowi*	167	163
Gray Thin-toed Gecko — *T. r. russowi*	171	167
Zarundy's Gray Thin-toed Gecko — *T. r. zarundyi*	172	167
Spiny-tailed Thin-toed Gecko — *T. spinicauda*	175	171
Lebanese Thin-toed Gecko — *T. amictopholis*	179	175
Bampur Thin-toed Gecko — *T. sagittifer*	180	176
Asia Minor Thin-toed Gecko — *T. heterocercus*	182	176
Asia Minor Thin-toed Gecko — *T. h. heterocercus*	184	180
Mardin Asia Minor Thin-toed Gecko — *T. h. mardinensis*	184	180
Subgenus Cyrtopodion — *Cyrtopodion*	185	180
Kachhi Thin-toed Gecko — *T. kachhensis*	185	181
Pakistani Thin-toed Gecko — *T. watsoni*	188	183
Rough Thin-toed Gecko — *T. scaber*	190	186
Salt Range Thin-toed Gecko — *T. montiumsalsorum*	192	188
Kashghar Thin-toed Gecko — *T. elongatus*	194	189
Agamuroid Thin-toed Gecko — *T. agamuroides*	196	192
Tibeto-Himalayan Group of *Tenuidactylus*	197	193
Tibetan Thin-toed Gecko — *T. tibetanus*	199	194
Chitral Thin-toed Gecko — *T. chitralensis*	201	197
Minton's Thin-toed Gecko — *T. mintoni*	202	197
Stoliczka's Thin-toed Gecko — *T. stoliczkai*	205	200
Kirman Thin-toed Gecko — *T. kirmanensis*	207	201
Genus Spider Geckos — *Agamura*	208	203
Persian Spider Gecko — *A. persica*	209	204
Farsian Spider Gecko — *A. gastropholis*	211	207
Misonne's Spider Gecko — *A. misonnei*	213	207
Kharan Spider Gecko — *A. femoralis*	215	210
Literature cited	217	212
Alphabetical index of Russian names of animals	227	—
Alphabetical index of Latin names of animals	229	231

* * *

{p. 3} FOREWORD

Gecko lizards form a reptilian group remarkable in many respects. First of all, they are one of the most ancient reptilian families, whose ancestral species can be traced back to the Mesozoic [238], and they exhibit a number of primitive traits [111]. Second, the vast majority of geckos are found in tropical climate zones [307]. Few species of these lizards have made their way into the Palearctic. Among the subjects of our research are the genera *Eublepharis, Teratoscincus, Crossobamon, Alsophylax, Bunopus, Tropiocolotes, Carinatogecko, Tenuidactylus,* and *Agamura,* which are found in the U.S.S.R. and the neighboring countries. These comprise the vast majority of Palearctic forms.

Geckos have many distinct differences from all other lizard species. They have a characteristic vertical pupil and unblinking eyes covered with a motionless transparent membrane, just like the snakes. They are capable of emitting resonant sounds. Primarily nocturnal, they, unlike most reptiles, lay eggs in a solid calcereous shell. The main adaptive radiation among geckos is along the path of digit modification; thus, many species are capable of moving along vertical surfaces.

Until recently, the taxonomy of the Palearctic geckos has not been explored thoroughly enough, although it appeared to be more or less successful. It was known that the experts did not always view the various taxa in the same way. Thus, rock geckos in the work of European herpetologists [307 and others] were known as the genus *Gymnodactylus,* whereas the American zoologists and, in recent years, many West German researchers treat them as the genus *Cyrtodactylus* [191, 205, *et al.* in text]. An especially large number of disputed classifications existed in the systematics of the small and the rare species, which at different times were classified as *Alsophylax, Bunopus, Tropiocolotes,* and *Microgecko.* Under close scrutiny it turned out that there exist disagreements with regard to the methods of determining the generic allocation of a number of gecko species. Their classification is founded on traits of questionable taxonomic value, the geographic variation of certain species has not been sufficiently studied, and descriptions of a number of forms are in need of revision [149]. The purpose of our work is the study of the taxonomic placement and phylogenetic relationships of geckos in the fauna of the U.S.S.R. and the neighboring countries, revision of their taxonomy, construction of well documented descriptions and the study of geographic variations of certain species, creation of accurate dichotomous keys, the study of ecology, identification of rare and endangered species, and recommendations for their protection. Our data on the rare species were included in the first [1977] and second [1984] editions of the *Red Book for the U.S.S.R.*

The detailed study and description of the species from the bordering areas is conditional, first of all, on the fact that it is impossible to review our species and understand their taxonomy without studying the genera in general. Second, the domestic literature lacks the descriptions and the identification keys of gecko species from neighboring countries, and even though specimens often end up in the collections of the museums in the U.S.S.R., their processing [i.e., identification] presents certain difficulties. Years of our experience in doing this work have shown that the discovery of species spreading from the south still occurs in several Soviet republics, and accurate classification is possible only if the corresponding reference literature is available.

Our work in gecko taxonomy revision has yielded significant and unexpected results. Peculiarity of the group of Palearctic rock geckos was discovered. This group was separated into a new genus *Tenuidactylus* [153]. The new genus *Carinatogecko* from Iran [151] was described. Improvements in taxonomy led to the creation of three new subgenera: *Mediodactylus* [150], *Asiocolotes* [40], and *Altiphylax* [56]. New species *Gymnodactylus turcmenicus* [144], *Tropiocolotes levitoni* [43], *G. mintoni* [44], *Alsophylax tadjikiensis* [38], and *A. tokobajevi* [56] were discovered, as well as the new subspecies *G. caspius insularis* [14], *Teratoscincus scincus rustamovi* [146], and *A. loricatus szczerbaki* [42]. Also, *A. spinicauda* and *A. tibetanus* have been placed in a different genus, species new to the U.S.S.R. (*Bunopus tuberculatus* and *G. longipes microlepis*) [48] were found, the precision of distribution boundaries for the known species was significantly improved, and new data on the ecology, reproduction, and habits were collected.

This work was done in the laboratories of the Zoological Museum at the Institute of Zoology and at the Central Museum of Natural History of the Academy of Sciences of Ukrainian SSR during the period of 1972-84. During this time the authors have done extensive field research in Crimea, Caucasus, Transcaucasia, and Middle and Central Asia. They have had the opportunity to become familiar with the collections of the largest {p. 4} museums in this country, Western Europe, and the U.S.A. Altogether, over 5,000 specimens of gecko lizards were studied, both from the collections of the ZIK (Kiev), as well as those which were kindly shared with us by the following: Dr. I. S. Darevsky, Mrs. L. N. Lebedinskaya, and N. L. Orlov (Zoological Institute, Leningrad - ZIL), Dr. V. F. Orlova and T.

I. Aleksandrovskaya (Zoological Museum of Moscow State University - ZMMU), Dr. L. Khozatsky (Leningrad State University - MLSU {error in original – MG}, Dr. Ya. D. Davlatov and Dr. T. Ya. Yatgarov (Institute of Zoology and Parasitology, Uzbek SSR Academy of Sciences, Tashkent - UIZP), Dr. S. A. Said-Aliyev (Institute of Zoology and Parasitology, Tadjik SSR Academy of Sciences, Dushanbe - DIZP), Dr. S. M. Shammakov and Dr. Ch. A. Atayev (Institute of Zoology, Turkmen SSR Academy of Sciences, Ashkhabad - IZT), Dr. Z. K. Brushko and Dr. R. A. Kubykin (Institute of Zoology, Kazakh SSR Academy of Sciences, Alma-Ata - AAIZ), Mr. V. K. Yeremchenko (Institute of Biology, Kirghiz SSR Academy of Sciences, Frunze - FIB), Mrs. L. M. Pisareva (Zoological Museum of Kiev State University - ZMKU), Mr. V. I. Vedmederya (Zoological Museum of Kharkov University - ZDKU), Dr. V. V. Neruchev (Pedagogical State Institute, Gorky - GPI), Dr. T. S. Sattorov (Tadjik SSR Pedagogical Institute, Dushanbe - DPI), Mr. V. P. Velikanov (Krasnovodsky State Preserve, - KSP), Dr. A. P. Markevich (collections made in Arabian Republic of Egypt). Also, we received some collections from foreign depositories: Dr. J. Eiselt, Dr. F. Tiedemann, and M. Haüpl (Naturhistorisches Museum Wien, Austria - NMW), Dr. E. Arnold, Miss. A. Grandison, and Mr. B. Clarke (British Museum of Natural History, London - BMNH), Dr. W. Böhme (Zoologisches Forschungsinstitut und Museum Alexander Koenig, Bonn, BRD - ZFMK), Dr. M. I. Bunni (Iraq Natural History Museum, Baghdad - NHMB), Dr. G. de Witte (Institut Royal des Sciences Naturales de Belgique, Brussels - BZ), Dr. O. Dely (Magyar Nemezeti Museum, Budapest - UNM), Dr. R. Mertens and Dr. K. Klemmer (Natur-Museum und Forschungs-Institut Senckenberg, Frankfurt-am-Main, BRD - SMF), Mr. H. Marx and Mr. A. Resetar (Field Museum of Natural History, Chicago, USA - CMNH {= FMNH - MG}), Dr. L. Capocaccia and Dr. M. A. Cherchi (Institute of Zoology - MIZG and Zoological Museum - MCZN, Genova, Italy), Dr. A. G. Kluge (Zoological Museum, University of Michigan, USA - UMMZ), Dr. A. E. Leviton (California Academy of Sciences, San Francisco, USA - CAS), Dr. S. A. Minton (private collection, Indianapolis, USA - SAM), Dr. Günther Peters (Institut für Spezielle Zoologie und Zoologisches Museum der Humboldt Universität zu Berlin - ZMB), Dr. M. Poggezi (Museum de "La Specola," Universita di Firenze, Italy - MZSF), Mr. J. P. Rosado (Museum of Comparative Zoology, Harvard University, Cambridge, Massachusetts., USA - MCZ), Dr. Z. Rocek (Prague National Museum, Czechoslovakia - PNM), Dr. R. Roux-Esteve (Museum National d'Histoire Naturelle, Paris, France - MNHP), Mr. G. Swinny (Royal Scottish Museum, Edinburgh, Scotland - RSM), Dr. W. R. Heyer, Mr. R. I. Crombie, and Dr. F. McCullough (Smithsonian Institution, Washington, USA - USNM), Dr. M. S. Hoogmoed (Rijksmuseum van Natuurlijke Historie, Leiden, Holland - RMNH), Dr. R. G. Zweifel (American Museum of Natural History, New York, USA - AMNH), Dr. L. Cederholm (Zoological Museum, Lund, Sweden - ZML), Mr. B. Schätti (Zoological Museum, Universität Zurich, Switzerland - ZMZ), Mr. J. Schmidtler (private collection, Munich, BRD - ZSM), Dr. K. Edelstam (Naturhistoriska Riksmuseet, Stockholm, Sweden - ZMS), and Dr. B. K. Tikader (Director of the Zoological Survey of India, Calcutta).

The gecko karyotype preparations were done by Mrs. V. V. Manilo. The invertebrates found in the digestive tract were identified by Dr. V. M. Yermolenko.

In the process of preparing this work for publication, valuable comments were received from Prof. I. S. Darevsky.

Mrs. N. P. Bondareva, Mr. Yu. N. Isayev, and Mr. A. A. Tokar have provided great assistance in running the experiments and processing the collected materials. Many of the museum staff assisted with collecting in the field. Mr. V. Yu. Rayevsky participated in preparing the illustrations.

To all the people mentioned above the authors express their sincerest thanks.

Glossary of Unfamiliar Geomorphic Terms
Glossary of Russian geomorphological terms used by the authors
that are not routinely encountered in Western herpetological literature
(from Bates and Jackson, *Glossary of Geology*. 1987. Amer. Geol. Inst., Alexandria, VA)[Eds.]

adyr   A term used in Turkmenistan for a part of a desert plain devoid of sands and having soft ground; in Kazakhstan, the term is used to describe a flat top of relict high ground or a mesa-like hill; also loosely applied in Central Asia for a low mountain , small hill, or eroded ridge with gentle slopes.
barkhan [also *barchan*]   An isolated crescent-shaped sand dune lying transverse to the direction of the prevailing wind.
chink [cliff]   Abrupt loess or sandstone slopes of some Middle Asian hills [not in Bates & Jackson].
loess [structures]   Man-made structures, buildings, walls, etc., made from bricks, similar to adobe bricks of the southwestern United States and Mexico, of local origin.
sor, shor   A dry salt lake.
takyr [also *takir*] A clay-silt playa.

{p. 5} GENERAL SECTION

BRIEF REVIEW OF INVESTIGATIONS OF GECKOS OF THE FAUNA OF THE USSR AND ADJOINING COUNTRIES

Geckos are so distinctive a group of lizards as to have been given a separate taxonomic group at the very dawn of herpetology (Laurenti, 1768; Latreille, 1825 [cit. from 111]; [111, 215], and others). The system, at the level of families, has proven to be the most stable [199, 200]. Its revision has become possible only recently, after the accumulation of the corresponding collections and the use of new characters. The first researcher to investigate this group from a non-standard vantage was G. Underwood, who used the character "pupil shape" to distinguish three traditionally recognized families — Eublepharidae, Sphaerodactylidae, and Gekkonidae. The last one was separated into two subfamilies: Gekkoninae and Diplodactilinae [sic]. A. G. Kluge [237] has determined that the pupil shape varies and cannot be used to distinguish subfamilies. This author grouped 82 genera of the world fauna into four subfamilies (Eublepharinae, Diplodactylinae, Gekkoninae, and Sphaerodactylinae) of the Gekkonidae family in a major work, dedicated to the higher taxonomic categories of geckos and to their evolution. This point of view was accepted by the majority of the authors of contemporary compendia.

However, if one familiarizes oneself with the work of A. G. Kluge in detail, it is easy to notice that the author, having objectively shown the differences between the aforementioned families, was subjective in evaluating their depth and in making the "family–subfamily" distinctions. Consequently, a most important taxonomic category of family is diagnosed quite imprecisely in his understanding and the differences between subfamilies exceed the level of differences between a number of families of other lizards. It is especially true of the distinctions between eublepharid lizards and geckos (14 characters) [237], which are more [numerous] than exist between such stable families as Agamidae and Iguanidae. Eublepharians and geckos differ, first of all, in the structure of the vertebra (in the first they are procoelous and in the second, amphicoelous). In the taxonomy of frogs, the structure of the vertebra is the foundation for separating the suborders (Opistocoela [sic], Procoela). It would seem that the structure of the suborder Sauria should be revised as a composite by using the corresponding criteria. We treat eublepharids and the true geckos as separate families. We recognize, however, the similarities between these groups and share the opinion of other researchers [111 and others], who unite them into one superfamily (or suborder) Gekkota.

* * *

Representatives of the genus *Eublepharis* (established by Gray in 1827, type species *E. hardwickii),* which inhabit southwestern Asia, have not attracted the attention of researchers until recent times. The species *Cyrtodactylus macularius* [190], which was described in the middle of the last century, was transferred into the indicated genus 30 years later [200] and this designation persists to this day. Only once A. M. Nikolsky [96] noted that the specimen from Turkmenia has a number of differences from the ones from India and expressed regret at the absence of comparative materials in his possession. G. Boulenger [202] and M. K. Mikhailovsky [76] were the first to study the {p. 6} eublepharid lizard from Kopetdagh and classified it as the Indian species *E. macularius*. This viewpoint was accepted by the herpetologists who followed [23, 112, 123, *et al.*].

Doubt was cast on the independence of an earlier described species from Hyderabad (West Pakistan), *E. fasciolatus* Günther [1864], as far back as the middle of the last century. Later, this species was synonymized with *E. macularius* by M. Smith [291]. Recently, S. Anderson and A. Leviton [171] have determined that in southwestern Iran there exists a species, *E. angramainyu*, distinct from *E. macularius*. All this served as a stimulus for further attempts to revise the taxonomy of these lizards. These [attempts] were not always successful. In our opinion, an example of just such an attempt is the identification of a new species *E. gracilis* by A. R. Börner [195]. He described the lizard on the basis of a single live specimen of unknown origin, exhibited in the Cologne Zoo. The description is based on small nondiagnostic characters, such as coloration and pattern, which are highly variable.

We have determined that changes in coloration among the eublepharids depend on the physiological condition of the animal, particularly on hormonal factors. Incidentally, the differences between *E. fasciolatus* and *E. macularius* were also in coloration, although S. Anderson and A. Leviton [171] have expressed doubts, insufficiently founded in our opinion, about their identification. We consider the former to be the junior synonym of the latter. Subsequently, A. Börner [196, 197], this time using materials from collections, but the same methodology, described a new species, *E. afghanicus* from the area of Kabul-Jalalabad and three new forms, *E. macularius fuscus* from around Bombay, *E. m. smithi* from the vicinity of Delhi [India], and *E. m. montanus* from the vicinity of Karachi [Pakistan]. He also distinguishes two more, as yet undescribed, subspecies from the town of Nushki and Zhob region (W. Pakistan), as well as *E. m. fasciolatus* [225], beside the nominative form (from the vicinity of Lahore, northeastern Pakistan).

This kind of "splitting" has only complicated and confused the taxonomy of eublepharians (the habitats of some of Börner's forms are 150 km apart). We consider it unfounded and not very reliable. Until recently, the eublepharids of the U.S.S.R. fauna were also regarded as the species *E. macularius*. Several years ago Prof. I. S. Darevsky went to Iran, observed live eublepharians there, and began to suspect flaws in their taxonomy. He analyzed the existing materials and became convinced that there exists a distinct species in Turkmenia, which he named *E. turcmenicus* in 1978. We have had the opportunity to become familiar in detail with live and preserved specimens of this species, and of *E. macularius*, and were also convinced that they do, in fact, differ in a number of scale characters, and especially in coloration and pattern. These differences are greater than those between other forms. All this would indicate that there is a pressing need for a special revision of eublepharian taxonomy. The result of this may well be the recognition of only one or two species in southwestern Asia with several subspecies.

Until such a study is done, it seems possible to us to distinguish three species in this group: *E. macularius, E. angramainyu*, and *E. turcmenicus*. The most complete data on the ecology and behavior of *E. turcmenicus* are contained in the article by N. N. Szczerbak [145].

* * *

The skink gecko has been known (under the name *Stenodactylus scincus*) since 1858, thanks to the description by H. Schlegel, for which he used the materials from Prebalkhashye region. In 1863, A. Strauch had distinguished the genus *Teratoscincus*. The species *T. keyzerlingii* [sic], which he described from the collections from Iran, was of this type. Subsequently, G. Boulenger [200] identified these species. However, S. A. Chernov [125] believed the geckos from the republics of Middle Asia {The name "Middle Asia" is used by Russian naturalists to denote the part of Asian territories of the former U.S.S.R. republics such as: Kazakhstan, Tadjkistan, Turkmenistan, and Uzbekistan including Karakalpakistan. Another, equivalent designation for this area in Russian-language scientific literature is "Middle Asia and Kazakhstan" – MG} to be a part of the nominate group but *T. scincus keyzerlingii* [sic] to be common in Iran and be distinguishable by a larger number of scales around the body (31-36 instead of 28-32) and by numerous (42-50 instead of 30-48) granulated scales between the eyes. *T[eratoscincus] zarudnyi* from Iran, which was described by A. M. Nikolsky [87], S. A. Chernov considered to be a junior synonym of *T. s. keyzerlingii* [sic] [112, 125]. Chernov had also studied {**p. 7**} the holotype *T. roborowskii* Bedriaga 1907, from the Oasis Sa-chou, Peoples Republic of China, which differs from specimens from Middle Asia only in details of coloration and is recognized as a subspecies of *T. scincus*. Thus, in accordance with the current understanding [17, 307], *T. scincus* from the deserts of Central Asia and Prebalkhashye region is attributed to the nominative form.

Altogether, the genus encompasses four species: *T. przewalskii* Strauch, 1887 in Central Asia, *T. bedriagai* Nik., 1899 and *T. microlepis* Nik., 1899 in Iran [and the previously cited *T. scincus*]. The currently existing descriptions and conclusions on taxonomy are based on the study of a small number of specimens. The geographic variation of *T. scincus* has not been studied. The study of the population samples of this species from the U.S.S.R. [145], which we did in 1979, has shown them to be nonhomogeneous. The most isolated population, from the Fergan Valley, was described as a subspecies, *T. scincus rustamovi* Szczerbak.

* * *

The history of the study of fringe-toed geckos begins with the description of the type species under the name *Gymnodactylus eversmanni* Wiegmann, 1834 (at first Wiegmann placed it in the genus of thin-toed geckos). This discrepancy was noted by Strauch [300 {error in original – MG}], who distinguished a special genus *Ptenodactylus* in 1887 but mistakenly used a name that had already been applied to the genus of iguanids. O. Boettger [193] corrected the error a year later and called this genus *Crossobamon*. For many years the genus was considered to be monotypic. However, in 1967 A. G. Kluge [237] attributed three more South Asian species to this genus (*Stenodactylus orientalis* Blanford, 1876, *S. lumsdeni* Boulenger, 1877, and *S. maynardi* Smith, 1933). Since then, there has been no revision of the given group but the author of the compendium on the geckos of the genus *Stenodactylus* [176], nevertheless, confirmed that attributing these species to the genus of fringe-toed geckos was correct.

The materials on the ecology of *C. eversmanni* are widely represented in the works of Soviet zoologists [22, 23, 98, 107, 125, 130, *et al.*].

* * *

The beginnings of the study of the geckos of the genus *Alsophylax* are associated with the name of P.- S. Pallas [258], who described a new lizard species "*Lacerta pipiens*" on the basis of the material he collected near the mountain Bolshoi Bogdo in the area of Lake Baskunchak (currently in the Astrakhan region). Subsequently, this species was attributed to the genus *Ascalabotes* [247]. E. Eichwald [213] moved it into the genus *Gymnodactylus* and L. Fitzinger [215] separated it into an independent monotypic genus *Alsophylax*. G. Boulenger [200] united it with the genus *Bunopus* Blanford, 1874 (type species *B. tuberculatus*, Baluchistan). This led to significant changes in the original generic diagnoses, because they were characterized by mutually exclusive characteristics: smooth (*Alsophylax*) or tuberculated (*Bunopus*) subdigital lamellae. In addition to this, G. Boulenger united another species, *Gymnodactylus microtis* Blanford, with the P.-S. Pallas' species, without confirming his point of view, however, with comparative analysis. The resulting confusion remains to this day because the vast majority of the researchers, who continued the work, accepted G. Boulenger's point of view.

The founder of the collections at the St. Petersburg Zoological Museum, A. Strauch [300], simultaneously described three new species of the genus *Alsophylax*: *A. przewalski* (western China), *A. loricatus* (Fergan Valley), and *A. spinicauda* (northern Iran). Unfortunately, Strauch's very precise and detailed descriptions were not accompanied by some unifying generic characteristic. *Bunopus blanfordi (*Egypt) was also described in this work.

All the species that were included at the time in the genus *Alsophylax* had come from Middle and Southwest Asia and adjoining areas. Nevertheless, G. Boulenger [202] had attributed his new species from Tibet, *A. tibetanus*, to this genus.

{p. 8} J. Anderson [165] described the species *B. spatalurus*, which was also characterized by the key diagnostic characters of the genus *Bunopus* – the tuberculated subdigital lamellae, basing the description on collections from southern Arabia. For some reason, this species escaped the experts' attention and for more than 60 years was not mentioned in the herpetological literature.

The first attempt at dividing the genus *Alsophylax* was made by A. M. Nikolsky [93], who described *B. crassicauda* from northern Iran. He noticed one of the characters of the original diagnosis of the genus *Bunopus*, the spiny lateral digital scales. However, his opinion about the heterogeneity of this genus found no support among the contemporary herpetologists. In 1907, A. M. Nikolsky described *A. laevis* from southern Turkmenia, which is closely related to *A. pipiens*. S. A. Chernov [121] reduced this designation to a synonym of the squeaky pygmy gecko without critical analysis (this led to subsequent confusion in some taxonomic issues with the genus *Tropiocolotes*). The smooth pygmy gecko was restored to the status of a species by O. P. Bogdanov [19, 20].

The next attempt to restore the genus *Bunopus* was made by A. Leviton and S. Anderson [243], who suggested its separation from the genus *Alsophylax* on the basis of the presence of dorsal tubercles. R. Mertens [257] noted the lack of foundation for such an approach. Taking this into account, these authors returned to the "subdigital lamellae, the free distal margins of which are denticulate" character but added one more, "a denticulation on some or all of the ventral scales" [244, p. 166] and included the species *B. tuberculatus, B. blanfordi, B. crassicauda,* and *B. abudhabi* (described in the same work and found in eastern Arabia) in the genus *Bunopus*.

However, Yu. K. Gorelov et al. [48] discovered occasional serrated abdominal scales on the type species of the genus *Alsophylax*, *A. pipiens*, and objected to the diagnosis in this form. In addition, these authors noted that A. Leviton and S. Anderson did not indicate to which genus they attribute the tuberculated species *A. loricatus*. This omission allowed Gorelov *et al.* to agree with the opinion of R. Mertens [248] and consider the genus *Bunopus* (the species *B. tuberculatus* and *B. loricatus* found in the U.S.S.R. fauna) as just a subgenus of *Alsophylax* (the species *A. pipiens, A. spinicauda,* and *A. laevis* found in the U.S.S.R. fauna).

The presence of well-defined longitudinal and transverse dorsal tubercles (*Bunopus*), or their absence or random placement (*Alsophylax*), should have been used as the criterion for separating the subgenera. On the basis of these data, A. Bannikov *et al.* [17] limited the genus to 13 species, without defining the size of each subgenus. However, with these diagnostic characters each subgenus should have included the species with the smooth and {word omitted in original text – MG} tuberculated subdigital lamellae. This was still in disagreement with their original diagnoses. In one of our publications [150] we clarified this issue and suggested our own version of diagnoses for the genera *Alsophylax* and *Bunopus*, based on the detailed data from series and including the original descriptions. However, a recent discovery in Tjan-Shan of a new species, *A. tokobajevi* [56], which differs from other species of the genus in a number of morphological characters, has resulted in the recognition of a new subgenus *Altiphylax* and refinement of the generic diagnosis.

The inclusion of a new species, *B. aspratilis* [170] from southwestern Iran, in the genus *Bunopus* did not result in experts' objections. E. Arnold [175] restored the species *B. spatalurus*, which was mistakenly redescribed as *Trachydactylus jolensis* [227] and synonymized with *T. spatalurus* [244]. He emphasized that this is justified, at least until the revision of the entire genus is done. In a subsequent work [177] he reduced the species *B. blanfordi* and *B. abudhabi* to junior synonyms of *B. tuberculatus*.

The taxonomy of the genus *Tropiocolotes* is no less complicated and muddled. Thus, the genus *Tropiocolotes*, as described by W. Peters [276], originally included only one species, *T. tripolitanus* known from Libya. G. Boulenger [204] included *Gymnodactylus steudneri*, which was described by W. Peters [275] from northern Sudan, in this genus. F. Steindachner [294] described *T. nattereri* from Sinai, which was later accepted as a junior synonym of the preceeding species. By 1947 [249] it became clear that the area inhabited by the species of this genus covers North Africa to the north of the 10th parallel {p. 9} and that the type species is ubiquitous in this area and forms four subspecies. The second species is found only in the northeastern part of the continent. The common characters for these species were the keeled subdigital lamellae, as well as keels on some or all body scales, absence of dorsal tubercles, small size, and a number of other, less significant characters.

Shortly thereafter, one more species, *T. scortecci* [208], found in the southern Arabian Peninsula, was included in this genus. Its diagnosis did not contradict the generic diagnosis.

However, between the end of the [19]50s and the beginning of the [19]70s a question arose whether to include the only species of the genus *Microgecko*, *M. helenae*, in the genus *Tropiocolotes*. This species was described by A. M. Nikolsky [94] on the basis of several specimens collected by N. Zarudny in southwestern Iran. A number of characters, such as smooth subdigital lamellae and smooth scales, contradicted the diagnosis of *Tropiocolotes*, whereas other characters, such as a certain positioning of some of the head shields, the absence of dorsal tubercles, and small dimensions, supported it. A number of researchers [167, 256, 258, 262, 263, 307] came out in favor of including *Microgecko* in the genus of dwarf geckos (*Tropiocolotes*). Still others defended the independence of the genus *Microgecko* [168, 223, 302]. As it later turned out, almost all the authors were basing their conclusions on erroneously identified materials. It was shown that a species mistakenly identified as *T. helenae* was a closely related species, *T. persicus*, which was originally described by A. M. Nikolsky [94] as *Alsophylax* and erroneously identified [168] as *Bunopus*. S. Minton *et al.* [265] proved that *Microgecko* belongs to the genus *Tropiocolotes* and distinguished three subspecific forms of *T. persicus*. Simultaneously, they described *T. heteropholis* from Iraq, one of whose main characters, heterogeneity of the dorsal scales, contradicted the generic diagnosis.

During this time, the genus *Tropiocolotes* grew by two more South Asian {here the authors imply the territory to the south of Middle Asia, which, in fact, is Southwest Asia - MG} species with smooth subdigital lamellae and homogeneous dorsal scales, *T. depressus* from Pakistan [264] and *T. latifi* from central Iran [246]. The above authors noted that, in their opinion, the genus *Tropiocolotes* contains two groups. *T*[*enuidactylus*] *helenae*, *T. persicus*, *T. depressus*, and *T. latifi* they attributed to the south {= Southwest - eds.} Asian group of species ("*helenae*" – complex). The other species (*T. tripolitanus*, *T. steudneri*, *T. nattereri*, *T. scortecci*) were attributed to another, North African group. However, there was no taxonomic status given to these groups.

The list of the subspecific forms of the species from the genus *Tropiocolotes* also grew by two names. T. Papenfuss [271] described *T. k. apoklomax* from Mali and J. Schmidtler *et al.* [286] described *T. helenae fasciatus* from western Iran. The above authors considered the species' types to be lost and did not compare their specimens to them.

Thus, currently it is known that the distribution area of the genus *Tropiocolotes* includes North Africa and Southwest Asia but the intrageneric and subspecific relationships and a number of other questions in the systematics of the genus remain unanswered.

The data relevant to the species of the genera under discussion are contained in many publications [3-7, 15, 19-25, 59-61, 106-108, 123-125, 130, 131, 216, 250-252, *et al.*]. Their characteristic peculiarity is in the fact that the domestic literature in some measure treats various aspects of the ecology of the discussed geckos, whereas in the sources from abroad such data are almost completely absent. Consequently, the existing point of view, which considers *Bunopus* to be a subgenus of the genus *Alsophylax*, is not based on the results of special study and requires re-examination. The intrageneric composition of the genera *Alsophylax*, *Bunopus*, and *Tropiocolotes*, their intra- and interspecific relationships, as well as the geographic variation of the species included in these genera, have had almost no study. Due to the poor knowledge of the species from the aforementioned groups, there is no single view on the taxonomic value of the diagnostic characters at various levels.

In studying the systematic position of the geckos of the genera *Bunopus* Blanf. and *Tropiocolotes* Peters, it became apparent that these taxa erroneously contain species which have only convergent similarity with their other representatives. On the basis of series of specimens, the analysis of scale characters, including the degree and limits of their variation, has allowed for a more precise definition of the contents of the listed genera and, following the resulting taxonomic re-arrangement, for greater precision of generic {**p. 10**} boundaries. As mentioned above, the genus *Tropiocolotes* contains the species *T. heteropholis* [169].

Heterogeneous scalation is one of the main characters of this species, which, by the way, is reflected in the species name. In this case, this heterogeneity is determined by the presence on the back of 12 longitudinal rows of large scales. This character, as well as highly defined keeling of the subdigital lamellae and dorsal scales, has convinced the authors of this species' description to include it specifically in the genus *Tropiocolotes*. It is, however, distinguished from other species of the genus by the heterogeneous dorsal scales. Our examination of this species' type resulted in the conclusion that its placement in the genus *Tropiocolotes* is not justified. Subsequently, a new species from Iran was described [161], *Bunopus aspratilis*, which was also characterized by pronounced keeling of all the scales, including the large scales of the head. In addition, this species had distinct dorsal and tail tubercles in combination with the keeled subdigital lamellae, which prompted the author to attribute the new species to the genus *Bunopus*. He noted that this species is close to *T. heteropholis* but did not attach particular importance to this. After studying the types of the indicated species, we concluded that they possess a number of peculiarities and it is possible to distinguish them as a separate genus, *Carinatogecko* Golubev & Szczerbak, 1981.

* * *

Until recently the largest number of gecko species were unified into the genus of thin-toed geckos, *Gymnodactylus* Spix, 1825, which was described on the basis of the type species *G. geckoides* from the New World. Two years later, Gray published a diagnosis of the genus *Cyrtodactylus* (type species *C. pulchellus*) based on the materials from Southeast Asia (Singapore), which differed from the previous genus in the order of positioning of preanal pores. The unreliability of this character gave the catalog authors [96, 200, 212, 215, 300, *et al.*] the right [i.e., a reason] to retain the first name [*Gymnodactylus*] as valid for one or another of the species groups that are [now] included within the genus.

In contemporary literature, there is no universally accepted opinion on the content of the thin-toed gecko genus (*Gymnodactylus* – *Cyrtodactylus*). European herpetologists [257, 307, *et al.*], whose opinion is shared by the Soviet experts [17], accept the thin-toed gecko genus (*Gymnodactylus*), which covers the species distributed cosmopolitically. G. Underwood [305] has split the genus *Gymnodactylus* (type species – the South American *G. geckoides*) into two genera on the basis of the eye pupil structure and some other characters. For the Old World species, he proposed the genus name *Cyrtodactylus* {error in original – MG} (type species – the southern Asian *C. pulchellus*), which was accepted by American herpetologists [170, 245, 263, *et al.*]. It also figures in the works by the European authors from abroad [*i.e.*, outside the USSR] [194].

We [153] subscribed to the opinion of H. Wermuth, the author of the world gecko checklist [307 {error in original – MG}], and looked at the group of Eurasian thin-toed geckos as the subgenus *Cyrtodactylus*. Simultaneously, the generic and subgeneric descriptions for the geckos of the U.S.S.R. fauna were made more precise. This decision, it seemed, was quite satisfactory but did not withstand the test of time. A new gecko species, which possesses the main characters of the genera *Gymnodactylus* and *Alsophylax* {*fide* Szczerbak – MG}, was discovered in 1983 in central Tjan-Shan. At the time, the diagnosis of the first genus contained the following main characters: (1) the digits are thin and crooked, as the last two or three phalanxes are laterally compressed and joined together at an angle; (2) the digit undersides are covered with one longitudinal row of subdigital lamellae; (3) the body is covered by homogeneous granulate scales, frequently interspersed with larger keeled scales (tubercles) that form more or less regular longitudinal rows, abdominal scales larger than lateral ones); (4) the males usually have preanal and femoral pores or only preanal ones.

Our in depth review of the U.S.S.R. gecko fauna revealed a number of surprises. Cursory acquaintance with the representatives of the genus *Alsophylax* from the U.S.S.R. fauna has shown all the characters, with the exception of the first one (distortion of the digits {digits angularly bent - MG} {**p. 11**}), to be characteristic of this genus as well. The species *A. spinicauda* and *A. tibetanus* have bent digits, just as the thin-toed geckos do. The study of Eurasian thin-toed geckos has shown their heterogeneity. Two groups of species, Mediterranean and Middle Asian, are clearly distinguishable here. *G[ymnodactylus] kotschyi, G. russowi* and some others should be attributed to the first group and *G. caspius, G. fedschenkoi,* and *G. longipes* — to the second. The group of Mediterranean species is characterized by the less distorted [*i.e.*, angularly bent] toes, which brings them closer to some southern lowland geckos included in the group *Bunopus*.

The study of thin-toed geckos revealed differences between these groups in some other traits. Thus, the Mediterranean thin-toed geckos differ from the Middle Asian ones, as we understand it, in the absence of femoral pores (only preanal pores are present) and in a number of other scale characters. These data allowed us to separate the group of Mediterranean thin-toed geckos into the subgenus *Mediodactylus* Szczerbak et Golubev, 1977[1]. Our revision of the genus *Alsophylax* has also shown that the species *A. spinicauda* is closer to representatives of the subgenus *Mediodactylus* and should be moved into the genus of thin-toed geckos. This would stabilize the diagnosis of the indicated genus and pygmy geckos {*Alsophylax* – MG}. On the basis of the totality of characters, we also moved *A. tibetanus* into the genus *Gymnodactylus* and supplemented its diagnosis with the character "presence on the lower surface of the tail of a row of enlarged scales or scales."

In returning to the 1983 discovery in Tjan-Shan, it is necessary to emphasize that the hindlimbs of the new gecko species have the same bent toes as do some thin-toed geckos. On the basis of the majority of characters, it was included in the genus *Alsophylax* as a special subgenus *Altiphylax* but, at the same time, the uselessness of the traditional description of the genus *Gymnodactylus* became apparent (the diagnosis of the genus *Alsophylax* also requires major modification) {*fide* Szczerbak – MG}. Acquaintance with the type species *Cyrtodactylus pulchellus* from Singapore has shown that there is a number of fairly significant differences between it and the group of Palearctic species of thin-toed geckos (in the former the basal phalanges are significantly widened and covered on the underside by a row of broad lamellae, distal phalanges are distinctly narrow; in the latter, the toes are uniformly thin over their length).

The structure of the digits, especially the positioning and form of the subdigital lamellae, are so consistent for the different species as to be the decisive factor in the systematics of geckos. It is on the basis of these very characters that taxonomists distinguish the groups of naked-toed {*Cyrtodactylus* – MG}, leaf-toed {*Hemidactylus* – MG}, least {*Sphaerodactylus* - MG}, fan-toed {*Ptyodactylus* – MG}, and other geckos. The adaptation to movement along vertical surfaces is accomplished in entirely different ways. To this day nobody has paid attention to these differences in the structure of Palearctic and South Asian species, which do not allow for their inclusion in the same genus. It was, however, G. Underwood [305] who noted that the Caspian {*C. caspius* – MG} and Turkestan {*C. fedtschenkoi* – MG} geckos available to him did not fit his diagnostic criteria for the genus *Cyrtodactylus*.

A. Kluge [239] convincingly demonstrated that the application of the above genus to some tropical gecko species is groundless. He writes: "The remaining characters set forth by Underwood also fail to diagnose *Cyrtodactylus* either because of their individual and interspecific variation or because the common state is pleisomorphic. Thus, the available evidence indicates that *Cyrtodactylus* cannot be diagnosed as it is presently constituted taxonomically, and, furthermore, the variation in the size and shape of the enlarged dorsal tubercles and their keeling and the preanal pore arrangement suggest that the group may be polyphyletic." {Kluge, 1983:469 – MG}

The validity of the indicated group of Old World geckos became doubtful as the result of our work in the northern part of its habitat and A. Kluge's work in the south. This brought us to the thought of separating the Eurasian thin-toed geckos into a special genus *Tenuidactylus* (type species *T. caspius*) [153]. The content of the genus in this monograph is defined in accordance with this work. It is no accident that the Caspian thin-toed gecko is selected as the type species. First of all, it was the first one to be described among {p. 12} our thin-toed

---

[1] In a new catalog of European and Southwest Asian reptiles that was published in the U.S. [322], without any justification, this subgenus is raised to the level of a genus, the presence in this region of the genus *Gymnodactylus*, as well as *Cyrtodactylus*, is acknowledged, and other errors are included.

geckos; second, the species contains the entire suite of very well-defined characters associated with the whole group. According to preliminary data, the genus *Tenuidactylus* unites 18 species of Palearctic geckos in three subgenera. These are the nominate genus and the [genus] *Mediodactylus*, recently described by us, which contains four Mediterranean species, as well as *Cyrtopodion*, made up of six species from Southwest and Central Asia.

The separation of the species group from the Tibet-Himalayas region is, probably, not far off in the future. The genus *Cyrtodactylus*, as we see it, contains a large and varied group of tropical and primarily South Asian species (around 40). The presence of groupings within the genus of thin-toed geckos was noted by M. Smith [291 {error in original – MG}], even though his research concerned primarily the Indian geckos. Nevertheless, even a brief acquaintance with the Palearctic forms convinced this herpetologist of their peculiarity. The latter author's first group (of the type "*scaber*") matches closely the subgenus *Cyrtopodion* in contents and some traits characteristic to it. The last group is not as homogeneous and is made up of two groups. In many ways, the first one occupies an intermediate position between *Tenuidactylus* and *Mediodactylus* and includes four species (*T. kachhensis, T. scaber, T. watsoni*, and *T. montiumsalsorum*). The second is a small group (*T. agamuroides* and *T. elongatus*) of "protoagamuroid" geckos (in some characters they resemble the representatives of the genus *Agamura*). Nevertheless, they cannot be unified with agamuroid geckos, as they possess the main traits of Palearctic thin-toed geckos.

The process of describing all the species of this genus is, probably, not yet completed. The evidence for this is the description of the species *T. turcmenicus* Szczerbak, 1977, in Badkhys, an area well studied by zoologists. New discoveries in Iran, Afghanistan and, especially, in the Tibet-Himalayas region are to be expected.

As the work in karyology has demonstrated [73, 154, *et al.*], our arrangements of nomenclature and the placement of certain species in the genera *Alsophylax* and *Tenuidactylus* are convincingly supported by distinctive chromosomal structures.

* * *

A small genus of so-called spider geckos (*Agamura*) from Iranian Baluchistan, described in 1874 by the famous English naturalist W. Blanford [186 {error in original – MG}], unites two species, the type species *A. persica* and *A. femoralis* [211, 290], which inhabit desert areas of eastern Iran, Afghanistan, Pakistan, and northwestern India [49]. In 1874, the form *A. cruralis* Blanf. was described (listed under this name in Boulenger's Catalog [200]), but M. Smith showed it to be identical to *A. persica*. The species *A. femoralis*, known only from the type locality, is considered valid [307] but, probably, requires greater attention from systematists.

Initially, *A. persica* was attributed to the genus *Gymnodactylus*. However, its peculiar appearance and a number of characters in scalation so clearly distinguish it from the representatives of the latter genus that the new generic description was never questioned after its publication. Nevertheless, the genus *Agamura*, in our opinion, is close to the genus *Tenuidactylus*.

As we determined, the agamuroid lizard *Gymnodactylus gastropholis* Werner, from Iran, described in 1917 and subsequently synonymized with *G. agamuroides* Wettstein, 1951, strictly on the basis of similar appearance, corresponds, in fact, to the genus *Agamura* diagnosis, and we consider it to be in this genus.

The Belgian zoologist de Witte described a new genus *Rhinogecko*, from Iran (type species *R. misonei* [sic]) [320], which is characterized by very bulbous nasal shields, which form the nostril channels, and are noticeably elevated above the facial surface. In all other characters this species belongs to the genus *Agamura*. Bulbous nasal scales, although less developed, are found in a representative of the genus *Tenuidactylus* (*T. elongatus*). Therefore, the genus of the species "*misonei*" [sic] requires clarification.

{p. 13}* * *

Recently [323], the German herpetologist J. Frizsche, during a tourist visit to Ashkhabad, caught, supposedly on a building wall, a specimen of a species new to the U.S.S.R. fauna, *Hemidactylus turcicus*. We searched specifically for this gecko in the indicated region but without success. Most likely an individual of the Turkish half-toed gecko was accidentally imported and then caught or else this circumstance can be attributed to the category of museum error and inclusion of this species in the list of the U.S.S.R. fauna is not appropriate. Therefore, we do not consider the genus *Hemidactylus* here.

MATERIAL AND METHODS

In the foundation of this work are, primarily, the collections and observations by the authors during field research. We collected the material on geckos prior to the planned work on the given group and used it in creating the regional compendium [141 {error in original – MG}] in conjunction with collecting other reptiles [143]. Concurrently, we studied the taxonomy and ecology of the Mediterranean thin-toed gecko [138, 139] and made a number of new finds [147]. Also during this time, field work was regularly done in Crimea (1956-64, 1975, 1980, 1981, and 1982), as well as in Azerbaijan (1958), Georgia and Armenia (1963), Turkmenia (1963, 1964, 1966, 1970, and 1971), Uzbekistan (1964, 1968), Tadjikistan (1964), Kirgizia (1957, 1960, 1962, 1972), Kazakhstan (1973). Starting in 1974, we conducted extensive field work with the main purpose of collecting material on geckos.

The field research covered the areas of Caucasus and Transcaucasus (1974, 1975), Turkmenia (1975-84), Uzbekistan (1976), Tadjikistan (1976, 1980, 1981), Kazakhstan (1976), and Kirghizia (1984). Particularly extensive research in Middle Asia was done in 1976. A highway transect {by vehicle – MG} along the route Kiev – Baku – Krasnovodsk – Kushka – Kerki – Termez – Fergan Valley – Kyzylkum Desert – Ustyurt – Gur'ev – Astrakhan – Volgograd – Voronezh – Kiev covered more than 20,000 km. In those years, we crossed just the Karakum desert eight times south to north and back, as well as west to east. We used various methods to collect the animals. In the daytime, we most frequently collected them from their shelter by turning over rocks, and by digging out rodent and insect burrows, termite hills, and bases of dead shrubs. The active search allowed us to collect many species of lizards and snakes, which, until then, were thought to be rare and were found in collections as single specimens. At twilight and at night we examined rock and loess cliffs or barkhans (sandhills). On the plains (in the steppe and desert regions), we used the GAZ-66 truck with additional lights as well as several observers. From a slowly moving vehicle, a strip of up to 5 m could be seen well (even ants could be distinguished). This method was indispensable for collecting and counting small ground geckos (the speed of 4-5 km/hr, the distance noted by the speedometer). Some of the captured lizards were fixed in alcohol at the capture site and others were transferred to field terraria for later study in the laboratory.

Our extensive collections and observation of ecology, as well as the specimens received from many domestic museums and those from abroad, allowed us to study series, primarily made up of the population samples from various regions of the species range, from the conceptual position that we took of the polytypic species. This gave us the opportunity to compile sufficiently well-founded taxon descriptions (we acquainted ourselves with the type materials still in existence), to study the individual, age, and the especially important geographic variation, and to revise the systematic positions of the studied taxa. Altogether, we examined {**p. 14**} over five thousand specimens and used around 20 characters in compiling the descriptions and studying geographic variation.

We selected and used a number of new characters in addition to the traditional ones. The boundaries of their variation became clearer as we processed the materials. This allowed us to establish characters and combinations of characters from the generic to the subspecific level with high taxonomic value and low levels of variation and, for the first time, use them as diagnostic traits. This way the generic diagnoses lost the advantage of ease and brevity but gained a greater degree of reliability. We used the following characters:

1. Body length (**L** {= SVL}) (from snout to vent).
2. Length of unregenerated tail (**LCD**) (from vent to tip of tail). The above measurements were made with calipers. Characters 3-9 were measured with the binocular microscope MBS-2 micrometer.
3. Head length (from front nostril edge to front ear edge).
4. Snout length (from the front nostril edge to front orbit edge).
5. Transverse eye diameter.
6. Maximum (vertical) diameter of ear opening.
7. Head width (behind eyes).
8. Head height (behind eyes).
9. Dorsal tubercle length (in one of two central rows in the middle of the back).

Some of these measurements were used to generate the following indices:
1. **L/LCD** {= body length/undamaged tail length – MG}.
2. Head height to width ratio (×100) {**HHW**}.
3. Ear opening to eye diameter ratio (×100) {**EED**}.
4. Dorsal tubercle to body length ratio (×1000) {**DTL**}.
{Acronyms in braces { } in 2-4 appear only in the English text – MG}

The number of scales and scales counted:
1. Scales across the head (interorbital; the scales on the ridge above the eyes were not counted) {*i.e.*, superciliary scales – MG}.
2. Scales along the head (only for skink geckos; from nasal scales to enlarged dorsal scales).
3. Supralabial scales separately on right and left (also taken into account the scale onto which the front edge of the orbit is projected).
4. Nasal shields.
5. Scales separating first nasal shields.
6. Infralabial shields.
7. Pairs of postmental scales or scales replacing them (in *Bunopus*, some *Teratoscincus*, and others).
8. Scales around dorsal tubercle (in one of two central rows in the middle of the back).
9. Longitudinal and transverse rows of dorsal tubercles.
10. Scales along underside of body from first pair of postmental scales to vent (if the boundary between the preanal scales and small granular scales between them and the vent was well defined, then these granular scales were not counted). {**GVA**}
11. Ventral scales across abdomen (at the middle of the body).
12. Scales around the middle of the body (only for the species without dorsal tubercles and sharp boundary between abdominal and lateral scales).
13. Subdigital lamellae on the 4th toe of the hindlimb on the right and left (in skink geckos – number of lateral fringe scales on the same toes).
14. Preanal (femoral) pores (in females of the subgenus *Tenuidactylus* the number of scales that may carry these pores was counted).

In addition, taken into account were the shape, size relationships, surface condition, and relative position of all the scales and scales and gecko body undersides. Color and pattern were described.

For the size characters (1, 2), scale characters (1, 6, 8, 10-14), and the indices (1-4), we computed the statistics: Min – max, $\overline{X} \pm S_{\overline{x}}$, V. The samples were subjected to the Student's *t*-test. Statistical treatment of characters and comparison was done using standard {**p. 15**} methods [100, 101] on the microcomputer *Electronica BZ-21* using a program developed for this purpose by L. I. Francevich [115].

Each account is preceded by the Russian and Latin names of the species. For some species, subgenera, and genera the Russian names[2] are used for the first time. The type locality follows. In some cases its precision was improved through the study of the type specimens. We determined the validity of the names and, where necessary, designated a lectotype in accordance with the procedural rules of the *International Code of Zoological Nomenclature* [74]. The karyotype data are given. The majority derive from our laboratory. The references in which a name was used for the first time, are listed chronologically in the list of synonyms. Descriptions are preceded by diagnoses (frequently these are new, devised by the authors). In examining distributions, we clarified the current range boundaries and made the first attempt to provide a distribution map, taking into account all the known locations, the list of which is contained in the legend. In the biotope study, we gave the descriptions of the soil, relief, and vegetation, as well as elevation above sea level, the exposure of the slope, and its angle. Quantitative data were acquired either by the common transect method (during walking or driving counts) or by doing an absolute count as we turned over rocks and dug out shelters. Recalculation of the numbers per 1 hectare is possible in both cases.

Response to temperature was studied by measuring the temperature of the substrate and air whenever an active lizard was found on the surface, as well as in the laboratory in the thermogradient apparatus (modified Herter's apparatus), which allowed us to determine the preferred temperature for the gecko sample being studied.

The data on the daily activity cycle were acquired during field research (appearance from and retreat to shelter) and on the peak of activity — during counting and the study of the response to temperature. The seasonal activity cycle we ascertained during the direct search in nature and by comparing the collecting dates of the collection materials.

Shedding was recorded during collecting in nature and for lizards in captivity.

The feeding data were acquired by the stomach content analysis, as well as through observations in nature and in the terrarium. Because we studied feeding most actively in recent years, stomachs were dissected only when data in literature were limited or absent.

We preferred to obtain data on reproduction, clutch descriptions, and incubation when keeping geckos in the terrarium. Precise data on these topics are impossible to acquire in nature. Variation in the incubation period of specific clutches depends on the temperature regime throughout the entire period of development. In recent years, it has also become known that the incubation temperature can affect the sex ratio and even aspects of pattern. We consider the study of reproductive characteristics of reptiles under the laboratory conditions to be promising and timely.

Growth rate was studied primarily using statistical methods by comparing the body size of specimens captured at various times.

We acquired (along with the literature sources) data on the natural enemies of geckos by capturing predators in the lizards' biotopes and through examinations of the nests of predatory bird during the period of feeding the fledglings. The reasons for fluctuations in the numbers of lizards of interest to us were determined by comparing the human factors and their impact on the biotope.

We also paid detailed attention to the documentation of behavioral characteristics. The data acquired during captivity have an important place here, along with the observations from nature. Here we described the methods of locomotion, aggression, defense; reproductive behavior; and characteristics of sound communication. The materials on the summer and winter shelters were acquired mainly through study in nature.

The practical importance [of each species] was first and foremost determined from the viewpoint of preservation of the rare species' genofond {gene pool – MG}. In the necessary cases we give recommendations {**p. 16**} on protection. Judging by the food types, all geckos are insect eaters. They consume many harmful insects, have a certain place in biocenosis (many are common species), and are beneficial.

Data on ecology of many species from beyond the U.S.S.R. borders are extremely scanty even though we used all the available literature. Unfortunately, many herpetologists from

---

[2] The Russian names have been translated directly into English. These names often do not match the commonly used English colloquial names. We believe that it is more important for the translation to match the original text than to match a terminology that has neither regional nor international standing. [Eds.]

abroad do not give these issues enough attention. Hence, there is a certain lack of uniformity in the material in the species accounts.

The vast majority of the drawings and photographs are original. The color plates of many species we reproduce for the first time.

PHYLOGENETIC RELATIONSHIPS AND ZOOGEOGRAPHICAL ANALYSIS

The ancient origins of geckos (and their primitive nature, which, in the final analysis, is often not the same thing) has become a textbook notion. However, the study of the phylogeny of these lizards to this day is not afforded sufficient attention. This can be partly justified by the fact that the fragile bones of geckos are not preserved in the fossil record and paleontological specimens, as a rule, are not available. By using the comparative morphological method, we can, to a great extent, approach a solution to this problem.

The famous Soviet herpetologist P. V. Terentjev [111] summarized the work on the phylogeny of lizards and proposed a scheme of their relationships. At the base of the branch of the geckonid lizards, he placed representatives of the Jurassic lizards Ardeosauridae (fossilized remains of representatives of two genera are known from Bavaria, West Germany and Manchuria). This is the current view. Specific research into the evolution of gecko lizards (so far, perhaps, the only work of its kind) was done by A. Kluge [237]. He examined phylogenetic connections of gekkonids through comparisons of internal and external morphology (18 diagnostic indicators) on the basis of extensive material at the subfamily level, established their evolutionary levels, primitive and advanced characters, and discussed possible parallels.

According to his data, primitive characters are considered to be the presence of real eyelids (their absence is considered to be an advanced character; in parentheses below we state the advanced condition of the character); development of premaxillary bones from two ossification centers (one center); postcranial endolymphatic apparatus lacks calcified sac (the sac present); escutcheon scales absent (present) {original in error – MG}; ability to vocalize (no ability); lay two eggs (one); supratemporal bone present (absent); number of sclerotic ossicles in eye greater than 14 (equal or fewer); cloacal sacs and bones present (absent); angular bone is noted (absent); splenial present (absent); single frontal bone (paired); nasal bones paired (fused into one); single parietal bone (paired); lumbar vertebrae procoelous (amphicoelous); hypohyal-ceratohyal joined (absent) {original in error – MG}; second visceral arch complete (variously incomplete) {original in error – MG}; squamosal bone present (absent).

The majority of the preceding characters can be accepted as primitive, the value of such characters as "amphicoelous-opisthocoelous vertebrae" or "clutch contains one or two eggs" is questionable. Using statistics, the author computes the number of one or the other kind of characters for each of the groups under study and places them on a certain evolutionary level in accordance with this. Even though the evaluation of characters contains some subjectivity, the conclusions overall are of interest and correspond to reality. The comparative morphological scalation analysis we have done allowed us to distinguish the following alternative character values (primitive – advanced) {fide Szczerbak – MG}: the head and body scales small (large); scalation has weak differentiation or the scales are {p. 17} homogeneous (well differentiated); the presence of only preanal pores (femoral pores also present); basal phalanges of the digits widened (not widened); segmentation of the tail not pronounced (clearly pronounced); a dark transverse band on the back of the head (band absent); the species range is more southern (more northern; exceptions are frequent here).

P. V. Terentjev [111] places the family of scalefoots (Pygopodidae) closer to the ardeosaurs, then the family of plate-bellied night lizards (Xantusiidae), flat-tailed geckos (Uroplatidae), and, finally, eublepharians (Eublepharidae) and the family geckos (Gekkonidae). In our opinion, eublepharians should more correctly be treated as a family but quite a primitive one, whose place on the phylogenetic branch is lower than that of the highly specialized family of scalefoots. Plate-bellied night lizards occupy a sort of intermediate position between the geckos and skinks [49], probably closer to the latter. The final resolution of this issue requires special research. And, finally, the flat-tailed geckos are currently treated as a genus of the gecko family. Thus, the phylogeny of gekkonids should appear as a branch from ardeosaurs to eublepharians, from which geckos branch off, and at the beginning of which the scalefoots branch off.

A. Kluge [237] grouped all the geckos into four subfamilies. At the base of this branch he places the subfamily of eublepharians (in our understanding it is a family), then the subfamily of Australian spike-tailed geckos (Diplodatylinae) branch off from the main trunk, and at the end this branch splits and leads to the real geckos and the subfamily of round-toed geckos (Sphaerodactylinae), this last group occupying the highest evolutionary level as the most advanced branch. Thus, with a small correction, we accept the phylogeny of A. Kluge on the subfamily level. His views are in agreement with zoogeography: Southeast Asia as the center of origin for the circumglobal family of eublepharians (individual species are found, in addition, in Africa and America. They probably got there across the Bering Isthmus in the Paleocene, which again speaks to the ancient origins of this group).

It can be assumed with a fair degree of certainty that all the geckos evolved during the late Jurassic and early Cretaceous periods in Southeast Asia. A relatively primitive subfamily is tied to the Australian region and New Zealand (autochthon of the Australian region). Their ancestor evolved in Southeast Asia from the eublepharian ancestors sometime in the late Mesozoic period and penetrated into Australia through the Indo-Australian islands. Probably, the geckos that migrated west to Africa also came from eublepharian ancestors. Currently, Gekkoninae are the dominant group and within it can be observed up to three levels of evolutionary radiation: the two {p. 18} early levels led to their penetration into Africa, Madagascar, and Southwest Asia. The geckos of the New World evolved from the paleotropical forms whose ancestors inhabited Africa and the Mediterranean (later introduction by humans is not included). Thus, A. Kluge expresses the thought that Sphaerodactylinae of the New World came from African ancestors, which utilized the trans-Atlantic route.

Palearctic gecko genera and species phylogenetic diagrams are absent from the literature familiar to us. We make an attempt to present the relationships within this group by constructing dendrograms, which are derived from morphological research (Fig. 1, 2). Undoubtedly, the genus *Tenuidactylus* is connected in its evolution with the paleotropical genus *Cyrtodactylus* and takes up a central position. Its penetration into palearctic Asia through the Himalayas is unquestionable. The right branches (*Bunopus – Stenodactylus – Crossobamon*) {close phylogenetic relationship between the last two genera and *Cyrtodactylus* seems doubtful – MG} unite the genera, which adapted to psammophilous mode of

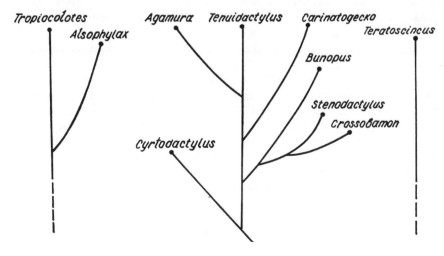

FIG. 1. Phylogenetic relationships among the gecko genera of the USSR and contiguous lands.

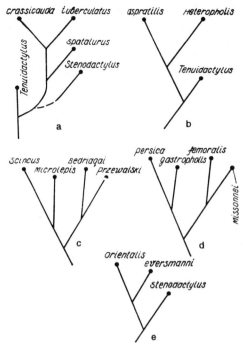

FIG. 2. Phylogenetic relationships among the species of the genera: a – *Bunopus*, b – *Carinatogecko*, c – *Teratoscincus*, d – *Agamura*, e – *Crossobamon*

life, and are characterized by the development of ribbed subdigital lamellae, crests on the digits, and by generally small scales but with areas of large scales. This is probably a progressive branch, because sand deserts are a more recent formation than tropical forests and rocks. The genus of skink geckos and the genera of northern lowland pygmy geckos {*Alsophylax* – MG} and dwarf geckos {*Tropiocolotes* – MG} developed independently, and, so far, it is difficult to picture their ancestral groups. It is possible that the genus of dwarf geckos is related to the African genus *Saurodactylus* (they are similar in small size and head scalation).

We proved the independence of the genera *Alsophylax* and *Bunopus*, which were previously considered to be connected. This position is also well supported by the zoogeographic data. The genus *Alsophylax* is endemic to the Turan Plain-Desert, Central Asian Desert, and Mountain Asian provinces [148, 301]. The genus *Bunopus* is characteristic of the Sahara-Arabian, Mediterranean, and Transitional Iran-Afghan provinces. The above allows us to draw not only the conclusion that these genera are independent but also that they are not closely related (no overlap in range). The phylogeny of the genus *Alsophylax* can be pictured as follows: the genus is divided into two parts — the relict subgenus *Altiphylax* with the species *A. tokobajevi* and the nominative subgenus. In the center of the latter are positioned the close species *A. laevis* and *A. pipiens*. They appear close due to the least differentiated scalation in the nasal area, and even though the latter species reaches the northern boundary of the species range, it should be considered to be the more ancient species, because the biotope of *A. laevis* — takyrs {areas with smooth hard clay surface, dry bottoms of former lakes or reservoirs measuring from several square meters to dozens of kilometers and having extemely sparse or no vegetation – MG}, which were formed by alluvial processes in the Quaternary period — are a more recent landscape formation. {The clay plains of the central Kyzylkum, where probably the oldest populations of *A. laevis* occur, are considerably older than the takyrs of the northern plain adjacent to Kopetdagh, which was penetrated by this species probably no earlier than mid- to late Pleistocene. However, the assertion "the older the landscape, the older the species inhabiting it" cannot be fully accepted as convincing because of the possible later speciation after penetration of an area by other species – MG}. The extensive range of *A. pipiens*, from the banks of the Volga to Mongolia, should also be taken into account here.

A new "smooth" species, *A. tadjikiensis*, had recently split from *A. laevis*. The Central Asian branch, the species *A. przewalskii*, is referred to the subgenus *Altiphylax*, whose range lies within the Central Asian Mountain region, and *A. loricatus*, based on the level of dorsal tubercle development, occupies a somewhat isolated position, but still close to *A. przewalskii*.

{p. 19} As can be seen from the morphological data, the genus *Tropiocolotes* is close to the genus *Alsophylax*, especially the subgenus *Asiocolotes*, which, probably, is the ancestral branch for the genus *Alsophylax* (Fig. 4). R. Mertens [256] considered the genus *Tropiocolotes* to be North African, which penetrated deep into western Asia (it served as

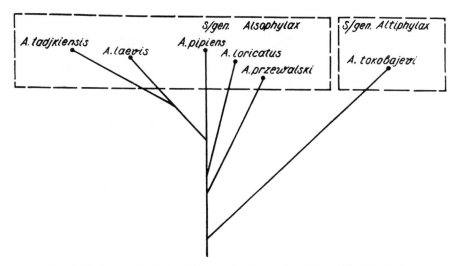

FIG. 3. Phylogenetic relationships among the species of the genus *Alsophylax*

the bridge for penetration of many North African reptiles onto the Asian continent). This opinion is in agreement with A. Kluge's data [237], who demonstrated on the basis of extensive osteological material that the spread of the subfamily Gekkoninae, accompanied by species formation, on all continents went from the tropics to the poles. Thus, the spread of reptiles from North Africa through Southwest and into Middle Asia appears to be fairly common. This direction coincides with the direction of variation of some scale characters within the genus *Tropiocolotes*.

In the branch *T. tripolitanus* – *T. steudneri* – *T. depressus* – *T. levitoni*, in the transition from species to species, there is a gradual increase in the number of preanal pores (correspondingly 0 – sometimes 2 – 2-5 – 6-7), subdigital lamellae (11-17 – 15-18 – 16-18 – 17-23), and in body size (up to 26 mm, in the eastern part of the range up to 27 – 32 – 45 mm); gradual formation of a row of plates on the lower surface of the tail (none – none – only on the last third – along the entire tail); and others. In some of the individuals of *T. levitoni* there appear a third nasal scale and, especially importantly, tubercles in the crest area. This species has a body pattern close to that of *Alsophylax* (among the species of the latter genus, *A. laevis* is closest to *Asiocolotes*). Current ranges of the indicated species are (in the same order): *T. tripolitanus* – North Africa; *T. steudneri* – eastern part of North Africa and Southwest Asia;

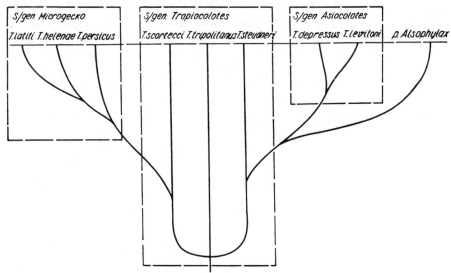

FIG. 4. Phylogenetic relationships of different species of *Tropiocolotes*

T. depressus – northeastern Baluchistan (Central Brahui Range); *T. levitoni* – southwestern Hindukush; *A. laevis* – southern part of Middle Asia. All this allows a suggestion that there existed several intermediate forms between the ancestors of *T. levitoni* and those of *A. laevis*.

In some cases, the extraordinarily similar scales of the nasal area in *Microgecko* and *Alsophylax* (*A. loricatus*) can, in our opinion, be explained by one of the propositions of N. I. Vavilov's law of homologous rows, according to which "not only the genetically close species but also genera exhibit similarities in the rows of hereditary variability" [32, p.24]. In this case, the path of character formation (specific pattern of nasal scales and scales) in the series *Tropiocolotes – Asiocolotes – Alsophylax* is considerably longer in the number of recent "intermediate" forms than in the series *Tropiocolotes – Microgecko*. This is likely caused by different rates of evolution in these gecko groups. R[obert] Mertens' opinion about the North African origins of the genus *Tropiocolotes* applies to an even greater degree to the nominative subgenus, which includes three species. {p. 20} Characteristics common to them are the keeled subdigital lamellae; generally undifferentiated nasal scales; fairly large dorsal scales; absence of subcaudal plates and well defined preanal pores present in all the males.

The existing deviations from the diagnosis make it possible to establish connections with other subgenera. Thus, some males of *T. steudneri* have two weakly defined pores. This allows us to suppose its greater closeness to *Asiocolotes* than that of the other species of the subgenus (other considerations are discussed above). In *T. scortecci*, the supranasal scales do not touch the nostrils but they are already somewhat enlarged and resemble the shape of those in *Microgecko*. Probably, the three species of the nominative subgenus evolved from a single common ancestor (similar size of the flattened scales and variations in the degree of their keeling). In all *Microgecko*, the supranasal scales are not only enlarged but also touch the nostrils, unlike those in *T. scortecci*. Two additional nasal scales and one to two pairs of large postsupranasal scales constitute required subgeneric characters (very rarely a deviation from the norm can be observed: the first additional scale is pushed aside by a second one and by the supranasal scale). The combination of these characters with the smooth subdigital lamellae (no more than 15) and the absence of subcaudal plates completely separates the species of this subgenus from the other ones. If the long list of *T. persicus* synonyms, which developed in the last 25 years, is noted, then the closeness of these species becomes even more apparent.

In many scale characters, including the diagnostic ones, *T. persicus* and *T. helenae* overlap somewhat, which also speaks to their closeness. According to our data, these species can be reliably distinguished only by the number of supralabial scales and by the color of the regenerated tail tip (in the former it is white, in the latter – black). The absence of representative collections of *T. latifi* and the unreliability of its main diagnostic character (as our observations have demonstrated, the absence of postmental scales is very frequently a character at a level below that of the species or even subspecies, for example, as in *A. laevis*, *T. depressus*, *B. spatalurus*, and others) do not allow us to establish its species independence with confidence and its greater closeness to any of the species of the subgenus. However, in the small number of infralabial scales it is closer to *T. helenae*.

Along two directions, which lead from *T. tripolitanus* (*T. steudneri – T. depressus – T. levitoni* and *T. scortecci – T. helenae*), parallel variation of the quantitative characters can be observed: the reduction of the dorsal scale size (greater [number] of scales around midbody) and an increase in the number of scales along the lower body and across the head. The central position (Fig. 4) of the nominative subgenus species is determined by the less differentiated scalation, weak development or absence of preanal pores and commonality of distribution. Within the subgenus *Microgecko* there is a distinct tendency towards the loss of postmental scales: in *T. persicus* they are almost always in contact, in *T. helenae* almost always separate, in *T. latifi* totally absent. An increase in the number of pores and distinctiveness of the subcaudal plates, as well as geographical closeness of their ranges, can be observed in the species of the subgenus *Asiocolotes*.

The ranges of the genera *Tropiocolotes* and *Bunopus* partially coincide but are separated

1 — *Eublepharis macularius*, adult (photo by A. D. Bautin), 2 — *Eublepharis macularius*, juvenile (photo by V. Miropolsky), 3 — *Eublepharis turcmenicus* (photo by N. N. Szczerbak).

4

5

6

4 — *Teratoscincus scincus* (photo by N. N. Szczerbak), 5 — *Teratoscincus przewalskii* (photo by N. L. Orlov), 6 — *Crossobamon eversmanni* (photo by N. L. Orlov).

7 — *Stenodactylus stenodactylus* (photo by P. Kodym), 8 — *Alsophylax tokobajevi* (photo by N. N. Szczerbak), 9 — *Alsophylax pipiens* (photo by N. N. Szczerbak).

10 — *Alsophylax laevis* (photo by N. N. Szczerbak), 11 — *Alsophylax tadjikiensis* (photo by N. N. Szczerbak), 12 — *Alsophylax tadjikiensis* (photo by N. N. Szczerbak).

13 — *Alsophylax loricatus szczerbaki* (photo by N. L. Orlov), 14 — *Bunopus tuberculatus* (Turkmenistan) (photo by N. N. Szczerbak), 15 — *Bunopus tuberculatus* (Iran) (photo by Steven C. Anderson).

16 — *Tropiocolotes tripolitanus* (photo by P. Kodym), 17 — *Tenuidactylus caspius* (photo by N. N. Szczerbak), 18 — *Tenuidactylus turcmenicus* (photo by N. N. Szczerbak).

19 — *Tenuidactylus fedtschenkoi* (photo by N. N. Szczerbak), 20 — *Tenuidactylus longipes* (photo by N. N. Szczerbak), 21 — *Tenuidactylus elongatus* (photo by J. Robert Macey).

22 — *Tenuidactylus kotschyi danilewskii* (photo by N. N. Szczerbak), 23 — *Tenuidactylus russowi* (photo by N. L. Orlov), 24 — *Tenuidactylus spinicauda* (photo by N. N. Szczerbak).

by the vertical distribution along the biotopes (the transitional Iran-Afghan Province). Only one species, *B. tuberculatus*, is endemic to the Mediterranean province but representatives of the genus *Tropiocolotes* are absent here. The Arabian zone of the desert Sahara-Arabian Province is characterized by the abundance of forms of the genus *Bunopus* (two species, three subspecies). This area was probably the center of formation of the genus. Only in the extreme south of the region (Hadramaut) do the generic ranges coincide (*B. spatalurus* and *T. scortecci*). Consequently, the genus *Bunopus* is, on the one hand, close to the genus *Tenuidactylus* (the species *B. crassicauda* is close to the *Cyrtopodion* group in the absence or weak development of keeling and scalation of the tail's dorsal side), and on the other, to the genus *Stenodactylus* (*B. spatalurus* is characterized by the increase in the number of body scales, their reduction in size and by the reduced definition of preanal pores) (Fig. 2).

{p. 21} The phylogenetic branch *Carinatogecko* splits from the main trunk, which produces the *Mediodactylus* group, somewhere near *T. heterocercus* [45]. The connection between *C. aspratilis* and *T. heterocercus* is expressed in the greater keeling of the body scales in comparison to other species of the subgenus.

The phylogeny of the genus *Teratoscincus* (Fig. 2) is characterized, in our view, by more or less the same development of the branches. However, within the genus, the tendency towards an increase in body scale size and its advance toward the back of the neck can be observed in *T. microlepis* and *T. scincus*, while the branches *T. bedriagai* and *T. przewalskii* are close in this way. According to zoogeographic data, the oldest center of the desert fauna in our region is Central Asian and a younger one is Middle Asian and its derivative Iran-Afghan mountain-steppe centers [142]. Consequently, the oldest branch is *T. przewalskii*, the youngest one is *T. microlepis* and *T. bedriagai* {*fide* Szczerbak – MG}. The most specialized psammophilous form and, consequently, the most progressive is the *T. scincus* branch.

The phylogeny of the genus *Crossobamon* is also simple (Fig. 2).

The phylogeny of the genus *Tenuidactylus* we see as the most complex (Fig. 5). It has been established that this branch comes from the genus *Cyrtodactylus* and the group of Tibet-Himalayan species is closest to the latter (it is not an accident that they are in neighboring territories). The positioning of species in this initial group is done by taking into account the successive acquisition of the most characteristic traits of the genus *Tenuidactylus*. First of all, this involves the development of a segmented tail, increase in the size of dorsal and caudal tubercles, and [increase] in limb length (*chitralensis* – *stoliczkai* –

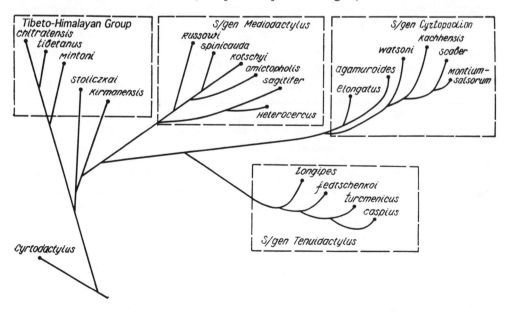

FIG. 5. Phylogenetic relationships among the species of the genus *Tenuidactylus*

kirmanensis). The species *T. fasciolatus*, *T. lawderanus* and *T. himalayanus* should, probably, be positioned on a branch with *T. chitralensis*, closer to *T. tibetanus*. The gap between *T. chitralensis* and *T. stoliczkai* is quite large, and it is conceivable that "intermediate links" may be found in the future. In this region, undoubtedly, there are as yet undescribed species. {This supposition has been confirmed. By 1995, eight new species from this region have been described [*Alsophylax boehmei* Szczerbak, 1991, *Cyrtodactylus battalensis* Khan, 1993, *Tenuidactylus medogensis* Zhao & Li, 1987, *T. indusoani* Khan, 1988, *T. rohtasfortai* Khan & Tasnim, 1990, *T. kohsulaimanai* Khan, 1991, *T. baturensis* Khan, 1992, and *T. fortmunzoi* Khan, 1993] and their generic and specific status are in need of review – MG}

The Mediterranean geckos are closest to the initial Tibet-Himalayan group (the territorial connection was made through the Iran-Afghan Province). They are close in the scalation of the tail, its slight bulbousness, the relatively small oval and oval-triangular dorsal tubercles. Within the subgenus, the "statistically average" *T. kotschyi* occupies the central position. The line *kotschyi – sagittifer – heterocercus – Carinatogecko* (*aspratilis – heteropholis*) is characterized by an increase in keeling with a subsequent reduction in body size. *T. russowi* is the only Middle Asian species that is found, as a rule, on the plain. It deviates from the other representatives of the subgenus in the number of chromosomes (44 instead of 42) [73]. *T. amicto-* {p. 22} *-pholis* and *T. spinicauda* are mountain species, which is a factor in the reduction in their overall size and in the homogeneity of dorsal scalation.

Representatives of the subgenus *Mesodactylus** have deviated further from the initial species than those of the *Mediodactylus* group. They have larger dorsal tubercles and the scales of the head are larger, but individual representatives (*T. agamuroides* and *T. elongatus*) bring the indicated subgenera closer. The last two species within the subgenus form a somewhat distinct group, which is close also to the genus *Agamura* in a number of characters. The following has been noted in both: thinner tail and elongated limbs, bulbous nasal scales, uniformity of tail scales, appearance on the lower [tail] surface of three scale whorls, increase in the size of scales with preanal pores. Increase in the size of the abdominal scales and dorsal tubercles is noted in the group *T. watsoni*, *T. kachensis*, *T. scaber*, and *T. montiumsalsorum*. The characteristic of this group's scalation on the supracaudal surface allows the supposition of the group's closeness to the genus *Bunopus*.

The nominative subgenus is, undoubtedly, closest to the subgenus *Mesodactylus** through the species *T. montiumsalsorum*. Their relatedness is evidenced by the appearance of embryonic femoral pores and by movement to the north. The central species of the subgenus *Tenuidactylus* is, probably, *T. fedtschenkoi*. The absence of specimen series for *T. montiumsalsorum* does not allow us to discover now which of species from the group under consideration is closest to it. Elongated limbs and some other characters (including the most southwestern range) allow the perception of similarity between *T. longipes* and representatives of the genus *Mesodactylus**.

Taking into account the facts that *T. caspius* is the most plastic species, has adapted to life on the plain, and has advanced furthest to the north and west from the initial forms, it should be considered the most advanced species of the group. If dorsal tubercles are compared (their size, keeling, extension towards the head), we get the series *caspius – turcmenicus – fedtschenkoi – longipes* in which, from left to right, dorsal tuberculation increases, the abdominal scales get larger, and the contact of the first pair of postmortal scales broadens. Doubtless, closeness exists also between *T. turcmenicus* and *T. caspius*, which is evidenced by many characters (until recently these species were not even distinguished). As a whole, a fairly compact group of species is observed within the subgenus. The most deviated among these is *T. longipes*.

And finally, there is the small but distinctive genus *Agamura*. Many characters {p. 23} reveal its closeness to the genus *Tenuidactylus*. In its subgenus (*Cyrtopodion*), there is a group of species (*T. agamuroides* and *T. elongatus*) that are carriers of "agamuroid" characters. Within the genus *Agamura* there is a noticeable strengthening of "agamuroid-

---

* {The authors had intended this to read *Cyrtopodion*; *Mesodactylus* is a junior subjective synonym of *Cyrtopodion* – MG.}

ness" from *A. persica* to *A. misonnei*; the nasal scales become more convex, the limbs elongate, the lower femoral scales enlarge, and the expression of pores is reduced. Within the genus, *A. persica* and *A. gastropholis* are, on the one hand, closely related and, on the other hand, *A. femoralis* and *A. misonnei* – are closer to one another (Fig. 2 {error in original – MG}). The following conclusions can be drawn as the result: 1. Our materials confirm A. Kluge's [237] deduction about the penetration of geckos from the south to the north and west in Asia (the more primitive initial group is *Cyrtodactylus* and the more progressive and specialized is the genus *Tenuidactylus*). 2. It has been established that the transition of geckos from one zoogeographic category into another is accompanied by taxonomic restructuring (in Indo-Malay area – genus *Cyrtodactylus*, in Palearctic – genus *Tenuidactylus*, in Iran-Afghan Province – genus *Bunopus*, in Turan and Mountain Asian Provinces – genus *Alsophylax*, in Sahara-Arabian Province – genera *Stenodactylus* and *Tropiocolotes*, in Turan Desert Province – genera *Crossobamon*, *Alsophylax*, and *Teratoscincus*, even though a number of them – *Stenodactylus*, *Tropiocolotes*, *Crossobamon*, and *Teratoscincus* – can be found in Iran-Afghan transitional province; in Middle Asian mountain region – *A. tadjikensis*, in Central Asian mountain region – *A. tokobajevi*, et al.).

Key to the Genera

1 (2) Separate eyelids present . . . . *Eublepharis* (Fam. EUBLEPHARIDAE)
2 (1) Separate eyelids absent, eyes covered with transparent membranes
. . . . (Fam. GEKKONIDAE)
3 (4) Digits distinctly broadened (except distal phalanges) . . . . *Hemidactylus*
4 (3) Digits become gradually thin from base to tip 5
5 (8) Digits covered below with numerous small scales 8
6 (7) Tail covered above with longitudinal row of large [finger]nail-like plates; abdominal scales large, flat, "scincoid" . . . . *Teratoscincus*
7 (6) Tail scales uniform, small; abdominal scales small, usually subconical
. . . . *Stenodactylus*
8 (5) Digits with one row of transverse subdigital lamellae 9
9 (16) Subdigital lamellae smooth 16
10 (13) Three (rarely, two) more-or-less congruent small convex nasal scales; tubercles usually present; also tubercles present along almost entire tail; digits usually angularly bent 13
11(12) Limbs very long and thin; adpressed forelimb reaches tip of snout at the wrist; length of foot with toes distinctly shorter than length of crus; as a rule, tail shorter than body, not fragile, becomes sharply thinner near its base . *Agamura*
12(11) Limbs usually of moderate length; adpressed forelimb reaches tip of snout or a little farther with tips of fingers; length of foot with toes almost equal to length of crus; as a rule, tail longer than body (if equal or shorter, it is thick, "fleshy"), fragile . . . . *Tenuidactylus*
13(10) Nasal scales flat, the first considerably larger than the other one or two, which can be absent altogether (if two to three more or less equal nasals present, then no dorsal tubercles); tubercles on tail almost always absent 14
14(15) No dorsal tubercles; females (and, often, males) lack preanal pores; upper labial scales curve up onto the upper surface of snout so that each one is not in a single plane 15
15(14) Dorsal tubercles almost always present; females usually with preanal pores (if dorsal tubercles absent, then both females and males with preanal pores; if females lack preanal pores, then dorsal tubercles distinctly expressed); supralabial scales are in the same plane and drop straight to the edge of mouth . . . . *Alsophylax*
16 (9) Subdigital lamellae spinous or keeled 17

17(20) No distinctly expressed dorsal tubercles (if present, they are barely distinguished from small dorsal scales and are irregularly scattered) 20

18(19) Dorsal tubercles large, flat; nostril in contact with two flat nasal scales, which do not differ in size from other scales of snout . . . . *Tropiocolotes*

19(18) Dorsal scales small, granulated; nostril in contact with three distinctly convex nasal scales, which are considerably larger than other scales of snout . . . . *Stenodactylus*

20(17) Dorsal tubercles expressed more or less distinctly and arranged into rows . . . 21

{p. 24} 21(22) Lateral digit scales form distinctly expressed fringes . . . . *Crossobamon*

22(21) Lateral digit scales spinous or keeled 23

23(24) Caudal tubercles in each semicircle of one segment do not contact each other; first nasal scale larger than others; height of the first supralabial scale from nostril to the edge of mouth distinctly shorter than its width along the edge of mouth . . . . *Carinatogecko*

24(23) Caudal tubercles in each semicircle of one segment are in contact with each other along the entire lateral edge; nasal scales congruent; height of the first supralabial scale from nostril to the edge of mouth distinctly greater (if smaller, then insignificantly) than its width along the edge of mouth . . . . *Bunopus*

{p. 25} SYSTEMATIC SECTION

Family EUBLEPHARIDAE

Vertebrae procoeleous. Unpaired parietal bone. Premaxillary bone formed by two centers of ossification. Supratemporal bone often present, angular bone rarely absent. Neck lacks calcified endolymphatic sacs. Real movable eyelids present. Eggs have a soft shell.

Two genera are known from Asia, two from Africa, and one from southern part of North and Central America.

Genus Fat-tailed Geckos — *Eublepharis* Gray, 1827*

Type species: *E. hardwickii* Gray.
1827 - *Eublepharis* Gray, Philos. Mag., London (2), vol. 2, p. 56.

DIAGNOSIS. Digits short, straight, cylindrical, without lateral fringe, with single row of transverse lamellae beneath. Body covered with small scales mixed with separate larger cone-like scales. Postmentals large. In majority of species axillary cavities transformed into deep skin pockets. Males with distinct preanal pores. Tail distinctly shorter {0.5-0.75 – eds.} than body and head (1.5 to 2 times {as stated in original – eds.}), middle part dilated, pointed end.

Eight species are known, six of them distributed in Southwest Asia and northern India, two on Hainan Island, Îles de Norway {also known as Xuy Nong Chao – MG} in Tonkin Gulf and the Ryukyus. One species in the U.S.S.R. fauna.

This group of Palearctic, Southwest Asian species differ from each other only in details of color and pattern, and small number of stable characters of scalation. All this makes possible its description by choosing characters that are general for the examined species, *E. macularius*, *E. angramainyu*, and *E. turcmenicus*.

GROUP OF SOUTHWESTERN EUBLEPHARIANS

DESCRIPTION. Large lizards with body length up to 160 mm; head massive, ovoid, slightly depressed, sharply delimited from neck; all body scales, except subdigital lamellae, smooth. Nostril opens at the rear of nasal scale and surrounded by wide rostral, first supralabial, large and flat supranasal and six to eight small nasal scales. Supranasal scale quadrangular with uneven upper edge. Supranasal scales separated by one to three large scales. Scales of snout large, flat, polygonal, heterogeneous, becoming distinctly smaller in preocular region, where roundish tubercles can be found; they become more and more convex nearer to neck. Several eyelash-like scales in rear part of upper eyelid.

Supralabials small, becoming gradually smaller toward the back; height of the first labial from nostril to the mouth edge is equal to its width along {p. 26} the mouth edge; ear opening large, oval. Chin scales pentagonal, its width along the mouth edge is twice the height; first pair of large postmentals usually contact each other or separated by one to two scales; one or two pairs of smaller postmentals can follow; several sharply enlarged gulars located below them; lower gular scales small, slightly convex. Infralabial scales wide, becoming gradually smaller toward the rear.

---

* {L. Grismer (1987, Acta Herpetol. Sinica: 6:455-456 ; 1988, Ph.D. Thesis, Stanford Univ., pp. 455-456; 1994, Zool. Sci. 11:319-335) referred four eublepharid species, *E. hardwickii*, *E. macularius*, *E. turcmenicus*, and *E. angramainyu* to the genus *Eublepharis*; two other species, *E. lichtenfelderi* and *E. kurowiae*, were placed in the resurrected genus *Goniurosaurus* Barbour. – MG.}

Dorsal scales small, polygonal-roundish, slightly larger on each side, where the border with abdominal scales not clearly defined; the length of a dorsal tubercle in two central longitudinal rows on the middle of the back smaller than the interspaces between these tubercles; size of the tubercles increases towards flanks, their diameter becomes equal to or greater than interspaces. Conical or spherical dorsal tubercles form about 20 transverse and about 40 longitudinal rows, not very distinct. Abdominal scales large, polygonal and roundish, slightly tiled patterned; row of preanal pores angularly bent, sometimes interrupted by nonperforated scales (more often in the middle of the row). Caudal segments covered above with rows of small rectangular roundish scales intermixed with whorls of conical or spherical tubercles (up to 10), which, as a rule, do not contact each other; the segments covered below with three to four rows of flat polygonal scales, which gradually increase in size from first to last row and from flanks to midbelly but do not reach size of real scales. Regenerated tail covered with rectangular and roundish scales arranged in more or less regular rows only on upper surface; tubercles absent.

Limbs of moderate length; fingers of adpressed forelimb reach to the middle of interspace between eye and nostril, and toes – to elbow or shoulder; forelimb covered with uniform, flat scales, smaller on lower surfaces; separate larger convex scales, analogous dorsal tubercles, occur on forearm; numerous distinct conical tubercles scattered on the upper surfaces of hindlimbs.

Key to the Southwest Asian Species of the Genus Eublepharis

1 (2) Subdigital lamellae smooth . . . . *E. angramainyu*
2 (1) Subdigital lamellae tuberculated 3
3 (4) Tubercles on subdigital lamellae feebly keeled; eight or fewer preanal pores
. . . . *E. turcmenicus*
4 (3) Tubercles on subdigital lamellae strongly keeled; eight or more preanal pores
. . . . *E. macularius*

SPOTTED FAT-TAILED GECKO — *EUBLEPHARIS MACULARIUS* (BLYTH, 1854)

Type locality: Salt Range, Punjab, northwestern India.
Karyotype: 2n = 38, all chromosomes acrocentric, NF = 38.
1854 - *Cyrtodactylus macularius* Blyth, Jour. Asiat. Soc. Bengal, vol. 23, p. 737.
1864 - *Eublepharis fasciolatus* Günther, Ann. Mag. nat. Hist., London (3), 14, p. 429 (Type locality: Sind, Hyderabad).
1871 - *Eublepharis macularius* John Anderson, Proc. Zool. Soc. London, p. 163.*

DIAGNOSIS. Subdigital lamellae distinctly tuberculated; eight to 11 preanal pores.
HOLOTYPE. Kept in the Indian Museum, Calcutta, India [245].
DEFINITION (Fig. 6) (from 5 specimens in ZIK). L adult 92.3-128 mm (according to literature, up to 150 mm); males larger than females; L/LCD (n = 3) 1.35-1.53; HHW (n = 5) 54-68 (61.6 ± 2.56); EED (n = 5) 0.78-1.0 (0.91 ± 0.04); supralabials (n = 10) 8-12 (9.6 ± 0.37); infralabials (n = 10) 8-10 (9.0 ± 0.21); scales across head (n = 5) 24-33 (28.0 ± 1.87); DTL (n = 5) 9.0-11.0 (9.80 ± 0.37); scales around {p. 27} dorsal tubercle (n = 5) 12-13 (12.8 ± 0.2); GVA (n = 4) 153-165 (158.5 ± 2.5); ventral scales across midbody (n = 4) 24-26 (24.75 ± 0.48); preanal pores (n = 3 {not in original – MG}) 8-11 (according to literature, 11-17) subdigital lamellae (n = 9) 14-21 (17.44 ± 0.69); additional nasal scales: five in one case, six not found, seven in one case, eight in three cases; supranasal scales separated by one (two cases), two (one case), or three (two cases) scales. Postmental scales in first pair contact each other (four cases) or separated by single scale (one case); in two cases these scales are separated from infralabials by a row of small scales; first pair of postmentals followed by one to two pairs of much less distinct scales; five to six supralabials reach front edge of orbit.

---

* {Add to synonymy: 1980 – *Cyrtodactylus madarensis* Sharma (see Das, 1992, Asiatic Herp. Res. 4:55-56.) – MG}

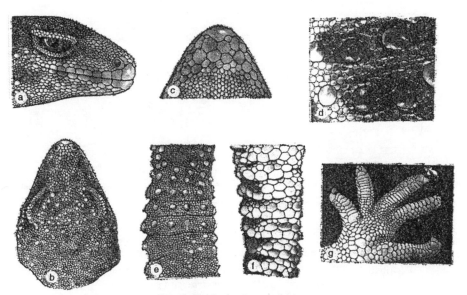

FIG. 6. *Eublepharis macularius*
a – head from side; b – head from above; c – chin; d – dorsum; e – tail from above;
f – tail from below; g – toes from below (ZIK SR 2763, captive born)

COLOR AND PATTERN. Main upper surface color lemon-yellow, brightest on the body. Combination of numerous dark brown spots and light thin interspaces creates a reticulate pattern on the head; these spots become larger on body, arranged in longitudinal rows, sometimes merging, more or less narrow stripe along midline lacks them; tail spots can merge creating transverse bands. Limbs also covered above with spots. Lower surfaces of head and tail with less distinct, washed out small spots; all other lower surfaces are white. Dark body spots in specimen from Tatta (Pakistan) arranged in dark longitudinal stripes such that three light strip interspaces are formed. Also, this specimen has three distinct light transverse bands on shoulder, midbody, and lumbar regions.

SEXUAL DIMORPHISM AND AGE VARIATION. Males larger than females. Upper surfaces in young geckos flesh-colored; black-brown "hood" covers upper part of head from nostrils to the mid-ear projection; in the occipital region of this hood there is a light bird-track-shaped pattern with three tips directed toward the neck; solid black-brown stripe between ear-openings on the neck; the area between this stripe and the hood is white; three wide distinct transverse brown bands on the body; four to five similar bands cover tail; {**p. 28**} longitudinal brown stripe on upper thigh; barely distinguishable stripe on upper arm. Abdominal surface pinkish without spots.

DISTRIBUTION (Fig. 7). From eastern Afghanistan southward throughout Pakistan to Baluchistan and eastward to western India; in mountains reaches 2500 m elevation. The discovery of two specimens of this genus (probably this species) by N. A. Zarudny [60] moves the western boundary of its range considerably.

LEGEND to Fig. 7.
Afghanistan: 1 - 2 {[more than 800 km (from Kopetdagh, Turkmenistan) to SE within Afghanistan territory] omitted in original – MG}[50]; 3 - near Gulbahar, 65 km N of Kabul [318]; 4 - Charikar, 35°05′N, 69°10′E [209]; 5 - Jalalabad, 34°26′N, 70°25′E [172]; 6 - 5 to 10 miles of Nimla, 34°19′ to 21′N, 70°10′ to 15′E [209]; 7 - Kandahar [318].
Pakistan: 1 - Zhob Dist.: Cherat [263]; 2 - Salt Range, Punjab (Terra typica) [258]; 3 - Chiniot [235]; 4 - Lahore [258]; 5 - Fort Sandeman [263]; 6 - 21 miles west of Hindubagh [263]; 7 - Ahmadpur Sial [235]; 8 - Quetta Dist.: 16 miles NE of Quetta [263]; 9 - Sibi Dist.: Bolan Pass near Mach [263]; 10 - Nushki [258]; 11 - Las Bela Dist.: 16 miles N of Bela [263]; 12 - Thar Parkar [258]; 13 - Hyderabad, Sind [225]; 14 - 3 miles of Hab Chowki [263]; 15 - Tatta [258]; 16 - Karachi Dist.: Hill Park [263];

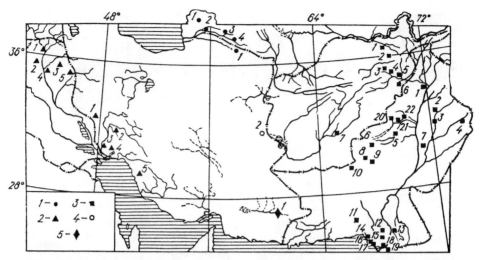

FIG. 7. Distribution of *Eublepharis turcmenicus* (1); *E. angramainyu* (2); *E. macularius* (3); *E.* sp. (Zarudny, 1904) (4); *Tenuidactylus sagitiffer* (5)

17 - Malir Cantonment [263]; 18 - Pir Pahto [258]; 19 - Batoro; Jati [258]; 20 - Wana [232]; 21 - Manzai [232]; 22 - Ladha [232].

GEOGRAPHIC VARIATION. Variability of pattern and some characters of scalation that are very important in species diagnoses, as well as a poorly studied distribution (especially in western and northwestern parts of the area), do not allow for final conclusion about intraspecific systematics of this species. Subgenera descriptions by Börner [197] are not persuasive and need to be revised on the basis of study of a series of specimens.

ECOLOGY. Even though *E. macularius* has been successfully bred by terrarists in recent years, data concerning its natural mode of life are practically absent. This is complicated by the very recent use of *E. macularius* as a name for several other Southwest Asian species. We had the opportunity to keep both *E. macularius* and *E. turcmenicus* in captivity and can state that they are very similar in behavior.

HABITAT. Inhabits rocky terrain. N. A. Zarudny [60 {error in original – MG}] found specimens, probably of this species, on the plain ". . . on clay-gravel soil covered by sand and abounding in bushes of *Zygophyllum*."

RESPONSE TO TEMPERATURE. In captivity, preferred soil temperature[1] 20-24°C [258].

{p. 29} DAILY ACTIVITY. Nocturnal. In nature, was found "at sunset" [60].

SEASONAL ACTIVITY. According to collection data [197], the earliest recorded find in western Pakistan was made on Feb. 15, 1976, the latest – on Dec. 19, 1960. In laboratory, does without hibernation..

SHEDDING. In captivity, shedding observed in March and April.

FEEDING. In terrarium, eats crickets, cockroaches, and even baby mice.

REPRODUCTION. The clutches of *E. macularius* in captivity were noted on Apr. 7 and 8, 1984. The egg dimensions are (n = 6) 24.4-30.5 × 14.5-15.5 mm, mass 2-2.1 g. Lays two eggs. Incubation period in our terrarium lasted 61-63 days (t = 26-29°C, humidity 75-100%). Two lizards hatched on 27 and 29 Sept. 1984. Their measurements were L 49 mm, LCD 31-32.5 mm. According to G. Emst's data [214], one female laid four clutches with four week intervals during one year and their incubation took 42 to 56 days. Eggs were laid in wet sand.

GROWTH RATE. According to our observations in terrarium, the increase in body length during the first two months (Sept., Oct.) was 10 mm (L 5.8-5.9 mm), during the next (Dec.) month another 10 mm (L 70-71 mm); then the rate increased, in the next month (Jan.) they added almost 20 mm (L 90-92 mm).

---

[1] Temperatures in this monograph are in degrees Celsius.

NATURAL ENEMIES. Can be eaten by foxes [277].

ASPECTS OF BEHAVIOR. Less timid than *E. turcmenicus*; can take food from tweezers; summer feeding pause absent. Males are aggressive towards other males; when fighting, can tear off tail or piece of skin from body and eat them. There is a case noted of mating between a male *E. macularius* and a female *E. turcmenicus*. Unfortunately, the clutch died.

IRANIAN FAT-TAILED GECKO — *EUBLEPHARIS ANGRAMAINYU*[1] ANDERSON & LEVITON, 1966

Type locality: between Masjid-i-Suleiman and Batwand, Khuzistan Prov., Iran.
Karyotype unknown.

1963 - *Eublepharis macularius*, S. C. Anderson, Proc. Calif. Acad. Sci., ser. 4, vol. 31, p. 435.
1966 - *Eublepharis angramainyu* S. C. Anderson, and A. Leviton, Occas. Papers Calif. Acad. Sci., no. 53, p. 1.

DIAGNOSIS. Subdigital lamellae smooth; 11-13 preanal pores arranged in one row.

HOLOTYPE [171] CAS 86384, adult male. Caught along the old road between Masjid-i-Suleiman and Batwand, Khuzestan Prov., Iran, coll. S. C. Anderson; May 20, 1958.

$L = 147$ mm; $LCD = 100$ mm. Supranasal scales separated by single, almost pentagonal supranasal scale, the width of which is more than its length; five to seven small additional nasal scales; 10 supra- and infralabial scales; ear large, its length 2.5 times greater than its width; pentagonal chin scale followed by four rows of enlarged scales; dorsal tubercles on flanks of body almost contact each other; 24 scales across abdomen; 13 preanal pores; 24 smooth subdigital lamellae.

LIFE COLOR AND PATTERN (in alcohol). Light continuous stripe from neck to the base of tail along spine; on both sides it is bordered by transverse marks as well as by dark punctuated stripes which can merge in longitudinal direction; black and white reticulate pattern on head; horseshoe-shaped dark or light band on the neck absent; numerous spots on the limbs; tail covered with numerous irregular dark transverse marks, wider than the light interspaces. Abdomen white, yellowish-brown.

{p. 30} In comparison with holotype, the paratypes have some differences: 11-13 feebly marked [preanal] pores in females; no. 86333 has an additional row of four pores behind the row of 13 pores. Three dark transverse dorsal bands in juveniles around rear part of the neck and shoulder, on midbody and on lumbar region. Size extremes: L males 142-154 mm; LCD 97-100 mm; L females 126-137 mm; LCD 86-90 mm; L juv. 90 mm, tail regenerated.

DISTRIBUTION. Western foothills of Zagros and north of Mesopotamian Plain in Iran and Iraq (Fig. 7). Observed between elevations of 300 and 900 m [200].

LEGEND to Fig. 7.
Iraq [269]: 1 - Nineveh Prov.: Ruins of Nineveh (about 300 m elevation) near Al Mawsil; 2 - Nineveh Prov.: Al-Hadhr; 3 - Kirkuk Prov.: Chalga village south of Chemchemal; 4 - Nineveh Prov.: Lesser Zab River bank near point of junction with Tigris River; 5 - Sulaymaniyah Prov.: Mera-di village and vicinity, along the highway between Darbandi-Khan and As-Sulaymaniyah.

Iran: 1 - Khuzestan: Dehloran: Ali-Khosh (32°41′N, 47°16′E) [303 {error in original – MG}]; 2 - Khuzestan Prov.: old road between Masjid-i-Suleiman and Batwand [168]; 3 - 5 - [50 {localities plotted in accordance with Fig. 3 on p. 207 – MG}].

ECOLOGY. Not studied. It is known that these lizards are nocturnal and inhabit stony hills and ruins of Nineviah on the Mesopotamian Plain. In Iran, egg laying occurs from end of May to the beginning of June [59]. In the Iranian Khuzestan, they can be found most often one hour before midnight; they catch crickets from among stones, and also eat scorpions, solpugids, large spiders and beetles, small geckos of the same and other species [277].

TURKMENIAN FAT-TAILED GECKO — *EUBLEPHARIS TURCMENICUS* DAREVSKY, 1978

Type locality: Turkmenia: Bakharden (Kopetdagh).
Karyotype: not studied.

---

[1] *angramainyu* - "Spirit of Darkness."

1890 [error in original – MG} - *Eublepharis macularius* Boulenger, Ann. Mag. Nat. Hist., [ser. 6], vol. 6, p. 352.
1978 - *Eublepharis turcmenicus* Darevsky, Tr. Zool. Inst., Leningrad, vol. 61, p. 204.

DIAGNOSIS. Subdigital lamellae slightly tuberculated; five to eight preanal pores.

HOLOTYPE [50]. ZIL 10103, adult male; near Bakharden (Kopetdagh), April 17, 1903, coll. I. V. Vasilyeva.

L = 125 mm; LCD = 80 mm. Supranasal scales separated by two scales; 10 supra- and infralabials; two large postmentals touch each other behind mental, followed by nine enlarged polygonal gular scales; each postmental separated from the first infralabial by small granules; 32 scales across abdomen; five preanal pores (the row is twice interrupted by several scales lacking pores). Tubercles on subdigital lamellae weakly developed.

DEFINITION (Fig. 8) (from 4 specimens in ZIK collection). L adult (n = 4) 111.9-129.3 mm; LCD 64.0 mm; males larger than females; L/LCD (n = 2) 1.94-2.61; HHW (n = 3) 0.59-0.68; EED (n = 3) 0.75-0.86; supralabials (n = 8) 9-11 (9.75 ± 0.25); infralabials (n = 5) 10-11 (10.6±0.24); scales across head (n = 3) 26-30; DTL (n = 4) 8.06-10.05 (9.16 ± 0.43); scales around dorsal tubercle (n = 4) 10-13 (11.75 ± 0.63); GVA (n = 4) 148-156 (152.25 ± 1.93); ventral scales across midbody (n = 4) 20-22 (21.25 ± 0.48); preanal pores (n = 2) 8; subdigital lamellae (n = 7) 20-23 (21.43 ± 0.37); seven (two cases), six (one case), or eight (one case) additional nasal scales; supranasals separated by one to three scales; first pair of postmentals form a long suture or only point of contact behind mental; these scales separated from infralabials by a row of scales and followed by one to two much less distinct scales; six supralabials reach orbit. Mass to 62 g.

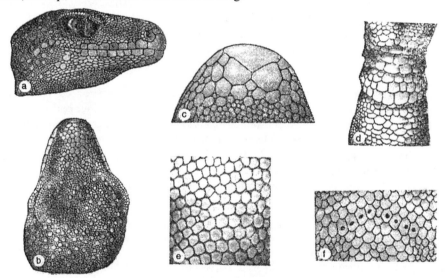

FIG. 8. *Eublepharis turcmenicus*
a, head from the side; b, head from above; c, chin; d, tail from below; e, belly;
f, preanal pores (ZIK Re 8/I; Kopetdagh, Chandyr Gorge).

LIFE COLOR AND PATTERN. Background color of body ocherous with lilac-brown spots and dots; {p. 31} on head these dots thickly arranged, almost without interspaces, bordering light longitudinal body stripe from neck to tail base; separate dots between them sometimes forming thin transverse lines; three wide light transverse bands on shoulder, midbody, and lumbar regions. Tail and limbs covered with dots and spots in disorder. Lower surfaces white, except regenerated part of tail, which is covered with a pattern of dark speckles.

SEXUAL DIMORPHISM AND AGE VARIATION. Maximum body length is 143 mm in males and 135 mm in females. Young specimens have the same pattern as young *E. macularius*.

DISTRIBUTION. Kopetdagh and Turkmen-Khorasan Mountains {the mountain crests between the Kopetdagh Mountains and Dasht-i-Kavir Desert are labeled the Turkmen-Khorasan Mountains in the Russian literature – MG} (Fig. 7).

LEGEND to Fig. 7.

Turkmenistan: 1 - W Turkmenistan: Kyurendagh ridge: Kemal [23] and Danata [145 {error in original – MG}] Wells; 2 - Karakala Dist.: Chandyr Gorge [145]; Damghan and Shikhindere Wells between Sumbar and Chandyr Rivers [109]; 3 - Arvaz River valley near Bakharden [23]; 4 - Chuly and Phiryuza [13].

Iran: 1 - Turkmen Khorasan Mountains N of Meshed [50]; 2 - Chakh-i-Ziru [60].

SYSTEMATICS AND GEOGRAPHIC VARIATION. Material studied does not show any geographic variation.

HABITAT AND QUANTITATIVE DATA. (Fig. 9). In Kopetdagh, this lizard occurs on stone foothills, occupying slopes of hills of crumbled schists and fragments of rocks covered with bushes of wormwood and ephedra, and sometimes with separate trees of *Paliurus spinachristi*. Based on observations of four individuals, it turned out that they prefer northern and north-western solar exposure, slope about 30°, height (from the foot of a hill) 100 to 120 m, altitude up to 812 m. As a rule, colonies of *Microtus afghanus* {MG}, or burrows of *Meriones persica* {MG} or *Ochotona rufescens* {MG} were found near the collecting sites of these lizards. Two individuals were captured 35-37 m from each other on the plot 2000 m^2. This species is rare and can be found sporadically. Only 12 specimens are known from the Soviet Union.

FIG. 9. Biotope of *Tenuidactylus spinicauda* and *Eublepharis turcmenicus* (a barren stone-gravel hillside in the vicinity of Danata Village, western Kopetdagh, Turkmenia)

RESPONSE TO TEMPERATURE. An active lizard was observed in the spring with temperature of 18°C; they preferred 30°C in terrarium.

{p. 32} DAILY ACTIVITY. Nocturnal. One specimen was caught near Chuly around 2400 hrs. (S. Shammakov, pers. commun.); in terrarium, leaves shelter after twilight.

SEASONAL ACTIVITY. Our earliest find of Apr. 22, 1977 does not clearly reflect the times of this species' appearance after hibernation. Probably this occurs at the beginning of April. It is possible that fat-tailed lizards disappear from the surface in summer (the last known find was made on July 22, 1973 [12]), because, in accordance with observations in captivity, they sharply reduce food intake. However, in terrarium they also did without hibernation.

SHEDDING. The male that was caught on May 12, 1984, shed on July 13, 1984. In captivity, another male shed five times during the year, with the longest interval from April to August. Shedding takes two days, then the skin is eaten.

FEEDING. In the feces from the lizard that was caught in May 1977 in Turkmenia, we found the remains of Heteroptera, Hymenoptera, a beetle, and a racerunner (probably, *Eremias strauchi*). In the stomach, the remains of a Myriapoda, a wood-louse, and Turkestan white ant were found. Perhaps they eat small rodents. Our specimens readily ate crickets, cockroaches, and larvae of mealworms {*Tenebrio molitor* – MG}.

REPRODUCTION. Sex ratio (n = 10) 1:1 in collections examined. This species lays two oval eggs in a soft leathery cover. According to a verbal report from Dr. Ch. Atayev, a female that was captured on July 16, 1972, laid eggs 38 × 19 and 40 × 19.5 mm (mass 7 and 7.5 g). In our terrarium, a female laid one egg 32 × 19.4 mm on July 20, 1984 and the second one on July 23, 1984 (30 × 19 mm). In the latter case, copulation occurred no later than the beginning of April (at that time the male was separated), and distinct signs of pregnancy were visible on May 20, 1984. In accordance with Dr. N. L. Orlov's information, the female, caught near Phiryuza in May, laid three, and then another two eggs during two weeks of July that year.

GROWTH RATE. Not studied. Known lifespan of an adult individual in captivity was as long as seven years (Dr. S. Shammakov's report). A young individual's {**p. 33**} dimensions at one year of age were: L = 71 mm, LCD = 32 mm, mass 5.1 g. (Dr. Ch. Atayev's report).

NATURAL ENEMIES. A cobra {*Naja oxiana* – MG} that was captured near Karakala, regurgitated one individual of this species.

ASPECTS OF BEHAVIOR. The Turkmenian fat-tailed gecko moves deliberately, on outstretched limbs and rarely runs. On stone screes, the lizard rocks to and fro before moving. It grabs its prey tightly, the strength perceptible even to a man, and tries to kill it with sturdily-built jaws (compressing them so that the eyes sink into orbits and the head becomes distinctly flattened). It revolves on its {longitudinal – MG} axis, as the glass-lizard {*Pseudopus apodus* – MG} does, if the prey resists. Drinks like a cat, drawing water with the wide tongue. Can dig burrows in soft soil (from observations in terrarium). In nature, probably uses rodent burrows because all lizards were captured under stones or slabs (up to 1m long) after the rains and heavy showers, which heavily moisten the soil. They hibernate in the same places. In nature and in the terrarium, they have a special place for a "toilet," where feces are collected. Such a "toilet" was found near Karakala under a slab of limestone near the lizard's shelter. *Eublepharis* is a very fearful, nervous animal, which handles captivity badly. It emits two sorts of sounds – bark-like or (when being handled, when agitated) thin and twanging, which can be reproduced as "y-a-zzz." Condition of the lizard's tail is an indicator of nourishment: it is wrinkled in a starved animal (such occur in nature, too) or smooth and plump in a well-nourished one. In nature, the animals are most often solitary. Two females fought after they were put into one terrarium and one of them broke the other's lower jaw. When copulating, the male grabbed the female's skin, starting at the base of the tail and moving up to the neck, and, while holding her in this position, approached the female's vent from one side with his hemipenis.

When blinded with light, a fat-tailed lizard does not stop but tries to hide.

PRACTICAL SIGNIFICANCE AND PROTECTION. In Kopetdagh, Turkmenian fat-tailed *Eublepharis* is found on the northern border of its range; probably the cold winters greatly reduce the numbers of the southern species. This peculiar lizard is of great scientific interest. To this day it has not been found in Turkmenian preserves (it is found near their boundaries). It is included in the *Red Book of the U.S.S.R.* It is advisable to work on breeding this species in captivity with subsequent release into the corresponding biotopes within the preserves.

FAMILY GEKKONIDAE

Vertebrae amphicoelous, having remnants of notochord. Parietal bone paired. Premaxillary bone formed from one center of ossification. Supratemporal arc and angular bone almost always absent. Calcified endolymphatic sacs present on neck. No movable eyelids (eyes covered with fused transparent eyelids). Eggs laid in rigid calcified shell.

The family's distribution encompasses the globe through the tropical, subtropical, and (partly) temperate zones between 50°N and 40°S (between 35°N and 48°S in the New World). Eighty-two genera are known.

Genus Skink Geckos — *Teratoscincus* Str., 1863

Type species: *Teratoscincus keyzerlingii* [*sic*] Str.
1863 - *Teratoscincus* Strauch, Bull. Acad. Sci. St. Petersbourg, vol. 4, p. 397.

DIAGNOSIS. Straight digits, not widened or compressed, covered below with numerous small spiny scales, and on each side – with fringe of long, flat scales; body covered with homogeneous, very large, as a rule, roundish rhomboid "scincoid" scales; head covered with {p. 34} small feebly convex scales largest on snout; last two thirds of upper surface of tail covered with a row of large fingernail-shape plates; small homogenous scales on lower tail surface; no preanal or femoral pores.

ADDITIONAL CHARACTERS. Head massive, body cylindrical, tail shorter than body. Pupil vertical with smooth edges. All body scales smooth; nasal scales slightly convex (three large and, often, one to two small subnasals); besides them, nostril contiguous to {p. 35} large quadrangular rostral scale; height of first supralabial scale from nostril to edge of mouth greater or equal to its width along the edge of mouth; ear opening large, oval; "supraorbital peak" well developed (Fig. 10, b). Mental scale large, quadrangular-trapezoid; one to three pairs of small postmental scales joining infralabials {error in original – MG} can be present on its sides; postmentals in the first pair never contact each other; gular scales small, slightly convex; dorsal tubercles absent; caudal segments not defined. Adpressed forelimb reaches snout with tips of fingers, hindlimb almost reaches axilla with toes; limbs covered with scincoid scales analogous to body scales, except for posterior surfaces of thigh, which is covered with small convex scales.

The genus includes four species generally distributed in Central, Middle and eastern part of Southwest Asia.

Key to the Species of the Genus Teratoscincus

1 (2) More than 80 scales across midbody . . . . *T. microlepis*
2 (1) Less than 50 scales across midbody 3
3 (4) Large dorsal scales end at neck level . . . . *T. scincus*
4 (3) Large dorsal scales end at shoulder level 5
5 (6) More than 40 scales across midbody . . . . *T. bedriagai*
6 (5) Less than 40 scales across midbody . . . . *T. przewalski*

COMMON SKINK GECKO — *TERATOSCINCUS SCINCUS* (SCHLEGEL, 1858)

Type locality: Ili River in "Turkestan."
Karyotype: 2n = 36, 2 metacentrics {error in original - MG}, 28 acrocentrics, 6 subtelocentrics, NF = 44.

1858 - *Stenodactylus scincus* Schlegel, Handl. Beoefen. Dierk., vol. 2, p. 16.
1863 - *Teratoscincus keyserlingii* Strauch, Bull. Acad. Sci. St. Petersbourg, vol. 4, p. 397. Terra typica: Seri Chah, Persia.
1885 - *Teratoscincus scincus*, Boulenger, Cat. Liz. Brit. Mus., vol. 1, p. 12.
1896 - *Teratoscincus zarudnyi* Nikolsky, Ann. Mus. Zool. Imp. Sci. St. Petersbourg, vol. 1, p. 370. Terra typica restricta: "Rume in Persia orientali."
1905 - *Teratoscincus roborowskii* Bedriaga, Ann. Mus. Zool. Imp. Sci. St. Petersbourg, vol. 10, p. {error in original – MG} 314. Terra typica: Oasis Sa-cheu.

DIAGNOSIS. Large dorsal scales end at neck level; 26-36 scales across midbody.
LECTOTYPE. Kept in the Leiden Museum [245 {error in original – MG}].
DEFINITION (Fig. 10) (from 216 specimens in ZIK, ZIL, MZSF). L adult males (n =

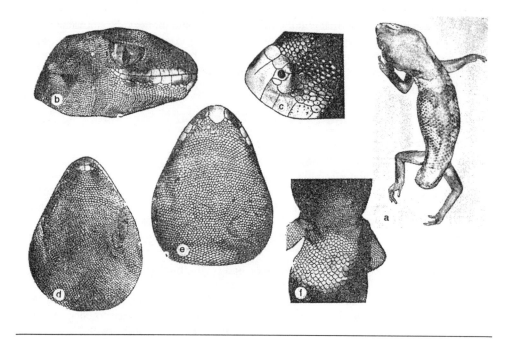

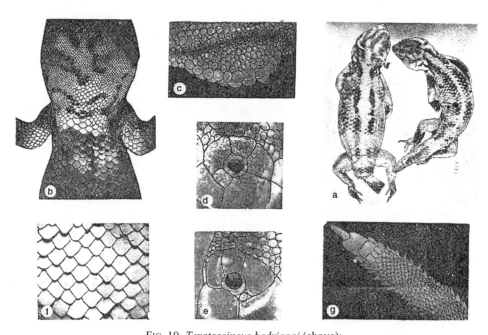

FIG. 10. *Teratoscincus bedriagai* (above):
a, general view from above; b, head from the side; c, snout from the side; d, head from above; e, head from below; f, dorsum (ZIL 9161; Khodji-i-du-Chagi, Iran).
Teratoscincus scincus (below):
a, general view from above; b, dorsum; c, upper eyelid; d-e, nasal scales; f, belly; g, toe from below (a, c, f: *T. keyzerlingii* [sic], MZSF 24152; 30°10′N, 60°50′E, Iran).

66) 50.9-92.8 (75.24 ± 1.34) mm; L adult females (n = 67) 50.6-116.2 (77.15 ± 1.57) mm; L/LCD (n = 122) 1.15-2.13 (1.49 ± 0.02); HHW (n = 30) 57-72 (65.27 ± 0.69); EED (n = 25) 60-110 (84.52 ± 2.60); supralabials (n = 416) 8-13 (10.57 ± 0.07); infralabials (n = 409) 8-12 (9.95 ± 0.05); scales across head (n = 154) 29-45 (36.07 ± 0.26); scales along head (n = 142) 50-80 (62.42 ± 0.47); GVA (n = 176) 84-113 (95.85 ± 0.44); scales across midbody

FIG. 11. Hatchling skink gecko and its egg shell (photo by A. D. Bautin).

(n = 111) 26-36 (30.62 ± 0.21); supracaudal plates (n = 140) 9-19 (15.36 ± 0.02); lateral scale fringes on 4th toe (n = 34) 20-30 (24.00 ± 0.35). One (85.4%) or two (8.3%) subnasal shields, 6% of specimens lack them, 0.3% have three of these scales (ratio of dimensions of subnasal scales to diameter of nostril is of importance for determining one of the subspecies); 1 (30.2%), 2 (62.7%) or 3 (7.1%) pairs of postmental scales. The generic description contains all other characters. Mass up to 30 g.

COLOR AND PATTERN variations are given in the descriptions of subspecies.*

SEXUAL DIMORPHISM AND AGE VARIATION. Females are not larger than males (no significant differences, $t = 0.95$). During reproduction cycle, male can be distinguished from female by swelling in postcloacal area. Juveniles have a more distinct pattern.

DISTRIBUTION. Middle Asia, eastern part of Southwest Asia (including eastern part of Saudi Arabian peninsula); a find is known from western part of Central Asia (Oasis Sa-cheu).

{p. 36} LEGEND to Fig. 12.

Turkmenia: 1 - Duzkir height [130]; 2 - Sarykamish (ZIK); 3 - ruins of fortress Izmykshir [ZIK {error in original - MG}]; 4 - Dikche and Yerburun Wells [130]; 5 - Kumsebshen [130]; 6 - W. Uzboi [102]; 7 - Uchtaghan Sands [130]; 8 - Doyakhatyn (ZIK); 9 - SW spurs of Chil-Mamed-Kum Desert (ZIL); 10 - 70 km W of Darvaza: Chaarly (ZIK); 11 - Darvaza [130]; near Serny Zavod; 40 km S of Serny Zavod; 30 km S of Darvaza (ZIK {omitted in original - MG}); 12 - Yaskha (ZIK); Kaplanly Well [130]; 13 - Uzun-Ada (Uzyn-Ada) [98]; 14 - Molla-Kara (ZIK); Djebel [80]; 15 - Akhcha-Kuyma [135]; 16 - Cheleken [130]; 17 - Kirpily Well [130]; 18 - Messerian Plain [130]; 19 - locality Kara-Bogaz, 40 km N of Kyzyl-Arvat (ZIL); 20 - 50 and 22 km NE of Kyzyl-Arvat; 35 km E of Kyzyl-Arvat (ZIK {omitted in original – MG}); 21 - 250 km N of Mary (ZIK); 22 - Yaradji Well [130]; 23 - Bugdaily village [130]; 24 - Farab (ZIL); 25 - Meshkhed (ZIK); Madau [133]; 26 - Lake Maloye Delily [130]; 27 near Bakharden (ZIK); Sinekly Well [130]; 28 - Porsykuyu Well; Karrykul [130]; 29 - Erbent Station [130]; Bakhardok (ZIK); 30 - near Chardjou (ZIK); 31 - N of Geok-Tepe (ZIK); 32 - near Ashkhabad; Annau Station; Lake Kurtlinskoye [130]; 33 - Repetek; 20 km N of Repetek; 37 km W of Repetek (ZIK); 34 - 100 km S of Chardjou [23]; 35 - Gyaurs [130]; 36 - Karabata [130]; 37 - Uch-Adji (ZIK); 38 - Kyzylcha-Baba [23]; 39 - Chekich village [133]; 40 - Kerkichy Station [23]; 41 - Dostluk settlement [133]; 42 - 20 km S of Yolotan settlement (ZIK); 43 - near Nychky (ZIK); 44 - Kelif [96]; 45 - Pul-i-Khatum [96]; 46 - near Tashkepry (ZIK); 47 - Kaghazly locality [130]; 48 - near Tahta-Bazar (ZIK); 49 - Manghyshlak: Ak-Tyube Sands [98].

Uzbekistan: 1 - Lake Sudochye [113]; 2 - SE of Nukus (ZMMGU); 3 - 150 km E of Nukus on road

---

* {See also Autumn K., and B. Han, 1989 – MG}

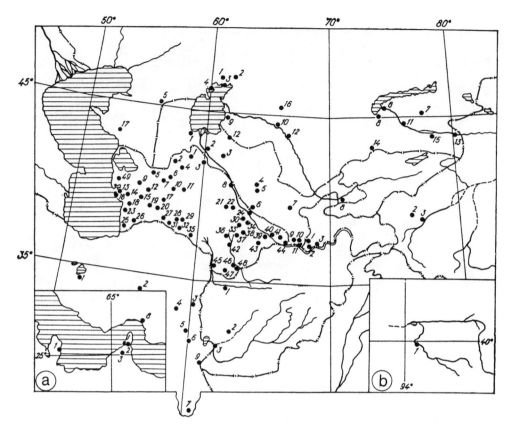

FIG. 12. Distribution of *T. scincus*.

to Uchkuduk: locality Balykbay-crus {error in original – MG} (ZIK); 4 - Kara-Katta (ZIK); 5 - Ayak-agytme Hollow (ZIK); 6 - Lower Zeravshan River: Alat and Karakul stations [22]; 7 - Samarkand [96]; 8 - near Kokand; Buvaidy Station [22]; 9 - 20 km of Shirabad on Kara-su River [22]; 10 - Jarkurghan Town (ZDKU); 11 - Termez [96]; 12 - 25 to 30 km N of Chabankazghan (Shabankazghan) (ZIK).

Tadjikistan: 1 - Shaartuz Dist.: Beshkent Valley (ZIK); 2 - {error in original – MG} - Chirchik settlement (ZIK); 3 - S border of Tigrovaya Balka Reserve (ZIK).

Kazakhstan: 1 - Malye Barsuky Sands near Koylibay (ZIL); 2 - 100 km NE Aralsk (ZIK); 3 - Aralsk antiplague Station (ZMMGU); 4 - Kara-tyup Peninsula [98]; 5 - Ustyurt: Sam Sands [84]; 6 - Ili River mouth (ZMMGU); 7 - between Ili and Karatal Rivers [98]; 8 - Lake Alakol (ZIK); 9 - Kuvan-Darja (Kuvan-Dzarma) [96]; 10 - Sol-Tobe (Solotobe) Station (ZIL); 11 - near Bakanas (ZIK); 12 - Bayghakum Station [96]; 13 - Dubunskaya Pereprava on Ili River (ZIK); 14 - Muyunkum: Akyr-Tyube Station (ZIL); 15 - Arasan Sands [98]; 16 - Aryskum Sands [98]; 17 - Asmatay-Matay [98].

China: 1 - Sa-cheu [300] (inset "b"); 2 - Yangihissar [189]; 3 - Yarkand [189].

Iran: 1 - Tehran Prov.: near Mirjaveh [216]; 2 - Chahardeh [194]; 3 - Ahangeran in Region of Zirkuh [88]; 4 - Rume [96]; 5 - Sarr Chah Hollow [59]; 6 - Hunik village [59]; 7 - Bampur River near Makhmudabad, 12 km SW of Iranshar settlement {error in original – MG} [256]; 8 - Minab [175](inset "a"); 9 - Ziya-i-Boloh and Ziya-i-Lahun, 30°0'N, 60°50'E (MZSF).

Qatar (inset "a"): 1 - Doha [175].

United Arab Emirates (inset "a"): 1 - Sharjah [175]; 2 - Ed-Dayd [175]; 3 - Jebel Fayah, 25°08'N, 55°50'E [175].

{p. 37} *Afghanistan*: 1 - Harirud River near Herat [{reference 193 in error in original; source unknown – MG}]; 2 - 48 km W of Dilaram, 32°15'N, 62°50'E [172 {error in original – MG}]; Herat and Kandagar Dist.: Fayah - Dilaram road [reference 172 in error in original; source unknown – MG}]; 3 - 16 km {omitted from original – MG} S of Qala-i-Kang, 31°05'N, 61°52'E [172].

SYSTEMATICS AND GEOGRAPHIC VARIATION. Within this widespread species one can recognize three forms. In contrast to other investigators [17, 112], we consider the use of color and pattern valuable for the differentiation of adults of the southern subspecies instead of the number of midbody scales and scales across the head, which seems to overlap with some other forms. When identifying the southwestern subspecies, the presence of a zone of intergradation between this and the nominate subspecies in southern Tadjikistan must be considered. The previously described *T. zarudnyi* [86] and *T. roborowskii* [182] do not have taxonomic significance.

Key to Subspecies of Teratoscincus scincus

1 (2) If subnasal scale present, then its diameter equal to or smaller than half of naris diameter . . . . *T. s. rustamowi*
2 (1) Diameter of one to two subnasal scales together equals more than half naris diameter 3
3 (4) Dorsum of adult specimens with transverse bands . . . . *T. s. scincus*
4 (3) Dorsum of adult specimens with longitudinal bands . . . . *T. s. keyserlingi*

COMMON SKINK GECKO — *TERATOSCINCUS SCINCUS SCINCUS* (SCHLEGEL, 1858)

Type locality: as in species.
1858 - *Stenodactylus scincus* Schlegel, Hand. Beoefen. Dierk., vol. 2, p. 16.
1905 - *Teratoscincus roborowskii* Bedriaga, Ann. Mus. zool. Acad. Sci. St. Petersbourg, vol. 10, p. 159.

DIAGNOSIS. One or more additional subnasal scales between first supralabial and naris; their diameter always more than half of naris diameter; dorsum in adults with transverse bands.

ADDITIONAL CHARACTERS (from 122 specimens in ZIK and ZIL). L adult males (n = 58) 52.4-92.8 (76.72 ± 1.49) mm; L adult females (n = 54) 50.6-93.1 (75.12 ± 1.48); GVA (n = 122) 81-113 (97.38 ± 0.51); scales around midbody (n = 80) 26-33 (29.71 ± 0.16); scales across head (n = 106) 29-45 (35.94 ± 0.34). The same type as in species.

LIFE COLOR AND PATTERN. Ground color grayish lemon-yellow. Head with gray-brown spots (sometimes with violet tint) and stripes forming irregular spotted pattern that encroaches on the throat. Seven to eight dark transverse bands from neck to base of tail; on both flanks, a thin dark stripe extends from shoulders to base of hindlimbs; every body scale, especially in zones of spots and stripes, bordered with thin white piping on its free (distal) end; some of these scales dark in their centers but colored with lilac-reddish or ochre colors on their periphery; tail sooty black above; each fingernail-shaped plate has thin white piping along free edge; no pattern on the limbs. Throat white with slight lemon shade; abdomen white; tail dark below; inner surfaces of pads pinkish.

DISTRIBUTION. Northern part of species area from Caspian Sea to Semirechye {southeastern Kazakhstan to the south and southeast of Lake Balkhash – MG}; southern border passes through eastern spurs of Turkmen-Khorasan Mountains {"Turkmen-Khorasan Mountains" is used in Russian geographical, geological, and biological literature to indicate the mountain region to the south of Kopetdagh and to the north of Dasht-i-Kavir – MG} and western spurs of Paropamisus. Does not penetrate Fergan Valley; forms a zone of intergradation with Rustamow's skink gecko in southern Tadjikistan. Subspecies determination of the geckos from Central Asia (Oasis Sa-cheu) needs clarification.

{p. 38} KEYSERLING'S SKINK GECKO — *TERATOSCINCUS SCINCUS KEYZERLINGII* [sic] STR.

Type locality: Seri-Chah, E Iran.
1863 - *Teratoscincus keyzerlingii* [sic] Strauch, Bull. Acad. Sci. St. Petersbourg, vol. 4, p. 397. Terra typica: Seri Tschah, Persia.
1896 - *Teratoscincus zarudnyi* Nikolsky, Ann. Zool. Mus., vol. 1, p. 60. Terra typica: "Rume in Persia orientali."
1949 - *Teratoscincus scincus keyzerlingii* [sic], Terentjev, Chernov, Opredelitel presmykayuschihsya i zemnovodnih, M[oscow]., Sov. Nauka, p. 128.

DIAGNOSIS. One or more additional subnasal scales present between first supralabial and naris; their diameter always larger than half of naris diameter; dorsum in adult specimens with transverse bands (Fig. 10).

ADDITIONAL CHARACTERS (from eight females in ZIL and MZSF). L adult females (n = 5) 62.3-116.2 (98.14 ± 9.43) mm; GVA (n = 6) 100-118 (108.83 ± 2.43); scales around midbody (n = 4) 31-35 (32.5 ± 0.96); scales across head (n = 5) 39-46 (43.20 ± 1.24). The other characters are contained in the species description.

LECTOTYPE. ZIL 2396. Adult female. "Seri Tschah, Keyserling, 1862"; L = 104.1 mm; tail broken; 42 scales across head; 13/12 supralabials; 13/12 infralabials; one pair of postmentals; GVA 118; 35 scales around midbody; 18 supracaudal plates.

COLOR AND PATTERN [263]. Color pattern of dorsum varies between yellow, orange and different shades of brown, alternating with light-gray blotches. Two wide longitudinal dark brown stripes usually present on body but do not extend onto tail; tail lacks distinct pattern. Flanks and abdomen light pink to white. Juveniles dark yellow to light orange; four to five sooty black transverse bands present on both body and tail.

DISTRIBUTION. Species range to south of Paropamisus: central and southwestern Iran, western Afghanistan, northwest Pakistan, eastern Arabian Peninsula.

RUSTAMOV'S SKINK GECKO — *TERATOSCINCUS SCINCUS RUSTAMOWI* SZCZERBAK

Type locality: sands near Kokand and Kayrakkum in Pherghan Valley.

1979 - *Teratoscincus scincus rustamowi* Szczerbak, in: Protection of nature in Turkmenistan, part 5, Ashkhabad, p. 129.

DIAGNOSIS. No additional subnasal scale between supralabial scale and naris; if present, then its diameter less than or equal to half of naris diameter; dorsum of adult specimens with transverse bands.

ADDITIONAL CHARACTERS. L adult males (n = 13) 61.7-77.5 (68.64 ± 1.50) mm; L adult females (n = 15) 61.9-81.8 (72.93 ± 1.94) mm; GVA (n = 37) 84-97 (91.0 ± 0.60); scales around midbody (n = 37) 28-33 (30.60 ± 0.35); scales across head (n = 37) 30-40 (35.40 ± 0.60). The other characters are contained in the species description. Color coincides with color of nominate subspecies.

HOLOTYPE. ZIK Re no. 9, near Kokand, May 8, 1976, adult male. L = 84.0 mm; LCD = 43.0 mm; 38 scales across head; 11/10 supralabials; 11 infralabials; three pairs of postmentals; 88 GVA; 31 scales around midbody; 15 supracaudal plates; no subnasal on the left side.

DISTRIBUTION. Pherghan Valley; forms a zone of intergradation with nominate subspecies in southern Tadjikistan.

HABITAT AND QUANTITATIVE DATA. Data on the skink gecko's places of habitation are found in many works [22, 23, 98, 107, 125, 130, 133, *et al.*] (Figs. 11, 12). There is a unanimous opinion among all the zoologists that this is a {p. 39} typical psammophilous species. As we were able to establish, it inhabits various places in sand deserts but in a number of regions is found on hard clay and saline takyr-like soils. In different locations, varying numbers are also observed.

In the central areas of the range (Karakum), the skink gecko prefers semi-packed and loose sands and is found here more frequently than elsewhere, up to 40 individuals per 1 km. In the northern (Kyzylkum) and eastern (southwestern Tadjikistan) parts of the range, it lives on hard and saline soils reaching, at times, densities of up to 20 individuals per 1 km of route.

RESPONSE TO TEMPERATURE. From the observations in Turkmenia, the lowest soil temperature (14°) at which active skink geckos were found we recorded on 29 April 1979 near Chairly village to the west of Darvaz. Usually, they are found on the surface when the soil temperature is 19-27° and the air temperature is 15.5-30°.*

As A. D. Bautin reported, skink geckos laid their eggs in a terrarium in a spot where

---

* {See also Cherlin, *et al.*, 1983, Ekol. Sverdl. 14(3):84-87. – MG}

FIG. 13. Biotope of *Crossobamon eversmanni, Teratoscincus scincus* and *Tenuidactylus russowi* (semi-stabilized dunes near Lake Yaskha, Turkmenia SSR)

the sand temperature was 34-36°. In the observations in captivity, incubation occurred at a temperature of 28-30° [117].

DAILY ACTIVITY CYCLE. Leaves cover, as a rule, after dark, around 2100 hrs. The greatest numbers of lizards were observed in the period from 2300 hrs to 0100 hrs after which their numbers on the surface decreased, although some individuals were active until dawn (0500 hrs). In the fall, in October, they were observed starting at 1940 hrs [133].

SEASONAL ACTIVITY CYCLE. The skink gecko in Turkmenia emerges from cover early — [it] was encountered on 15 March 1966 to the north of Bezmein [133]. The mass emergence of these lizards is observed in April. In Tadjikistan, the geckos of this species awaken from the winter hibernation at the end of March (18, 20, and 31 March 1959) [107]. In Kazakhstan, they appear in Kyzylkum in the second half of March, in Muyunkum at the beginning of April, in southern Prebalkhashye {Prebalkhashye is a large desert region to the west and southwest of Lake Balkhash – MG} in the first half of April [98].

{p. 40} In Turkmenia, [it] evidently disappears for the winter at the end of October or beginning of November (our last collections are dated 17 October 1982, Repetek Station). In Afghanistan, active geckos have been captured in the middle of November [172]. In Iran (Khorasan), N. A. Zarudny captured them on 12 November 1900 (old style) [96]. In southern Prebalkhashye, skink geckos disappear for the winter in the early days of September [98].

SHEDDING. Skink geckos shed no fewer than three times a season. Shedding was observed in mid-April, the end of May and early June, and in the second half of August [23, 96, 128]. Because geckos shed in their burrows, it is difficult to observe in nature. In captivity, shedding lasts 5-6 days. The animal removes and eats the shed skin.

FEEDING has been studied sufficiently well [22, 23, 98, 107, 130]. In Karakum, the skink gecko feeds primarily on beetles [130].* In the spring (n = 50) the rate of encountering beetles is 86.0%. In the summer (n = 87), it is 82.7%, and in the fall (n = 13), 46.1%. Among the Coleoptera, darkling beetles and June beetles predominate and are consumed by the gecko in especially large numbers in April and May. Hymenopterans, primarily ants, have a prominent place in the diet of geckos. Geckos can occasionally feed on other invertebrate species (cockroaches, termites, grasshoppers, isopods, solpugids). The contents of the diet for the lizards captured in Uzbekistan and Tadjikistan were not significantly different.

---

* {An excellent analysis of the trophic connections of this species in Karakums was recently published by Zh. Mishagina (1992) – MG}

Grasshoppers predominated in the stomachs of the geckos collected in Muyunkum [98]. In captivity, skink geckos reproduce normally when fed with mealworm larvae, crickets, and cockroaches. However, once or twice a week calcium glycerophosphate was given with the food and once a month "Tetravite" was added at the rate of one drop per adult lizard [117].

REPRODUCTION. The sex ratio in the samples studied is close to 1:1. Sexual maturity is reached when the body size is 68-70 mm and greater. Mating occurs in April. As was experimentally established by V. E. Frolov, one mating is enough for the female to lay two fertile clutches. One to two eggs are laid in June or July and, less frequently, in the beginning of August [22, 125]. A. D. Bautin got a clutch from females, caught in southern Tadjikistan, on 20 July 1976. In the laboratory, his geckos laid eggs in January, April, June, and August. A female in captivity [126] laid eggs on 16-22 March and 22 January; according to V. E. Frolov's data (in press) {Frolov, 1987 – MG}, two clutches were noted in the period from 20 March to 26 June with an interval of approximately three months. The parents came from Turkmenia. The eggs are white, almost spherical, with a shallow dent. Their dimensions are 17-21 × 16-19 mm with the mass of 2340-3050 mg [116]. According to A. D. Bautin's data, the females caught in southern Tadjikistan (n = 12) laid eggs with the average dimensions of 18.7 × 17.1 mm. The incubation lasts 60 days (unknown conditions) [13], 72-93 days with the temperature of 28-30° [116], and only 54 days with the temperature of 34° (A. D. Bautin's report). Hatchlings of the year were observed in Turkmenia 13 September through 18 October [133] and 15-18 August 1969 in Kazakhstan [98]

GROWTH RATE. The newly hatched geckos have the following dimensions: L = 37-38 mm, LCD = 22-25 mm, and the mass is 1770-1890 mg [116]. In captivity, they produced offspring in one year. After hatching in July of 1976, a young female reached sexual maturity (without hibernating) in six months, in January 1977, and laid eggs (A. D. Bautin's observations). These data show that geckos of the first generation can reproduce already at the end of the next season, but more often this probably happens at the age of two years. The maximum size is reached no earlier than at the age of five years.

ENEMIES. The skink gecko is eaten by *Diplomesodon pulchellum*, *Crocidura suaveolens*, saxaul jay *Podoces panderi*, *Eryx miliaris* and *E. tataricus*, *Coluber ravergieri*, *Crossobamon eversmanni* {scientific names substituted – MG} [22, 23, 130, 133, 107, 98]. On 28 April 1979 we found a dead gecko in the nest with {*Athene noctua* – MG} fledglings near the Kyzyl-Gaty Well (185 km to the north of Bakharden).

BEHAVIORAL ATTRIBUTES. The prevailing opinion that the skink gecko is a clumsy animal is incorrect. When illuminated with a bright light, the lizard flattens itself against the ground and stands still or slow- {p. 41} -ly tries to move away. We saw how quickly it can move during observations on moonlit nights. After leaving its burrow for awhile (15-20 min.), the gecko does not move away from cover and, when disturbed, immediately retreats to the burrow. In case of danger and in the absence of a burrow, it takes cover in bushes. The skink gecko moves on extended limbs without its body touching the substrate. The tail is usually slightly lifted. When hunting, it carefully approaches an insect, rises up some on its limbs and with a snapping motion grabs its prey from above, chews and swallows it.

Burrows are summer and winter shelter for this species. The gecko finds a small indentation, usually at the bottom of a sand mound or under a bush, and digs into it with the forefeet, throwing the soil under the belly and to the sides and pushing it further with the hindfeet. The ceiling is packed down with the head and the entrance is plugged with sand (A. D. Bautin's observations). Burrows can be 20-150 cm long, 10-60 cm deep, and 4-5 cm in diameter. The most complete data on skink gecko burrows come from K. P. Paraskiv [98]. As the author observed, in soft soils the geckos dig their own burrows but can occupy the burrows of other animals (large carabid [*Scarites* sp., dung beetle, *Dipus sagitta*, and *Allactaga elater* {MG}, *Meriones meridianus* {MG}). The gecko barely enters the passage of the host's burrow and from there digs its own burrow in the layer of moist sand, where it then takes cover. The gecko closes off the main passage with a sand or loess plug. Sometimes the plug converts the entrance into the chamber (nest) and has a passage in front of it.

The skink gecko has a voice. Its squeak resembles the sound of a metal plate struck on one edge. It has good hearing.

The geckos lead a solitary life. Very rarely two females can be found in one system of passages. Males and juvenile geckos live one to a burrow. At night, females and juveniles do not stray far from the permanent burrow. In contrast, males move a significant distance (up to 300 m) from the burrow and frequently abandon the burrow where they had spent the day. Every animal has a territory, where it hunts insects without incursions into another individual's area. The exceptions are the males, which move around during the mating period and, probably, the young in search of new territories. According to V. E. Frolov's data, in captivity males were observed fighting over cover only once. Pregnant females often wound the approaching males. Thus, the appearance of injured males is evidence of pregnancy in females and requires their separation in order to avoid further trauma. The skink gecko, unlike all the other gekkonid lizards of our fauna, have not only a very fragile tail but also very easily damaged scales, which regenerate.

When mating, the male presses the female to the ground with the snout, then curves its body, brings a leg under the female, and copulates (A. D. Bautin's observations in captivity). This is how he describes the egg laying process. The female becomes agitated 30-40 minutes prior to laying, begins to search for a good site, finds an indentation in the soil, deepens it, places the hindlimbs near the pit, and lowers the entire rear end of the body into it. One soft egg is laid, after which the female frequently begins to roll it with the hind feet. In 5-10 minutes the egg hardens and the female either buries it or leaves it on the surface. There is a known case of collective laying (5 eggs) in one burrow [22]. After hatching, the newborn gecko is active, begins making a burrow and takes food.

Besides the usual voice, the skink gecko is capable of producing a characteristic crackling sound, which is generated by rapid motion of the tail from the friction of fingernail-like scales on its surface. It is known [98] that the crackling emitted by the thrashing, automized tail is interpreted as a danger signal by other individuals.

PRACTICAL SIGNIFICANCE AND PROTECTION. Undoubtedly, the skink gecko is a useful animal, an important species ubiquitous in the sand deserts of Middle Asia and Kazakhstan, and it should be protected from senseless destruction. The extensive range and fairly large numbers of this species make any special protection measures {p. 42} unnecessary. It is protected in many preserves (Repetek in Turkmenia, Kyzylkum in Uzbekistan, Tigrovaya Balka in Tadjikistan, and elsewhere).

PRZEWALSKI'S SKINK GECKO — *TERATOSCINCUS PRZEWALSKII* STRAUCH, 1887

Type locality: Oasis Hami (western China).

Karyotype: not studied. {Gang and Zhang, 1992:34, report 2n = 36, 2 subtelocentrics, 34 telocentrics – MG}

1887 - *Teratoscincus przewalskii* Strauch, Mem. Acad. Sci. St. Petersbourg, ser. 7, vol. 35, no. 2, p. 71.

DIAGNOSIS. Large dorsal scales end at shoulder level; 31 to 38 scales across midbody.

LECTOTYPE. ZIL 6564, adult female. "Hami, 1879, N. M. Przewalsky." L = 89.7 mm; tail broken; 42 scales across head; 11/11 supralabials; 9/10 infralabials; one pair of postmental shields; 108 GVA; 35 scales across midbody; 14 supracaudal plates.

DESCRIPTION (Fig. 14) (from 66 specimens in ZIL collection). L adult males (n = 29) 51.7-93.4 (72.8 ± 2.4) mm; L adult females (n = 34) 52.6-93.0 (76.54 ± 2.17) mm; L/LCD (n = 48) 1.06-2.12 (1.58 ± 0.035); HHW (n = 31) 57-71 (61.97 ± 0.61); EED (n = 93) 72-92 (82.79 ± 1.12); supralabials (n = 132) 9-13 (10.42 ± 0.07); infralabials (n = 132) 7-12 (9.74 ± 0.09); scales across head (n = 56) 38-50 (42.71 ± 0.35); scales along head (n = 45) 97-112 (104.24 ± 0.61); GVA (n = 22) 102-115 (108.82 ± 0.73); scales around midbody (n = 56) 31-38 (34.71 ± 0.20) (37-39, according to [300]); supracaudal plates (n = 62) 8-16 (13.0 ± 0.22); lateral crests on 4th toe (n = 59) 19-25 (21.97 ± 0.18). Lower nasal scale always single, its diameter more than or equal to half of nostril diameter (95.5%); one (1.5%), two

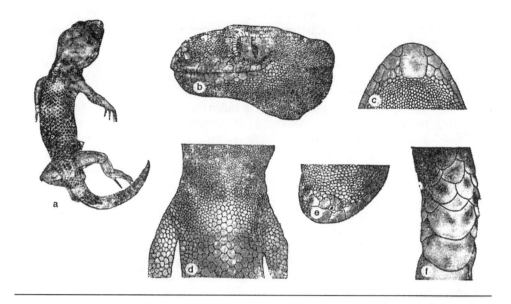

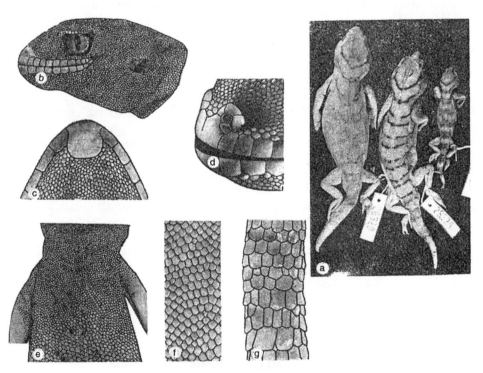

FIG. 14. *Teratoscincus przewalskii* (above):
a, general view from above; b, head from the side; c, chin; d, dorsum; e, snout from above; f, tail from above (ZIL 18272; southern Gobi, Mongolia)
Teratoscincus microlepis (below):
a, general view from above; b, head from the side; c, chin; d, nasal scales; e, dorsum; f, belly; g, tail from below (a – ZFMK 26328-330, b-g – ZFMK 26329; Tank-i-Graway, Baluchistan {error in original – MG})

(59.1%) or three (39.4%) pairs of postmental scales. All other characters are included in the description of the genus.

LIFE COLOR AND PATTERN. Main color among adults is gray-ochre, and yellowish-ochre for juveniles; lighter on front part of body. Wide transverse brown stripes on body (barely visible in adults and distinct, almost black in juveniles): five to {p. 43} six on dorsum, five on tail; they are faded on flanks. The transverse stripe on the shoulder is more distinct; arch-shaped dark brown stripe on the neck. Several dark spots on occiput, the same spots are on labial scales. Thighs and crus brownish-ochre with indistinct spotted pattern. Lower surfaces whitish or yellowish, except for brownish tail.

SEXUAL DIMORPHISM AND AGE VARIATION. Females are no larger than males ($t = 0.38$). Young specimens have very distinct pattern: broad dark brown bands with narrow yellowish interspaces.

DISTRIBUTION. Central and western parts of Gobi Desert. Western border runs along the line Djungar Gobi – Oases Hami and Sa-cheu, and eastern border through western parts of East-Gobi Aimags; from south and north, the distribution is limited by Mongolian Altai and Nan-Shan (Fig. 15 {error in original – MG}).

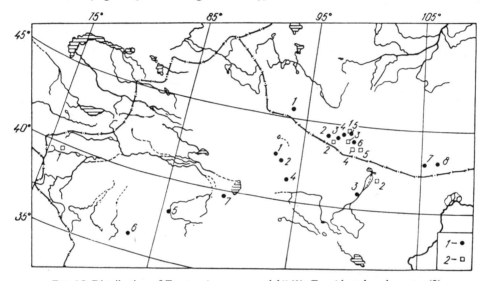

FIG. 15. Distribution of *Teratoscincus przewalskii* (1), *Tenuidactylus elongatus* (2)

LEGEND to Fig. 15.
Mongolia [78]: 1 - Altai-Somon; 2 - Sharahuls; 3 - Ehingol; 4 - Nogontsav; 5 - Ingheni-Hubrih-Bulak; 6 - Dzuldanay; 7 - Boruzhon-Gobi; 8 - Galbin-Gobi.
China: 1 - Oasis Hami [300]; 2 - Oasis Kufy [300]; 3 - Edzin-Gol River over Naryn; Tsagan-Tohoy; Beily [96]; 4 - from Sa-chou to Bugas (ZIL); 5 - Cherchen [96]; 6 - Nija [300]; 7 - Charlyk (Chaharlyk) [300].

SYSTEMATICS AND GEOGRAPHIC VARIATION. Within the collections studied no variation is found.

ECOLOGY largely unstudied. According to the observations in the Trans-Altai Gobi [31], it usually stays in sandy areas (here there are 15-20 individuals per hectare) but also occurs on the rocky plain – hammada, where 1-2 individuals per hectare can be found. Przhevalsky skink geckos dig their own burrows, although they will use rodent burrows as well. When pursued, they can climb onto bushes 80 cm in height. They possess stable individual territories. They feed mainly on darkling beetles, June beetles, [and] click beetles, and also large arachnids, such as scorpions and solpugids. In terraria they will attack small lizards.

They lay two white spherical eggs (16-17 × 15-16 mm) in the sand at the base of saxaul bushes {*Haloxylon sp.* – MG} at a depth of 30 cm. On 26 August 1982, the embryos were already formed (N. L Orlov's verbal report). Among the likely enemies is the long-eared hedgehog {*Erinaceus auritus* – MG}. The thrashing tail, discarded by the lizard, makes a loud rustling noise.

This is a common species, widely found in the deserts of Central Asia and protected in the Gobi Biosphere Preserve (PRM) {People's Republic of Mongolia – MG}.

SMALL-SCALED SKINK GECKO — *TERATOSCINCUS MICROLEPIS* NIKOLSKY, 1889

Type locality: Duz-ab in E. Kerman, Iran.
Karyotype: not studied.
1899 - *Teratoscincus microlepis* Nikolsky, Ann. Zool. Mus., no. 2, p. 145.

DIAGNOSIS. Dorsal scales small, gradually turn into neck scales; about 100 scales around midbody.

HOLOTYPE. ZIL 9164, adult male. "Duz-ab in Eastern Kerman, June 15, 1898, N. Zarudny." L = 53.2 mm; LCD = 34.0 mm; L/LCD = 1.56; 11 supra- and infralabial shields; no postmentals; 50 scales across head [88]; neck scales turn into dorsal ones gradually; 160 GVA; about 100 scales around midbody; about 10 supracaudal plates.

DEFINITION (Fig. 14) (from seven specimens in ZIL and ZFMK collections). L general (n = 6) 53.2-73.0 (64.42 ± 3.07); L/LCD (n = 5) 1.56-1.95 (1.78 ± 0.07); HHW (n = 5) 63-66 (64.0 ± 1.0); EED (n = 4) 53-89 (71.0 ± 9.90); supralabials (n = 14) 10-12 (11.07 ± 0.13); infralabials (n = 14) 9-11 (9.78 ± 0.24); scales across head (n = 7) 48-58 (50.57 ± 1.29); scales along head (n = 3) 135-154; GVA (n = 7) 150-188 (169.57 ± 4.67); scales around midbody (n = 7) 85-110 (96.14 ± 3.39); supracaudal plates (n = 6) {**p. 44**} 9-11 (10.0 ± 0.26); lateral crests on 4th toe (n = 7) 22-26 (24.29 + 0.52); one subnasal scale which is equal to nostril diameter, present (four cases) or absent (three cases); postmentals not distinct; 11 to 13 scales in contact with mental shield. All other characters are contained in description of the genus.

COLOR AND PATTERN [263]. Main color yellow to light brown; around six dark transverse bands, they can be ∧-sha- {**p. 45**} -ped, dark brown in adults and reddish in juveniles; five to six dark transverse bands on tail.

DISTRIBUTION. Kharan and Bempur Depressions, Dashte-Lut in southeastern Iran and West Pakistan; probably can be found in southern Afghanistan (Fig. 16 {error in original – MG}).

LEGEND to Fig. 16 {error in original – MG}.
Iran: 1 - Dashte-Lut near Kerman [263]; 2 - Duz-ab (Dobaz - Zagedan) (ZIL) [88]; 3 - Bempur: Rud-i-Bempur River (ZIL).
Pakistan: 1- Chagai Dist.: 3 km W Nushki; 14 to 17 km W Nushki (AMNH); 2 - S of Dalbandin (28°53′N, 64°26′E) [258]; 3 - Tank-i-Grawag (27°13′N, 63°25′E) [258]

GEOGRAPHIC VARIATION. None found.

ECOLOGY not studied {but see Anderson, 1993 – MG}. N. A. Zarudny [60] had reported that the specimen he obtained "stayed on highly saline saturated soils covered with a salt

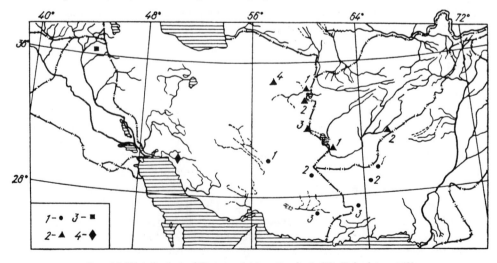

FIG. 16. Distribution of *Teratoscincus microlepis* (1), *T. bedriagai* (2), *Carinatogecko heteropholis* (3), *C. aspratilis* (4)

crust, which crunches underfoot." According to S. Minton's [263] observations, "Its habits seem to be much like those of [*T.*] *scincus*, but it is slower and not so prone to the irascible display of temper characteristic of [*T.*] *scincus*. Its skin is not so easily injured as that of [*T.*] *scincus*. An adult specimen kept several weeks in captivity refused all food except larvae of an unidentified large sand beetle."

BEDRIAGA'S SKINK GECKO — *TERATOSCINCUS BEDRIAGAI* NIKOLSKY, 1899

Type locality: Hodji-i-du-Chagi, E Iran.
Karyotype: not studied.
1899 - *Teratoscincus bedriagai* - Nikolsky, Ann. Zool. Mus., no. 2, p. 147.

DIAGNOSIS. Large dorsal scales end at shoulder level; 44-48 scales around midbody.

LECTOTYPE. ZIL 9161, adult male. "Hodji-i-du-Chahi, May 5, 1989, N. Zarudny." L = 63.0 mm; tail is broken; 32 scales across head; 11/10 supralabials; 10/9 infralabials; no postmentals; 103 GVA; 48 scales around midbody; 12 supracaudal plates.

DEFINITION (Fig. 10) (from 5 specimens in ZIL collection). L adult males (n = 4) 61.8-63.0 mm; L adult female (n = 1) 73.4 mm; broken tails; HHW (n = 5) 63-71 (67.40 ± 1.47); EED (n = 5) 86-93 (89.75 ± 1.49); supralabials (n = 10) 10-11 (10.70 ± 0.15); infralabials (n = 10) 9-10 {error in original – MG} (9.40 ± 0.16); scales across head (n = 5) 31-33 (32.0 ± 0.45); scales along head (n = 5) 69-90 (76.80 ± 3.62); GVA (n = 5) 93-103 (98.50 ± 2.10); scales around midbody (n = 5) 44-48 {**p. 46**} (46.0 ± 0.71) [88]; supracaudal plates (n = 5) 10-12 (11.0 ± 0.45); lateral fringe scales on 4th toe (n = 5) 17-21 (19.0 ± 0.84). Infranasal {error in original – MG} scale that equals nostril diameter, always present; postmentals not distinct. All other characters are contained in description of the genus.

COLOR AND PATTERN [88]. Body whitish above; oblique lines on neck, four broad dark transverse bands on back; white below.

DISTRIBUTION. Inhabits Iran along Iran-Afghan border and also southern Afghanistan (Fig. 16).

LEGEND to Fig. 16.
Iran: 1 - Niyaz-Abad, Haf Region [60]; 2 - Bamrud (Bemrud), Zirkuh Region [60]; 3 - Chah-i-Guyshe Wall ; Khodji-i-du-Chagi (Hodji-do-Chahi) [60]; 4 - 10 km E Doruneh [303].
Afghanistan: 1 - 16 km S of Qala-i-Kang, 30°58'N, 61°54'E [172]; 2 - Kandahar, 31° 36'N, 65°47'E [172].

ECOLOGY not studied {but see Anderson, 1993 – MG}. According to N. A. Zarudny's [60] observations, it was "found on gravel soil occasionally sprinkled with small road metal [= gravel] and covered in places with a thin layer of sand. Several individuals were observed . . . on highly saline, loose soil covered with a thin salt crust. Very common in the Sistan Depression, where it occurs in the untilled agricultural areas of this region, as well as, and especially outside of it, further south in the desert, overgrown with the tamarisk bushes. Here it stayed on clay and loamy soils, covered in places with small sand mounds." It is known [245] that this species reaches elevations of at least 4700 ft (1400 m).

Genus Fringe-toed Geckos — *Crossobamon* Boettger, 1888

Type species: *Gymnodactylus eversmanni* Wiegman [*sic*].
1887 - *Ptenodactylus* Strauch, Mem. Acad. Sci. St. Petersbourg, VII ser., vol. 35, no. 2, p. 64.
1888 - *Crossobamon* Boettger (nomen novum), Zool. Anz., vol. 11, p. 260.

DIAGNOSIS. Digits clawed, straight, gradually become thin at tips, with lateral fringe scales; covered below by one longitudinal row of transversely widened subdigital lamellae, feebly keeled or saw-toothed on free edge; all scales (except large scales on head and sometimes scales on abdomen) with more or less distinct keels; pupil vertical with serrated edges; in males nearly straight row of preanal pores, very frequently interrupted in the middle; small tubercles on back; no postmentals; tail tubercles or tail segments; tail thin, slender, not fragile.

ADDITIONAL CHARACTERS. Head and body slightly depressed. Rostral quadrangular; three slightly bulbous nasal scales, equal to each other; the first one can be separated from nostril by the second and rostral; first nasal scales contact each other or are separated by one to three scales; first supralabial scale higher than others, its height greater than width; scales on snout irregularly polygonal, heterogenous, become smaller toward neck; separate enlarged scales in loreal and supraocular areas; ear opening vertical, oval. Mental scale trapezoid with roundish and uneven posterior edge; infralabials become gradually smaller in size from first to last; gulars subconical, upper ones slightly larger than others.

Dorsal scales flat, rounded-quadrangular or rhomboid, no borders with abdominal scales observed; among these scales, flattened small roundish dorsal tubercles arranged in barely recognizable transverse and longitudinal rows or almost without order. Abdominal scales flat, rhomboid or triangular, slightly imbricate; preanal pore scales larger than others, not perforated in females; about 15 rows of small granular scales between row of preanal pores and vent. Tail covered with transverse whorls of scales slightly larger on lower surface. Adpressed forelimb reaches tip of snout by fingers, hindlimb reaches axilla by toes. Several tubercles, similar to dorsal ones, {**p. 47**} cover thigh and crus. Three to four rows of uniform scales on upper digit surface between lateral fringes; one wide upper scale contacts claw; one to two rows of small granules on each side of toe between subdigital lamellae and lateral fringes.

The genus contains two species distributed in Middle Asia, Kazakhstan, Iran, Afghanistan, and Pakistan.

Key to Species of the Genus Crossobamon

1 (2) Tail longer than body; males with five or more preanal pores . . . . . *C. eversmanni*
2 (1) Tail shorter than body; males with five or fewer preanal pores . . . . . *C. orientalis*

EVERSMANN'S FRINGE-TOED GECKO — *CROSSOBAMON EVERSMANNI* (WIEGMANN, 1834)

Type locality: Agytme (Kyzylkum).
Karyotype: $2n = 42$; 14 acrocentrics, 24 {error in original – MG} subtelocentrics, 2 submetacentrics, 2 metacentrics, NF = 70.

1834 - *Gymnodactylus eversmanni* Wiegmann, Herpetol. mexic., vol. 1, p. 19.
1887 - *Stenodactylus lumsdenii*, Boulenger, Cat. Liz. Brit. Mus., London, vol. 3, p. 479.
1888 - *Crossobamon eversmanni*, Boettger, Zool. Jb. Syst., Jena, vol. 3, p. 880.

DIAGNOSIS. Tail longer than body; five or more preanal pores in males.

TYPE SPECIMEN. Location unknown.

DEFINITION (Fig. 17) (from 308 specimens in ZIK, ZIL, ZMMGU, UMMZ, AMNH, and SAM collections). L adult males (n = 144) 34.0-52.2 (42.02 ± 0.36); L adult females (n = 129) 34.0-59.0 (45.92 ± 0.66); L/LCD (n = 276) 0.57-0.82 (0.66 ± 0.003); HHW (n = 301) 52-57 (54.37 {error in original – MG} ± 0.27); EED (n = 257) 0.27-0.71 (0.47 ± 0.0040); supralabials (n = 616) 10-16 (12.70 ± 0.05); infralabials (n = 616) 9-14 (10.82 ± 0.14); scales across head (n = 306) 22-37 (29.16 ± 0.28); DTL (n = 294) 5.45-14.1 (9.01 ± 0.089); scales around dorsal tubercle (n = 300) 8-12 (9.37 ± 0.04); GVA (n = 258) 165-291 (230.06 ± 1.63); preanal pores (n = 130) 5-12 (8.29 ± 0.15); subdigital lamellae (n = 611) 25-38 (32.08 ± 0.12).

LIFE COLOR AND PATTERN. Main upper surface color sandy pink. Along each side of body, a brown stripe extends from nostril through eye, above ear opening and forelimb, and in its last third can be split into separate marks; indistinct pattern of dark vermiculate stripes on head and limbs. Lower surfaces white. Other details of pattern are important in determining different forms.

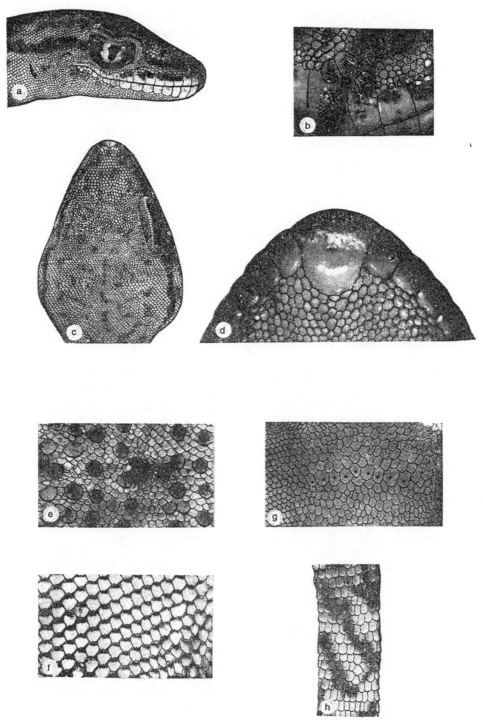

FIG. 17. *Crossobamon eversmanni*:
a, head from the side; b, nasal scales; c, head from above; d, chin; e, dorsum; f, belly; g, preanal pores; h, tail from above (ZIK SR 550; Turkmenia: Bakhardok)

{p. 48} SEXUAL DIMORPHISM AND AGE VARIATION. Females larger than males, they lack preanal pores. One year old juveniles have orange-lemon lower surface of tail.

SYSTEMATICS AND GEOGRAPHIC VARIATION. As our investigation of 10 specimens of the *maynardi* form showed, it quite fits {p. 49} in with the diagnosis of this species. The differences in color pattern between Baluchistan fringe-toed geckos and more northern specimens allow us to consider the form of *maynardi* (*lumsdenii* {error in original – MG}) as an independent subspecies. However, it is possible that transitional forms can be found because specimens from northern Iran have some differences from the nominate form but also are slightly similar to southern subspecies.

Key to the Subspecies of Crossobamon eversmanni

1(2) Thin vermiculate stripes along midline of dorsum and tail; 5-11 ($\bar{x}$ = 7.77) gular scales in contact with mental shield . . . . *C. e. eversmanni*

2(1) Two longitudinal stripes run along midline of dorsum and tail; space between them lacks transverse bars; 10-13 ($\bar{x}$ = 10.80) gular scales in contact with mental shield . . . . *C. e. lumsdenii*

EVERSMANN'S FRINGE-TOED GECKO — *CROSSOBAMON EVERSMANNI EVERSMANNI* (WIEGMANN, 1834)

Type locality: as in species.

DEFINITION (from 298 specimens). Ten to 12 thin short dark vermiculate stripes arranged along spinal column; dark stripes that extend from nostrils over eye and above ear usually end on the flanks in first third of body; tail with up to 30 transverse bars, on its sides can run together into longitudinal stripes; 5 to 11 (7.77 ± 0.07) gular scales in contact with mental shield. The other characters are in accordance with the description of the species.

GEOGRAPHICAL VARIATION. Investigation of 17 samples of the nominate subspecies has shown that eastern and southern populations (Beshkent Valley; right bank of Murghab River; Er-Oilon-Duz; Sistan) have a smaller number of scales across and along mid-belly {ventral scales across midbody and GVA – MG}. The border between these populations and others is Badkhyz Hills and Murghab River Valley. However, to the north, where the water barrier disappears, the variability of these characters becomes cline-like and probably marks the colonization pattern of Eversmann's fringe-toed gecko. The character "number of gular scales in contact with mental" varies similarly, but eastern populations have the highest values; these gradually decrease to the west and south. This latter phenomenon should make the diagnostics of subspecific forms easier.

DISTRIBUTION. Middle Asia, Kazakhstan; reaches Sistan (Iran) in the south (Fig. 18).

LEGEND to Fig. 18.

U.S.S.R., Kazakhstan: 1 - Irgyz (ZIL); 2 - sands near Aralsk [98]; 3 - Ak-Tyube Sands [98]; 4 - Kuvan-Darja dry bed [98]; 5 - Bayghakum Station near Julek (ZIL); 6 - Kendirly Bay [98]; 7 - Karai-Kuduk Well near Turkestan town [95]; 8 - Muyunkum Sands; right bank of Chu River: Chigil-tma Well [98]; 9 - bank of Talas River in 70 km N Djambul; Lake Kagharly-Kul (ZIL); 10 - Akhyrtobe Station [98]; 11 - Priaralskiye Karakumy Sands: Illyakuduk Well [98]; 12 - Ustyurt: Kyn-Tichkey spring [98]; 13 - Aryskum Sands [98]; 14-16 - south of Kysylkum Desert.

Turkmenia: 1 - SE Sarykamish (ZIK); 2 - Uch-Tahan Sands [130]; 3 - Shakhsenem; Merv (ZIL); 4 - W shore of Kara-Bogaz-Gol [130]; 5 - Takhta Dist.: 40 km S of Kommunar collective farm (ZIK); 6 - Chilmamedkum [130]; 7 - Kumsebshen [130]; 8 - Duzkyr Hills: 120 km SE Kunya-Urghench [130]; 9 - 70 km W Darvaza (our data); 10 - 100 to 150 km S of Kommunar collective farm (our data); 11 - Krasnovodsk (ZIL); 12 - Dardza Peninsula [110]; 13 - 185 km N Bakharden: Kyzyl-Katy Well (our data); 14 - 30 km S Darvaza (ZIK); 15 - Darghanata village (our data); 16 - Cheleken (ZIK); 17 - sands S of Djebel Station; Mollakara (ZIL); 18 - Yaskhan (ZIK); 19 - Kurtishbaba [130]; 20 - N of Ak-Molla, Unghuz, 136 km of Serny Zavod (our data); 21 -Perevalnaya Station [130]; 22 - Kirpily [130]; 23 - Doyakhatyn (ZIK); 24 - Bugdaily [130]; 25 - Akhcha-Kuyma Station [128]; 26 - 50 km NE of Kyzyl-Arvat (ZIL); 27 - Kultakyr Well [130]; 28 - 250 km N of Mary (ZIK); 29 - Madau [130]; 30 - Bakhardok (ZIK); 31 - 180 km N Bayram-Ali (our data); 32 - left bank of Amu-Darja on the road to Farab (our data); 33 - Farab Station (ZIL); 34 - 100 km NW of Repetek (our data); 35 - Repetek (ZIK); 36 - Karakala (ZIL); {p. 50} 37 - Bakharden [23]; 38 - N of Geok-Tepe [130]; 39 - 37 km W of Repetek (our data); 40 - Peski Station [23]; 41 - Uch-Adji (ZIK); 42 - 9 km NW of Gasan-Kuly [130]; 43 - Lake Maloye Delily [130]; 44 - Gyaurs Station; Annau village; Nysa village (ZIK); 45 -

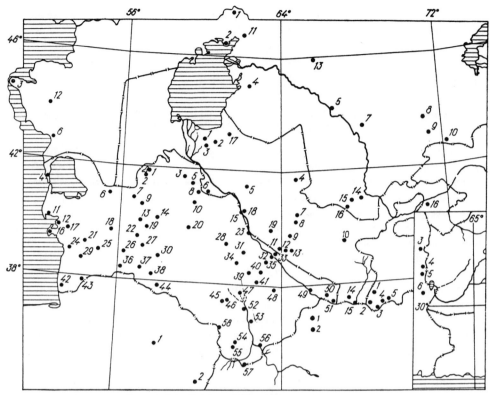

FIG. 18. Distribution of *Crossobamon eversmanni*

Dortkuyu [23]; 46 - Karabata (ZMMGU); 47 - Shakhsenem (ZIL); 48 - Djapar Well [130]; 49 - near Kerkichy [130]; 50 - Dostluk [130]; 51 - Kelif (ZIL); 52 - Yolotan (ZIK); 53 -Imambaba village (ZIK); 54 - Er-Oylon-Duz (ZIK); 55 - Kepely (ZIL); 56 - right bank of Murghab River near Takhta-Bazar (ZDKU); 57 - N of Kushka village [23]; 58 - Serakhs village (ZMMGU).

Uzbekistan: 1 - 25 to 30 km N of Chabankazghan (ZIK); 2 - Karakol River (ZIL); 3 - Chabankazghan (ZIK); 4 - Tamdy Well (ZIL); 5 - 100 km of Tartikul (Turtkul) (ZDKU); 6 - Khiva (ZIL); 7 - Ayakagytma and Tashkuduk (ZIK); 8 - 100 km N of Shafrikan: Kyzylkum Biological Station (ZMMGU); 9 - Shafrikan [22]; 10 - Samarkand (ZIL); 11 - Karakul Station [22]; 12 - Bukhara; Kagan; Kyzyl-Kuduk [22]; 13 - Uchkum Sands to W of Kagan [22]; 14 - Sands near Djarkurghan village (ZDKU); Surkhan-Darja River Valley and Aral-Payghambar Island [158]; 15 - 15 km N of Termez (ZIK); 16 - 18 km NW Kokand: Urghanchy [88]; 17 - Kemper-Tyube (ZMMGU); 18 - Kyzylrabat (ZMMGU); 19 - 20 km NW Gazly (ZMMGU).

Tadjikistan: 1 - Shaartuz Dist.: Beshkent Valley (ZIK); 2 - southern border of Tadjik and Uzbek SSR (ZIK); 3 - 8 km near Chirchik (ZIK); 4 - Tigrovaya Balka [88]; 5 - Abad collective farm, Karadum Sands (ZIL).

Iran: 1 - Mondehi and Feyzabad villages (in Sarr-Chah Valley) [59]; 2 - Mozhnabad and Fendukt [60]; 3 - Khouz in Zirkuh Region (ZIL); 4 - Tagh-i-Doroh lowland [60]; 5 - Chah-Isy Well in Seistan (ZIL); 6 - Khorasan Prov.: Tayabat, 30°47'N, 60°47'E (CAS).

Afghanistan: 1 - 20 km S of Andhoi (CAS); 2 - 30 km NW Shebergan (CAS).

LUMSDEN'S FRINGE-TOED GECKO — *CROSSOBAMON EVERSMANNI LUMSDENII* (BLGR., 1877) STAT. NOV.

Type locality: Baluchistan near the Afghan border.

1877 - *Stenodactylus lumsdenii* Boulenger, Cat. Liz. Brit. Mus., vol. III, p. 479 (between Nushki and Helmand).
1933 - *Stenodactylus maynardi* Smith, Rec. Ind. Mus., Calcutta, vol. 35, p. 18.
1967 - *Crossobamon maynardi*, Kluge, Bull. Amer. Mus. Nat. Hist., no. 4, vol. 135, art. 1, p. 23.

DIAGNOSIS. Two longitudinal stripes along midline of back and tail; space between these stripes lacks transverse bars; 10-13 ($\bar{x} = 10.80$) gular scales in contact with mental scales. All other characters in accordance with description of species (Fig. 20).

{p. 51} The type specimens are kept in the British Museum of Natural History (London) and in the Indian Museum (Calcutta) [245].

DISTRIBUTION. Baluchistan near Afghan-Pakistan and Afghan-Iran-Pakistan borders (Fig. 21).

LEGEND to Fig. 21.
Iran: 1 - Hormek [88].
Afghanistan: 1 - between Nushki (western Pakistan) and Helmand River - terra typica [245].
Pakistan: 1 - Nushki; 13 to 16 km NW of Nushki (AMNH); 2 - Kharan [258].

SYSTEMATICS AND GEOGRAPHIC VARIATION. The description of *maynardi*-form [290], despite the presence of insignificant differences, allows us, almost undoubtely, to synonymize it with the southern subspecies of Eversmann's fringe-toed gecko. We observe two nasal scales in a few specimens of both species of the genus (first nasal scale is separated from nostril by the second and rostral scales); besides that, A. M. Nikolsky [89], having identified a gecko from Hormuk as *lumsdenii*, noted the presence of three nasal scales, as well as more keeled subcaudal scales. Thus, variation of this character should be attributed to individual variation.

HABITAT AND QUANTITATIVE DATA (Fig. 13). This is a psammophilous species. It inhabits packed and semi-packed sands. As a rule, it is found along with another psammophile, the skink gecko; however, their abundance is different for various biotopes. The abundance of a lizard can be an indicator of its response to a given habitat. According to our observations (conducted in April-May; surveying after dusk for three hours), there is a distinct predominance of the fringe-toed gecko over the skink gecko in the extreme west of Turkmenia, in the vicinity of Cheleken settlement. Here they inhabit mounds of shells and salty-gray sand at the seashore with {*Salivornia* sp. – MG} vegetation. Along a 1 km route, two geckos were encountered and, altogether, the ratio was 13 fringe-toed to 20 skink geckos. In the vicinity of Lake Yaskhan, on clean wind-blown sands with kandym {*Calligonum* sp. – MG}, saxaul {*Haloxylon* sp. – MG}, sand acacia {*Ammodendron* sp. – MG} and selyn {*Stipagrostris* sp. – MG}, 31 skink geckos were caught for every 4 fringe-toed geckos. On the left bank of Amudarya, in the sand desert to the south of Doyakhatyn village, 25 skink geckos were noted for every 8 fringe-toed geckos.

FIG. 19. Three eggs of a Turkestan thin-toed gecko [*Tenuidactylus fedsckenkoi*] and one (smaller) of Eversmann's fringe-toed gecko [*Crossabamon eversmanni*] from Tadjikistan. (Photo by A. D. Bautin.)

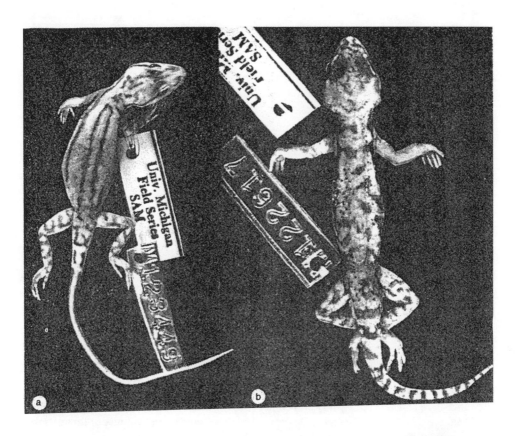
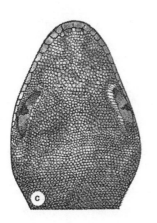
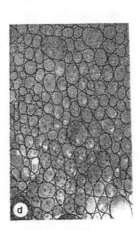

FIG. 20. *Crossobamon e. lumsdenii*:
a, general view from above.
C. orientalis:
b, general view from above; c, head from above; d, dorsum; e, toe from below; f, toe from above
(a, ZIK SR1450, near Nushki, Pakistan; all others, ZIK SR 1449, near Burra, Pakistan)

In a similar desert biotope, in the vicinity of Uch-Adzhy village (eastern Turkmenia), the ratio was 4 fringe-toed to 21 skink geckos. In the center of the Karakum Desert (250 km to the north of the Mary Town), 1 fringe-toed gecko was encountered for every 12 skink geckos. To the north of {p. 52} Ashkhabad, for 2 fringe-toed geckos 20 skink geckos were

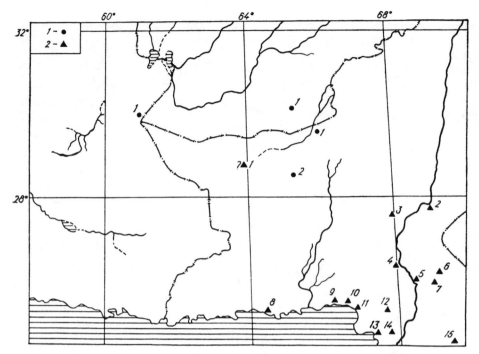

FIG. 21. Distribution of *Crossobamon e. lumsdenii* (1), *C. orientalis* (2)

found. In the Kyzylkum Desert, in the vicinity of Ayak-Agitme village in the hilly sands with tamarisk {*Tamarix* sp. – MG} and camel spine {*Alhagi pseudalhagi* – MG}, the ratio was 1 fringe-toed to 20 skink geckos. On the packed dusty sands in the vicinity of Imam-Baba village, on the right bank of Murghab River (southern Turkmenia), 9 fringe-toed geckos and not a single skink gecko were captured. On the stabilized sands with the {*Salsola* sp. – MG} bushes adjacent to sor {sor, shor = saline plain – MG}, in Er-Oylan-Duz and Namaksar Depressions (Badkhyz), only fringe-toed geckos are found.

In the vicinity of Gyaurs, on the loosely-packed and wind-blown sands with some loess mixed in (the vicinity of sands in the foothills of Kopetdagh), in an area 600 × 200 m we caught 15 geckos of the species of interest to us, that is one lizard was encountered on every 0.8 hectare. Consequently, even though the fringe-toed and skink geckos exist sympatrically side by side, they have differing requirements of the biotopes. The first prefers more or less packed sands with loess or salinated and the latter likes clean, weakly packed or blown sands. In Kazakhstan, *C. eversmanni* is found on packed and blown sands with occasional dzhuzghun {*Calligonum* sp. – MG}, teresken {*Ceratoides* sp. – MG} and {*Alhagi pseudalhagi* – MG} bushes [98]. From the middle of July to September, there is a sharp drop in the numbers of geckos from 15-25 individuals per 1 km to 5-15 can be observed [326].

RESPONSE TO TEMPERATURE. S. Shammakov [130] observed active geckos with air temperatures between 10° and 24°. In Tadjikistan (south of Babatagh), the geckos did not appear on the surface (burrow out of the soil) with a sand temperature of 17°. On 14 April 1977, in the vicinity of Yolotan the geckos were active with an air temperature of 22°, but when it dropped to 18° they disappeared. On 18 April 1978, very active geckos were observed in the sands south of Chirik village in the Shaartuz District of Tadjikistan S.S.R. with the air and soil temperatures at 19°. On 27 May 1982, in the vicinity of Gyaurs, when the air temperature was 27° [and] the wind began to blow and it started to drizzle, the geckos were encountered on the leeward side behind mounds and bushes. After the temperature fell to 18°, they disappeared. In terraria, fringe-toed geckos prefer a soil temperature around 30°.

DAILY ACTIVITY. They are nocturnal. In the spring they can occasionally be seen in the daytime (on 8 April 1977, one was caught in the Namaksar Depression, Badkhyz). We would usually catch them after dusk. In the spring time, as a rule, the geckos would disappear after it cooled, by 0130 hrs. From April to September they are active the entire night. In the middle of October they were observed only in the evening. On a cloudy night these lizards appear somewhat earlier than on a clear one. On a moonlit night their activity subsides.

SEASONAL ACTIVITY. The first discovery of this lizard after winter hibernation is known from Turkmenia on 15 March. Mass emergence is from the beginning of April, which is in agreement with our observations. In Tadjikistan, the first lizard was caught on 3 April and in Kazakhstan [98] the geckos are active at the end of April. In Turkmenia, they disappear for the winter in the second half of October. The last active individual was caught on 25 October 1967 [130]. On 10 October 1981, near Sultanbent, on the left bank of Murghab River, active lizards were no longer observed and one individual was dug out of its winter burrow. In Kazakhstan, in the Muyunkum Sands, the geckos were no longer encountered at the end of September.

SHEDDING occurred in an individual caught on 14 April 1977 near Yolotan (Turkmenia).

FEEDING. According to the data from Turkmenia [130], the geckos eat (especially in the spring) June beetles (rate of encounter 32.5%), ants (30.2%), and snout beetles, as well as butterfly caterpillars, spiders, termites, and crickets. Juvenile fringe-toed gecko individuals were found in their stomachs twice. In Uzbekistan [22 {error in original – MG}], the stomachs of these lizards had mostly beetles (rate of encounter 60% for June beetles, snout beetles), as well as spiders, homopterans, [and] butterfly caterpillars. In Tadjikistan [107], fringe-toed geckos eat primarily beetles (darkling, June, snout, leaf, click, wood-boring, [and] tiger beetles), as well as butterflies and their larvae, ants, flies, heteropterans, spiders, [and] praying mantis.

REPRODUCTION. The sex ratio in the collection materials is close to 1:1. Males reach sexual maturity when {p. 53} body size is greater than 35 mm, and the females at 42 mm. Courtship, judging from the lively gecko vocalizations, was noted on 14 April 1977 in the vicinity of Yolotan and on 13 April 1979 in the vicinity of Uch-Adzhy Station. O. P. Bogdanov [23] observed courtship on 14 May 1959, which continued until the beginning of June [326]. According to observations in Tadjikistan [107], courtship takes place in April, but egg laying is extended and lasts about two months. It is assumed that in Turkmenia two to three clutches are laid each season [326]. Clutches of one or two eggs are laid in Turkmenia from the end of May to the end of July [23, 130], in Uzbekistan from the beginning of June until 19 July [22], in Tadjikistan it has been observed in mid-June, maybe in the last ten days of the month [107]. K. P. Paraskiv writes that in the Ak-Tyube Sands on Mangyshlak he found a fresh clutch on 28 July 1947. Fully formed eggs in the oviducts of females, caught at the southern edge of the Muyunkum Sands, were found on 20 May 1949. There were scars in the ovaries of these females from the already-shed follicles and two enlarged follicles in each, which may indicate the possibility of repeated egg laying. The eggs are white, ovoid and can stick together during the laying process. The egg dimensions for the geckos from Turkmenia (n = 7) are 11.15-12.30 × 8.0-9.85 mm, with a mass of 0.458-0.848 g [281]. According to data from Uzbekistan, egg dimensions are 11-12 × 8-9 mm [62]. The largest eggs, extracted from the oviducts of females caught in Tadjikistan, are close in size to the ones mentioned above – 11.5 × 8 mm [107]. As confirmed in terraria, incubation lasts 45-53 days and averages 48 days. Newly hatched offspring have a body length of 21-26 mm, tail length of 33-41 mm, and mass of 0.365-0.501 g [281]. Hatchlings of the current year were encountered in Turkmenia, in the vicinity of Yolotan, at the end of July – beginning of August (first generation), and to the north of Gyaurs on 13-15 September (second generation) [22, 130]. At the same place, on 26 August 1947, an egg was dug up that contained a gecko ready to hatch [22].

GROWTH RATE. According to S. Shammakov's data [130], after hatching, the geckos grow 10-18 mm in 9-11 months and in two to three years reach their maximum size (average female L = 42, male L = 52 mm). The lizards of the first generation reach sexual maturity the year after they hatch, and individuals of the second generation reach it at the age of two.

ENEMIES. In Turkmenia, fringe-toed geckos are eaten by Karelin and red-striped racers

(*Coluber karelini, C. rhodorhachis*) , as well as the sand boa (*Eryx miliaris*) [130, 326]. On 22 June 1976, in the stomachs of owls {*Athene noctua* – MG} caught in the ruins of Dargan-Ata fortress, we found geckos (six in one, one in the other). In Uzbekistan, they may also be attacked by the Karelin racer [22]. In Tadjikistan, a gecko was found in the stomach of a maculated racer (*Spalerosophis diadema*) [107].

BEHAVIORAL ATTRIBUTES. They usually move slowly but can run quickly when disturbed. Having run for 5-6 m, they freeze. As reported by a number of authors [23, 98, 136], fringe-toed geckos climb on the branches of bushes and get food there [and] can jump from branch to branch for up to 10 cm. In searching for food, the fringe-toed gecko moves a considerable distance, up to 100-150 m, from its burrow, returning there by the morning. In the summer, it frequently burrows into sand to depths of 5 cm. It occupies the burrows of rodents, lizards, or it can dig its own. O. P. Bogdanov [23] reported that an egg was dug up from a depth of 15 cm.

On 10 October 1981, we dug out a hibernating individual from a mound under a kandym bush {*Calligonum* sp. – MG} from a depth of 60 cm. Winter burrows on the southern fixed sand slopes reached a length of 40-42 cm; the depth reached 27-30 cm. One individual was hibernating in the *Dipus sagitta* burrow. Summer burrows reached a length of 110 cm; depths did not exceed 20 cm and diameter of 1.5 cm [23]. From the Kazakhstan observations [98], the burrows of fringe-toed geckos are of simple design and consist of a straight passage that widens somewhat at the end in the zone of moist sand 30-40 cm from the surface. The length of the passage is not longer than 70 cm and is usually 40-50 cm long.

K. P. Paraskiv [98] repeatedly found these geckos in the same burrows with dung beetles and carabids. Using these beetles' passages, the gecko would dig its own side passage from the main one. The length of the side passage did not exceed 20-30 cm. During the day, the gecko plugs the burrow entrance or its side passage with a {p. 54} sand plug of up to 10 cm. Burrows can be positioned in close proximity. Three occupied burrows were found in an area of 1 m^2. As A. D. Bautin observed (written report), the fringe-toed gecko uses its tail when digging. It sticks the tail out of the burrow and begins to sweep in sand until the burrow [opening] is covered; it then pulls the tail back into the burrow. At that point, the tip of the tail works the most. This lizard's tail is not fragile, unlike in other species, and the tip, if broken, regenerates.

As we have established, during the mating period this gecko emits sounds that are lower than the thin-toed gecko's voice. They can be expressed as "yek...yek...yek."

When mating, the male grabs the female on the flank or in the area of the forelimb (thorax) and holds her down with the tail during copulation. For egg laying, the female digs a small pit and lays a pair of eggs within an interval of 5-6 minutes, during which she lies down to rest. The eggs are soft when first laid but harden quickly.

PRACTICAL SIGNIFICANCE AND PROTECTION. Judging by the nature of its diet, the fringe-toed gecko is a beneficial animal and occupies a significant place in the sand desert biocenosis. Because the desert areas inhabited by these lizards are fairly extensive and they are encountered there frequently, there is no need for any special protection programs for this species. The fringe-toed gecko, along with other animals, is protected in nine preserves (the most important among them are the Repetek and Krasnovodsk Preserves in Turkmenia; Baday-Tugay, Karakul, and Kyzylkum Preserves in Uzbekistan; Tigrovaya Balka in Tadjikistan).

EASTERN FRINGE-TOED GECKO — *CROSSOBAMON ORIENTALIS* (BLANFORD, 1876)

Type locality: Rohri and Shikarpur, Upper Sind (Pakistan).
Karyotype: not studied.

1876 - *Stenodactylus orientalis* Blanford, J. Asiat. Soc. Bengal, Calcutta (2), vol. 45, p. 21.
1884 - *Stenodactylus dunstervillei* Murray, Vertebr. Zool. Sind, p. 363.
1967 - *Crossobamon orientalis*, Kluge, Bull. Amer. Mus. Nat. Hist., vol. 135, art. 1, p. 23.

DIAGNOSIS. Tail shorter than body; five or fewer preanal pores in males.

TYPE SPECIMEN is kept in the Indian Museum, Calcutta.

DEFINITION (Fig. 20) (from 15 specimens in UMMZ, AMNH, and SAM). L adult males (n = 9) 29.5-47.8 (41.04 ± 1.67) mm; L adult females (n = 5) 27.0-56.5 (43.04 ± 5.33) mm; L/LCD (n = 6) 1.18-1.55 (1.31 ± 0.05); HHW (n = 15) 54-64 (53.73 {in error, error in original – MG} ± 0.81); EED (n = 15) 0.45-0.59 (0.53 ± 0.012); supralabials (n = 30) 9-14 (11.97 ± 0.23); infralabials (n = 30) 9-14 (11.40 ± 0.26); scales across head (n = 15) 20-26 (22.47 ± 0.48); DTL (n = 15) 6.80-12.70 (9.91 ± 0.42); scales around dorsal tubercle (n = 15) 7-9 (8.47 ± 0.16); GVA (n = 14) 185-225 (204.43 ± 3.17); preanal pores (n = 7) 2-5 (2.71 ± 0.42); subdigital lamellae (n = 29) 24-29 (26.55 ± 0.24).

LIFE COLOR AND PATTERN [263]. Main color of upper body surfaces brown to pale gray with three to five indistinct dark gray transverse bars; tail yellowish with distinct dark wide transverse bands. Abdomen white. The specimen from Las Bela (Pakistan) is noted for a unique pattern variation. Dark dorsal tubercles arranged in regular rows on an almost totally uniform black background.

SEXUAL DIMORPHISM AND AGE VARIATION. Females larger than males. Age variation not studied.

DISTRIBUTION. Sind (Pakistan) (Fig. 21).

LEGEND to Fig. 21.

Pakistan: 1 - Merui, Gaur Band, 28°57'N, 63°48'E [258]; 2 - Rohri, terra typica [258]; 3 - Sirla, southern Shahdadkot, 27°45'N, 67°55'E [258]; 4 - Hala [258]; 5 - 4 km S of Sehvan (AMNH); 6 - Sanghar Dist.: 18 km SW of Jamrao, sand hills {p. 55} near Jamrao Had (AMNH); 7 - Sanghar Dist.: 1 km W Burra; 8 km S of Sanghar (AMNH); 8 - Ormara [258]; 9 - Las Bela Dist.: Maini Hor Bay (AMNH); 10 - Sonmiani (as above); 11 - Naka Khari (as above); 12 - Dabu Dist.: Thana Bula Khan (as above); 13 - Tatta Dist.: 3 km E of Gharo (as above); 14 - Tatta; Jungshahi, 16 km NW of Tatta [258]; 15 - Diplo [258].

ECOLOGY not studied.

Genus Pygmy Geckos — *Alsophylax* Fitz., 1843

Type species: *Lacerta pipiens* Pallas.
1843 - *Alsophylax* Fitzinger, Syst. Reptilium, Fasc. 1, p. 90.

DIAGNOSIS. Nostril surrounded by pentagonal rostral, first supralabial, large and flat supranasal scales and often one or two additional small nasal scales; supralabial scales are in the same plane and drop straight to the edge of mouth; height of the first supralabial scale from nostril to the edge of mouth distinctly less than its width along the edge of mouth. Pupil vertical with serrated edges. No ciliary growths in the upper part of orbit. All small body scales flat; in addition to scales, dorsum can also be covered by tubercles (sometimes keeled); these tubercles can extend onto base of tail (no further than $1/3$). Caudal segments indistinct on upper side or feebly distinct; tail covered below with longitudinal row of plates (usually two per segment). No more than 13 preanal pores, females may lack them. Digits straight or slightly angularly bent, cylindrical, clawed, covered beneath with a row of flat lamellae; lateral digital scales do not form fringes. Small geckos, L to 46 mm.

ADDITIONAL CHARACTERS. Scales on snout large, granular, roundish, become gradually smaller toward the neck. 2.5-4 scales reach the front edge of orbit. Ear opening small, roundish. Upper gular scales enlarged; gulars roundish, slightly granular. Dorsum covered with granular scales that become larger on the flanks, where they gradually turn into abdominal ones. Up to 12 transverse rows of tubercles can be present on dorsum. Abdomen covered with large flat imbricate scales. Preanal pores usually arranged ∧-like. Tail thick, cylindrical, becomes abruptly thin in the last quarter; covered with whorls of large flat scales, larger on the sides; regenerated tail begins with swelling, covered with flat scales in no order.

Upper surfaces of forelimbs covered with scales that are similar to dorsal scales, lower surfaces – to abdominal scales; besides, upper crusal surface has tubercles, usually larger than dorsal ones. Fourth toe longer than others, covered above with one, and from sides — with two scale rows; base of claw covered with the upper and the lower lamellae; this "sheath" is in contact with one scale from above and from each side. Subdigital lamellae almost equal in size.

The genus includes six species, divided into two subgenera, which inhabit Kazakhstan, Middle and Central Asia.

Key to Species and Subspecies of the Genus Alsophylax

1 (4) One enlarged nasal scale, no additional nasal scales 4
2 (3) No dorsal tubercles . . . . *A. laevis* Nik.
3 (2) Dorsal tubercles present . . . . *A. pipiens* (Pall.)
4 (1) Besides enlarged nasal scale, many smaller scales contact nostril 5
5(10) A single additional nasal scale 10
6 (7) More than seven preanal pores . . . . *A. l. loricatus* Str.
7 (6) Less than seven preanal pores 8
8 (9) No dorsal tubercles . . . . *A. tadjikiensis* Golubev
9 (8) Dorsal tubercles present . . . . *A. przewalskii* Str.
10 (5) Two additional nasal scales 11
11(12) More than seven preanal pores . . . . *A. tokobajevi* Jeriomtschenko et Szczerbak
12(11) Less than seven preanal pores . . . . *A. loricatus szczerbaki* Golubev et Sattorov

{p. 56} Subgenus *Alsophylax* Fitz., 1843

Type species: *A. pipiens*.

DIAGNOSIS. Digits straight; preanal pores in both sexes; the length of forelimb is 25% of body length, adpressed forelimb reaches front edge of orbit by fingers (except *A. przewalskii*), adpressed hindlimb reaches elbow; caudal segments indistinguishable; caudal tubercles are not on sides of tail flanks. Subgenus includes five species.

SQUEAKY PYGMY GECKO — *ALSOPHYLAX PIPIENS* (PALL., 1811)

Type locality: Bolshoj Bogdo Mount (Astrachan Dist., R.S.F.S.R.).
Karyotype: 2n = 36; 30 acrocentrics, 4 subtelocentrics, 2 submetacentrics, NF = 42.

1811 - *Lacerta pipiens* Pallas, Zoogr. Rosso-Asiatica, vol. 3, p. 27 (terra typica: "Berg Bogdo, Kirgizensteppe").
1823 - *Ascalabotes pipiens*, Lichtenstein (in Eversmann), Reise von Orenburg nach Bukhara, Berlin, p. 145.
1831 - *Gymnodactylus pipiens*, Eichwald, Zoologia specialis, Vilno, vol. 3, p. 181.
1843 - *Alsophylax pipiens*, Fitzinger, Syst. Rept., p. 90.
1856 - *Gymnodactylus atropunctatus* Lichtenstein, Nomencl. Rept. Amph. Mus. Zool. Berlin, p. 6 (terra typica: "Tartarei").
1885 - *Alsophylax microtis*, Boulenger, Cat. Liz. Brit. Mus., London, vol. 1, p. 19 (err. syn.).

DIAGNOSIS. Usually many small roundish dorsal tubercles (sometimes, with slight trace of keels) in disorder; 8-13 preanal pores; one large supranasal scale, no additional small nasal scales; 4-7 dark wide transverse bands on body; lower surface of tail white.

NEOTYPE {erroneous designation, see Bauer and Günther, 1992; type specimen in Berlin Museum – MG}. ZIL 3600, adult male. "Mt. Bogdo magnus, 1872, leg. Becker." L = 34.3 mm; regenerated tail; only large supranasals, in contact with each other, present; distinct postsupranasals separated from each other by small scale; 9/10 supralabials; 6 infralabials; 13 scales across head; 11 preanal pores; 103 GVA; 18/19 subdigital lamellae; dorsal tubercles in more or less regular rows; every such tubercle in the middle of the back surrounded by 8-9 scales.

DEFINITION (Fig. 22) (from 690 specimens in ZIK, ZIL, ZMMGU, AAIZ, UIZP, UNM {error in original – MG}, NMW, MNHP, FMNH, USNM, and UMMZ). L adult males (n = 230) 25.0-38.0 (31.88 ± 0.21) mm; L adult females (n = 269) 25.0-41.6 (33.71 ± 0.26) mm; L/LCD (n = 188) 0.80-0.99 (0.880 ± 0.004); HHW (n = 420) 53-74 (62.27 ± 0.18);

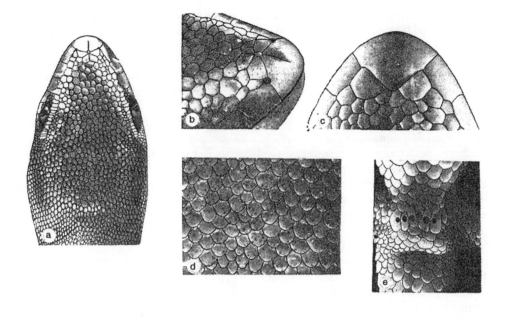
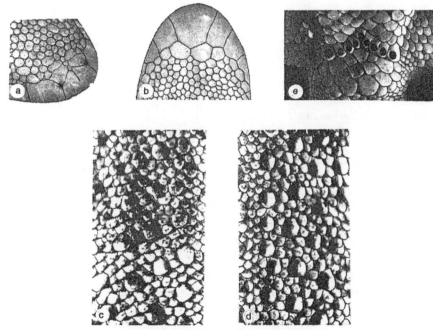

FIG. 22. *Alsophylax tadjikiensis* (top):
a, head from above; b, head from the side; c, chin; d, belly; e, preanal pores
(ZIK SR 942; near Sumbula, Tadjikistan).
Alsophylax pipiens (bottom):
a, snout from above; b, chin; c-d, dorsum; e, preanal pores
(a-c ZIL 3599, Bogdo, RSFSR; d-e BNM MG/8, Bayn-Dzon, Mongolia).

EED (n = 360) 6-29 (18.57 ± 0.20); supralabials (n = 1305) 5-9 (7.03 ± 0.02); infralabials (n = 1249) 4-7 (5.49 ± 0.02); scales across head (n = 636) 11-20 (14.76 ± 0.06); DTL (n = 571) 9-16 (12.38 ± 0.07); scales around dorsal tubercle (n = 286) 7-11 (8.59 ± 0.05); GVA (n = 515) 84-112 (96.54 ± 0.23); preanal pores (n = 646) 8-13 (9.44 ± 0.04); subdigital lamellae (n = 1148) 13-22 (17.35 ± 0.04). Mass up to 1.5 g. Nostril between rostral, first supralabial and large supranasal scales; in some specimens (up to 18%) from eastern part of range one additional small scale can be present; distinct (27.7%), indistinct (29.7%) or none (42.6%) postsupranasal {error in original – MG} scales. Mental scale quadrangular to pentagonal (74%), hexagonal (16.5%) or trapezoid (9.4%). None to two pairs of postmentals. Dorsal tubercles roundish, flat or with traces of keels, sometimes arranged in weakly distinguished rows.

LIFE COLOR AND PATTERN. Sandy ochre above. Dark brown stripe extends on each side of head from first supralabial through eye and far above ear opening. These stripes can come together on neck forming a horseshoe-shaped pattern. Between {p. 57} nostril and eye, they can be edged with white; indistinct dark brown pattern present on upper side of snout between these white edges. Four to seven dark brown bands across body from neck to lumbar, wider than interspaces. {p. 58} In the middle of the back they can be interrupted and shifted relative to each other along longitudinal axis. On upper surface of tail 11 wide transverse bands of the same color as dorsal bands. Limbs also covered above with indistinct transverse stripes. Abdominal surface white.

SEXUAL DIMORPHISM AND AGE VARIATION. According to our data, sexual dimorphism of the squeaky pygmy gecko becomes apparent in larger maximal and median sizes of females in comparison with males; preanal pores are less developed in females than in males and can almost completely disappear with age. Some body ratios are subject to age variation. Juveniles have shorter tails (in hatchlings it is shorter than the body) (Fig. 23) {Fig. 23, as well as Figs. 25, 33, 41, 47, 55, 62, and 75 contain only raw data and, consequently, cannot be used as proof of age variation in the character L/LCD – MG}. Besides, juveniles have a larger diameter of the eye compared to the body length. These parameters become more uniform in adult geckos. In this case, we took into consideration only the data from adult specimens.

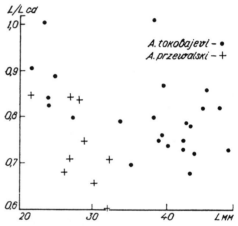

FIG. 23. Age variation of L/LCD index in *Alsophylax przewalski* and *A. tokobajevi*

DISTRIBUTION (Fig. 27 and 28). The Lower Volga, Ustyurt, Kazakh Knoll country {territory of Central Kazakhstan to the north of Balkhash and Betpakdala Desert, to the east of Aral Sea and to the west of Enisei River – MG}, Semirechye, Djungaria, southern Gobi and adjoining territories of China; southeastern boundary of the range {in China – MG} is not determined more precisely.

LEGEND to Figs. 27, 28.

Kazakhstan and Middle Asia: 1 - Astrachan Dist.: Bogdo Mountain (ZIL); 2 - Kirghiz Steppe of Bukeyevsky Orda [96]; 3 - Orenburg Territory: Almas-tau Mountain [58, 94]; 4 - 4 km W of settlement Emba [151]; 5 - Karazhar near Emba [90]; 6 - Guryev Dist.: Jiltau Ridge [65]; 7 - Donguz-tau (ZIL); 8 - nw shore of Aral Sea: Kaska-Jol locality [98]; 9 -Barsakelmes Island [16]; 10 - Jaslyk [29]; 11 - Kirk-kiz [29]; 12 - Sor-Barsakelmes (ZIK); 13 - chink {abrupt loess or sandstone slopes of some Middle Asian hills, especially of Ustyurt Plateau - MG} near Lake Sudochye (ZIK); 14 - NW shore of Lake Sarykamish: spit Kipillar-kir [34]; 15 - 15 km of Butentau Hills: Kunya-Darja old bed [23]; 16 - W Kyzylkum: Chabankazghan Well (ZMMGU; 17 - N Kyzylkum: 80 km NE of Chabankazghan Well (ZMMGU; 18 - Kum-Kala (ZMMGU; 19 - between Yerembet and Tupbugut settlements (ZMMGU; 20 - 21 km W of Tupbugud (ZMMGU; 21 - 50 to 70 km SW of Djusaly Station (ZMMGU; 22 - between Syrdarja and Janadarja (ZMMGU; 23 - Djusaly Station [98]; 24 - lower Sarysu River: Kok-Aladjar [98]; 25 - Uch-choku Hills [61]; 26 - Irgiz River [96]; 27 - upper Turgay River: Tort-mola Well [98]; 28 - Turgay: right bank of Djuligaly: Djilanchik (R. Kubykin, verbal report); 29 - Ulytau Mountain (ZIL); 30 - Djezdy settlement (ZIK); 31 - S of Bassagha settlement: Aktau Mountain (ZIL); 32 - Akchatau Mountain (ZIK); 33 - Bet-Pak-Dala (ZIL); 34 - Burubaytal [123]; 35 - Myn-Aral locality

[96]; 36 - Korajun-Tyubek cape [96]; 37 - Tas-aral Island [96]; 38 - between Ili and Karatal Rivers [137]; 39 - near Shaukar Peninsula [85]; 40 - E shore of Lake Balkhash (ZMMGU; 41 - Tumarcha Well [137]; 42 - Golova {p. 59} Bakanas (ZIL); 43 - Bartogoy near Turayghyr (ZIK); 44 - Turaygyr on Chilik River (ZIK); 45 - right bank of Charyn River Wof Chundja village (ZIK); 46 - foothills of Ketmen in the middle channel of Piyastyk River (R. Kubykin, verbal report); 47 - foothills of Djungarski Alatay: Katu-tau Mountain (R. Kubykin, verbal report); 48 - Panfilov (Djarkent) [96]; 49 - China: Chark-ukyur [3]; 50 - middle channel of Takhta River (R. Kubykin, verbal report); 51 - Rghaity (Dr. R. Kubykin, verbal report); 52 - Sredny Island [70]; 53 - Kishkene-Araltobe Island [70]; 54 - between Emelsu and Rghaity Rivers [137]; 55 - 60 km W of Zaissan Town [151]; 56 - Janghiztobe Station (R. Kubykin, verbal report); 57 - vicinity of Semipalatinsk: near the bridge across Mukur River (ZIK); 58 - Iman-Kara Mountain (V. V. Neruchev, written report); 59 - Besbay Mountain (V. V. Neruchev, written report); 60 - Ak-Tologay Plateau {p. 60} (V. V. Neruchev, written report); 61 - Ak-Keregeshe Plateau (collection of Ye. Yu. Kudakin); 62 - Kara-Kalpak S.S.R.: Takhta-Kupir Dist.: 30 km N of Chabankazghan settlement (ZIK); 63 - Kazakh S.S.R.: Kzyl-Orda Dist.: Kekrely settlement (ZIK); 64 - 40 km NW of Kekrely settlement: remnants of old fortress Chirik-Robat (ZIK).

Central Asia: Mongolia: 1 - 5 km N of Uench Town (BNM {error in original – MG}; 2 - E. Djungaria [18]; 3 - Bur-Nur (BNM); 4 - Khatan-Khairkhan-ula [78]; 5 - Somon-Khairkhan-ula [78]; 6 - Dzun-Mod (BNM); 7 - Uldziiyt-gol [78]; 8 - Nogon-Tsav [78]; 9 - Yaman-us [78]; 10 - Tsagan-Bogdo Ridge (ZIK); Shara-Huls [78]; 11 - Ehin Gol [78]; 12 - Naryn-Bulak (ZMMGU); 13 - Torin-Huduk (BNM); 14 - Haychin-Ula [78]; 15 - Djirgalantu-Huduk Well [78]; 16 - Nemegetu Ridge [15]; 17 - Muhor (BNM); 18 - Obot-Hural (BNM); 19 - Noyon-Bogdo (ZMMGU); 20 - 25 km N of Bulgan-Somon (BNM); 21 - Shovon-Huver [78]; 22 - Tevsh-Hayrkhan [78]; 23 - 58 km WSW Boyan-Dalay (BNM); 24 - Suhaytin-Zhuren-Tologoy, S slope of Zolon-Ula [77]; 25 - Dzuramtay [78]; 26 - Huhe Ridge [15].

China: 27 -Ninsia Dist. [278]; 28 - Gansu Dist. [118]; 29 - Sa-cheu Oasis [18]; 30 - 50 km N of Sevrey-Ula: near Dzurgan-Huduk {Mongolia – MG} (ZIK); [the following] have not been found on the maps: Mongolia: near Barun-Boyan Ridge (ZMMGU); Mongolia: Ucrus-Hundey locality (ZMMGU); Werow (?Nerow)-Bulag (BNM); Bajn-Dzon (?Dzak) (BNM).

SYSTEMATICS AND GEOGRAPHIC VARIATION. Thirteen samples that covered the entire range of this species were compared by 21 characters. Very weak differences between lizards to the west and east of the Aral Sea can be ascertained. Western specimens are slightly smaller, and less defined dorsal tubercles, which are arranged in rows; the averages of quantitative characters also have weak differences. However, in the southern part of the range (Lake Sudochye; N Kyzylkum; Lake Balkhash) these differences disappear. Only in the extreme eastern part of the range (in Mongolia) does the frequency of such characters as the presence of small additional nasal scales and slight keeling of dorsal tubercles increase. As it is known, these characters are unique, particularly to a related species, *A. przewalskii*, that lives in direct proximity with Mongolian populations of this species.

HABITATS AND QUANTITATIVE DATA. Our observations in the vicinity of the Emba settlement [143], in the mountain massives Ulytau and Akchatau, in the vicinity of Semipalatinsk (central and northeastern Kazakhstan) and the data of a number of authors [3-5, 15, 22, 23, 29, 34, 35, 55, 70, 77, 98, 112, 118, 125, Neruchev and Chkhivadze – written report], allow us to provide a unified description of the squeaky pygmy gecko habitats. This species is a typical denizen of semi-deserts and, in places, deeply penetrates the steppe zone, where it still occurs in the areas of semi-desert-like environment. It is encountered primarily on the slopes of slight elevations (up to 150-350 m in Middle Asia and Kazakhstan and 600-1550 m in Djungaria and Gobi), which are covered with large and small disintegrated rock, in the areas of weathering with disintegrated slabs, and less frequently, on clay-gravel plains. There are known isolated cases of this species' penetration onto the edge of dunes (Gobi Desert; Golovushkin, verbal report). All the habitats are characterized by scanty grass vegetation made up of boyalych {*Salsola arbuscula* – MG}, tasbiyurghun {*Nanophyton erinaceum* – MG}, teresken {*Ceratoides* sp. – MG}, sagebrush {*Artemisia* sp. – MG}, *Salsola* sp. {MG}, and short grasses. The references by several authors to seeing this species on the trunks of saxaul {*Haloxylon* sp. – MG} [64, 67, 79, 80, 121] relate, undoubtedly, to the gray thin-toed gecko {*Tenuidactylus russowii* – MG}.

Our observations, conducted in 1978 in several places in the Kazakh Knoll country, demonstrate that the squeaky pygmy gecko is a common species and can reach very high numbers. Near a bridge across Mukur River (eastern Kazakhstan), in areas especially rich in cover (disintegrating slopes of schist hills), the density of the squeaky pygmy gecko reaches 10 individuals per 1 m². Slightly lower density has been noted in the Ulytau and Akchatau Mountains, 7 and 3 individuals per 1 m² respectively.

RESPONSE TO TEMPERATURE. Active geckos have been noted at temperatures of 21-25°.

DAILY ACTIVITY CYCLE. At the end of July and in the early days of August 1979, in the Kazakh Knoll country, we observed the emergence of this gecko onto the surface after sunset, around 2030-2100 hrs. The maximum numbers were observed during 1.5 to 2 hours after appearance. The lizards were active until sunrise. During daylight hours they take cover. They have been noted on the surface in cloudy weather [98].

{p. 61} SEASONAL ACTIVITY CYCLE. In southern Prebalkhashye, they appear on the surface after hibernation at the end of March, in the beginning of April. In southeastern Ustyurt, the geckos enter winter shelters on 3 November [98].

SHEDDING. In the Akchatau Mountain massif, apparently repeat shedding was observed among adults and the current year's hatchlings on 3-17 August 1938 [4]. In the Sarykamish Depression, shedding individuals were captured in June and September [34]. According to our observations, neonate lizards shed several hours after hatching.

FEEDING. The main food of the squeaky pygmy gecko are insects (orthopterans, beetles, hymenopterans, flies and their larvae). In addition, it eats arachnids, including small scorpions [4, 17, 22, 98, 132]. According to our data (stomachs of 25 lizards collected in the vicinity of Semipalatinsk on 10 August 1978 have been analyzed), the highest encounter rates were for butterflies (22%), flies (14.3%), hymenopterans and ants (9.1 and 7.8% respectively), as well as spiders (6.5%). In the stomachs of geckos from the Akhcha-Tau Range (n = 25) that were caught on 19 August 1978, we most frequently encountered heteropterans (29.1%), ants (20%), flies (12.8%), and spiders (12.8%). Altogether, in the stomachs examined, no fewer than 20 species of invertebrates were found.

REPRODUCTION. The geckos reach sexual maturity at body lengths of 25-28 mm. The sex ratio in the collected materials is close to 1:1. Mating, judging by the most enlarged gonads, occurs at the end of April and in May [70]. Gravid females were found in May and June [15, 34, 123, 137]. The female lays one to two eggs. The following dates of egg laying are known: on 29 May 1983, from the vicinity of Chabankazghan (first laying), and 10 August, from the vicinity of Semipalatinsk (second laying; our data). On 15 August 1969, eggs with developing embryos 13 mm long were found in the Gobi [77]. On 26 August 1982, 20 eggs were found with the dimensions 9-10 × 6-7 (9.65 ± 0.52 × 6.82 ± 0.1) mm (N. L. Orlov's written report). The egg dimensions from Kazakhstan were (n = 30) 8.0-11.0 × 5.0-7.0 (9.72± 0.14 × 6.81± 0.84) mm. We noted the appearance of hatchlings from clutches found in the vicinity of Semipalatinsk on 16-28 August 1978. According to data from the literature [17], young geckos begin appearing in July.

GROWTH RATE. Hatchlings have body lengths of 16.0-18.0 mm (the body dimensions vary within these limits for the individuals caught in nature and hatched in the laboratory). By the end of the activity season, they reach the size (without the tail) of 24 mm. Consequently, individuals from the first generation already can reproduce the next year. Several age groups are characteristic of this species' populations (Fig. 24) and the senior group contains individuals of the ages three and higher. {See also Fig. 25, not otherwise referenced – MG.}

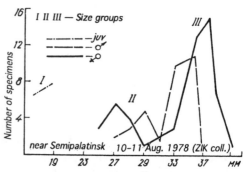

FIG. 24. Size groups of *Alsophylax pipiens*

ENEMIES, [AND] REASONS FOR DECREASE IN NUMBERS. The magpie has been noted

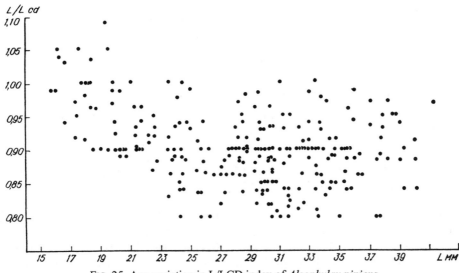

FIG. 25. Age variation in L/LCD index of *Alsophylax pipiens*

among the enemies of the squeaky pygmy gecko [3]. O. P. Bogdanov [27] determined that the numbers of *A. pipiens* decreased 600-700 fold in the 20 years from 1956 to 1976, which was caused by the drying of Aral Sea and movement of the shoreline {**p. 62**} away from rocks by 5-6 km, which (in turn) led to sharp reduction in the numbers of invertebrates. These lizards can also be displaced from their biotopes by plowing for cultivation.

BEHAVIORAL ATTRIBUTES. During the period of activity this lizard moves over the ground in short runs, lifting itself high above the substrate and bending the body. It can easily move along vertical surfaces, which, however, it avoids. It does not move far from cover. [The lizards] hunt for insects near the entrances to crevices and burrows, in the spaces under rocks, and chase each other. An agitated pygmy gecko emits single "metallic" tsk-type sounds that can be heard not only at night but also during the day, which is why they are often mistaken for bird calls. The vocal apparatus is especially well developed in the throat of the squeaky pygmy gecko in comparison with other geckos of the U.S.S.R. fauna. Over a small section of the throat opening, the walls are thickened and bulging [98].

The spaces under rocks and slabs, cracks in the soil, empty spaces around saxaul roots, burrows of vertebrates, and cracks in artificial stone structures [98] (our observations) can serve as the summer shelters for the pygmy geckos of this species. They spend the winter in burrows, cracks, [and] hollows of trees. There is a known case of wintering in a hole in a stone wall of a cattle enclosure [98]. The mating behavior of pygmy geckos remains unstudied due to their secretive way of life.

In schist outcrops, in the vicinity of Mukur River, at the end of July and beginning of August, according to our observations, the squeaky pygmy geckos lay their eggs in the cracks and hollows on the slopes with southeastern exposure at the depth of 5-20 cm. The eggs are laid singly and in groups, 5-10 of them per shelter from different females. These shelters are used for egg laying continuously for several years. Thus, in a disintegrating knoll slope we were able to discover a hollow at the depth of 13 cm with the dimensions of 30 × 50 cm in which had accumulated no fewer than 250-300 shell remnants from undeveloped eggs laid in different years. In the same space there were also fresh clutches. N. L. Orlov (written report) found 20 eggs (collective clutch) in the southern Gobi in a sand mound at the base of a saxaul plant {*Haloxylon* sp. – MG} on slopes with southern and eastern exposure at the depth of 30 cm.

PRACTICAL IMPORTANCE AND PROTECTION. The squeaky pygmy gecko is a beneficial species that, as a rule, inhabits areas not used by humans. However, there are exceptions to this. Thus, in a type locality region, on Mount Bolshoi Bogdo (Astrakhan region), its habitat is being destroyed due to the extraction of rock for gravel production. In this area, the squeaky pygmy gecko is in urgent need of protection measures. It must be included in the *Red Book for R.S.F.S.R.*, because Mount Bogdo is the only area in Europe inhabited by this species. Due to the small range and shrinking of the habitat of the pygmy gecko in Turkmenia, this species is included in the *Red Book for T.S.S.R.* [104].

PRZEWALSKI'S PYGMY GECKO — *ALSOPHYLAX PRZEWALSKII* STR., 1887

Type locality: Lower Tarim River; Hami and Cherchen Oases; Cherchen River; Lake Lobnor.
Karyotype: not studied.

1875 - *Gymnodactylus microtis* Blanford, J. Asiat. Soc. Bengal, Calcutta, vol. 44, part 2, p. 193 ("Nomen oblitum," Golubev, 1984).
1885 - "*Alsophylax microtis* (Blanf.)," Boulenger, Cat. Liz. Brit. Mus., London, vol. 1, p. 19 (nomen oblitum).
1887 - *Alsophylax przewalskii* Strauch, Mem. Acad. Sci. St. Petersburg, (7), no. 2, p. 55.

DIAGNOSIS. Large oval slightly keeled dorsal tubercles arranged in longitudinal and transverse rows; five to six preanal pores; additional nasal scale besides large supranasal scale; no broad transverse bands on body; tail white below.

LECTOTYPE. ZIL 5144, adult male. Tarim River (2500), 1878, N. M. Przewalsky. L = 29.2 mm; regenerated tail; large supranasal scale and small additional nasal scale; supranasals in contact with each other; pair of large flat postsupranasal scales follows supranasal; postsupranasals {p. 63} separated from each other by small scale. Eleven scales across head. 8/8 supralabials, 6/6 infralabials. 92 GVA; six preanal pores; 22/21 subdigital lamellae.

DEFINITION (Fig. 26) (from 31 specimen in ZIL, NMW, MNHP, UMMZ, USNM, and MCZ). L adult males (n = 12) 25.4-31.0 (28.85 ± 0.49) mm; L adult females (n = 17) 25.5-33.8 (30.89 ± 0.59) mm; L/LCD (summarized for both males and females) (n = 7) 0.66-0.85 (0.743 ± 0.028); HHW (n = 18) 63-71 (67.89 ± 0.66); EED (n = 23) 19-29 (24.83 ± 0.60); DTL (n = 29) 17-22 (18.59 ± 0.28); supralabials (n = 60) 7-10 (8.40 ± 0.08); infralabials (n = 60) 5-7 (5.83 ± 0.07); scales across head (n = 30) 9-13 (11.10 ± 0.15); scales around dorsal tubercle (n = 30) 8-11 (9.53 ± 0.15); GVA (n = 26) 84-98 (92.69 ± 0.74); preanal pores (n = 30) 5-6 (5.30± 0.09); subdigital lamellae (n = 54) 16-22 (19.63 ± 0.20).

Nostril between rostral, first supralabial, large supranasal scales and additional small nasal scale; supranasal scales contact each other (93.3%) or separated by a scale; postsupranasal scales defined distinctly (19.4%), weakly (38.7%) or not defined (41.9%); if defined, then they contact each other (44.4%), or separated by one to two (27.8% in each case) scales. Mental pentagonal, two to three pairs of postmentals; postmentals in first pair in contact with each other. Large oval slightly keeled dorsal tubercles arranged in regular longitudinal and transverse rows. Tail scalation has some differences compared to other representatives of genus: large oval weakly keeled tubercles at base of tail gradually disappear on first third of tail and do not extend onto its upper lateral surface. Longitudinal row of large plates absent or weakly defined and form combinations on single segment: (close to base of tail) 1 = 1 = 1 or 2 = 2 = 2, (in the middle) 1 = 2 or 1 = 1, (on the end) only 1 = 1.

COLOR AND PATTERN (in alcohol). Body sandy ochre above. Dark brown stripes three to five scales wide run from the middle of first to third supralabial scale through eye, above ear opening and along entire body on both flanks and then onto lateral surfaces at base of tail. Each such stripe edged along its upper edge by a light stripe 1.5 to 2 times narrower than the first one. This light stripe in the area between rostral scale and front edge of orbit, as well as on the last third of tail, is once again edged by thinner {p. 64} dark brown stripe that extends to the end of tail, merging at base of tail with the one from other side. No transverse stripes or bands on body and tail. Lower surfaces white.

SEXUAL DIMORPHISM AND AGE VARIATION. According to our data, females in comparison to males have less defined preanal pores; females are larger than males. There were not enough specimens in our sample to study the age variation (Fig. 23).

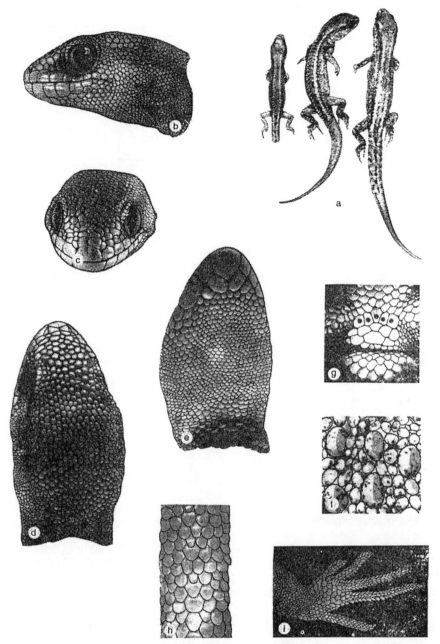

FIG. 26. *Alsophylax przewalskii*:
a, general view from above; b, head from the side; c, snout from front; d, head from above; e, head from below; f, dorsum; g, preanal pores; h, tail from below; i, hindfoot from above (b-e, ZIL 15690, Bugas to Lyukchun, China; others, ZIL 7030, Cherchen River, China).

DISTRIBUTION. Kashgharia; reaches Hami Oasis in eastern part of its range (Fig. 28).

{p. 65} LEGEND to Fig. 28.
1 - Chilan-su (ZIL); 2 - Kashkhgar [189]; 3 - Yarkand (MCZ); 4 - Yangihissar [189]; 5 - Khotan (ZIL); 6 - Nijadarya (ZIL); 7 - Kara-Sai (ZIL); 8 - Cherchen Oasis (ZIL); 9 - Lower Cherchen River (ZIL); 10 - Lower Tarim (ZIL); 11 - Lake Lobnor (ZIL); 12 - Lyukchun (NMW); 13 - from Bugas to Lyukchun (ZIL); 14 - Hami Oasis (ZIL).

SYSTEMATICS AND GEOGRAPHIC VARIATION. Based on the materials studied, no geographic variation was found.

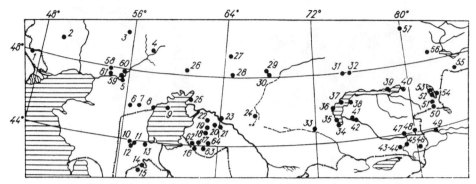

FIG. 27. Distribution of *Alsophylax pipiens* in Middle Asia and Kazakhstan

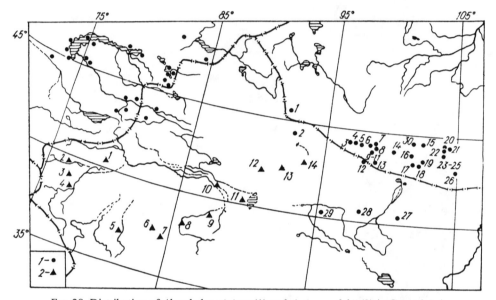

FIG. 28. Distribution of *Alsophylax pipiens* (1) and *A. przewalskii* (2) in Central Asia.

ECOLOGY not studied. From W. Blanford's work [187, 189] it follows that in western Kashgharia the Przewalski's pygmy gecko is found in inhabited and abandoned buildings. The habitat elevation for this species is 700-1400 m. Judging from the collectors' route descriptions, the lizards are active from April to October.

ARMORED PYGMY GECKO — *ALSOPHYLAX LORICATUS* STR., 1887

Type locality: Murzarabat, Mogoltau (northern Tadjikistan: Fergan Valley).
Karyotype: 2n = 36; 30 acrocentrics, 4 subtelocentrics, 2 submetacentrics, NF = 42.
1887 - *Alsophylax loricatus* Strauch, Mem. Acad. Sci. St. Petersb., ser. 7, vol. 35, no. 2, p. 59.

DIAGNOSIS. Large oval or triangular, distinctly keeled, dorsal tubercles arranged in strongly defined longitudinal and transverse rows; 8-12 preanal pores; besides large supranasal scale, one to two additional small nasal scales present; no broad transverse bands on body; white lower caudal surface.

LECTOTYPE. ZIL 4197. Adult male. Kishlak {a type of Middle Asian village – MG} Murza-Rabat, 1870, coll. Kushakevich. L = 28.8 mm; LCD = 41.8 mm; L/LCD = 0.69; large supranasal scale and one to two small additional nasal scales present; two supranasal scales in contact with each other followed by pair of postsupranasals, also in contact with each other; ten scales across head; 7/8 supralabials; 5 infralabials; two pairs of postmentals follow pentagonal mental, scales of first pair contact each other; 13 scales around dorsal tubercle; 91 GVA; 10 preanal pores; 18/19 subdigital lamellae.

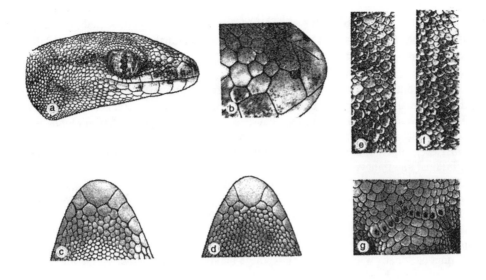

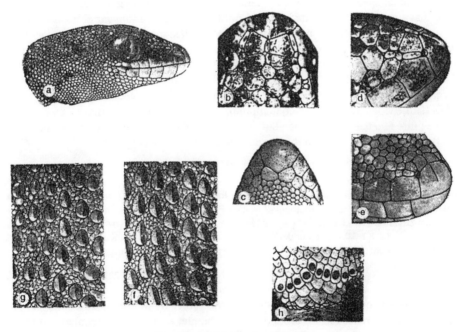

FIG. 29. *Alsophylax laevis* (top):
a, head from the side; b, nasal scales; c-d, chin; e-f, dorsum; g, preanal pores
(d, ZIK SR 1438, Kyzylkum; others, ZIK SR 2618, Meshkehd-i-Misrian, Turkmenia.
Alsophylax loricatus (bottom):
a, head from the side; b, snout from above; c, chin; d-e, snout from the side; f-g, dorsum;
h, preanal pores (c, e-f, ZIK SR 2725, Khiva, Uzbekistan, *A. l. szczerbaki*;
others, ZIK SR 940, Kanibadam, Tadjikistan, *A. l. loricatus*).

DEFINITION (Fig. 29) (from 229 specimens in ZIK, ZIL, DPI, DIZP, IZT, UIZP, CAS, UMMZ, USNM, and NMW). L adult males (n = 70) 22.0-30.8 (26.43 ± 0.28) mm; L adult females (n = 83) 22.3-32.8 (28.38 ± 0.29) mm; L/LCD (summarized for adult males and females) (n = 66) 0.64-0.86 (0.755 ± 0.007); HHW (n = 164) 57-75 (65.04 ± 0.33); EED (n = 108) 17-33 (23.01 ± 0.39); DTL (n = 110) 18-28 (22.77 ± 0.19); supralabials (n = 458)

6-9 (7.07 ± 0.03); infralabials (n = 458) 3-6 (4.96 ± 0.02); scales across head (n = 227) 8-15 (11.22 ± 0.09); GVA (n = 183) 82-101 (90.80 ± 0.30); preanal pores (n = 192) 8-12 (9.52 ± 0.06); {p. 66} subdigital lamellae (n = 403) 16-22 (18.46 ± 0.06); scales around dorsal tubercle (n = 227) 10-15 (12.25 ± 0.08).

Nostril between rostral, first supralabial, large supranasal scales, and one to two additional small nasal scales; behind supranasals, which are in contact with each other (99.1%), a couple of postsupranasals that also contact each other (94.6%) almost always present (96.6%). Mental pentagonal; one to two pairs of postmentals, scales of the first pair contact each other. Large oval or triangular keeled dorsal tubercles arranged in distinct longitudinal and transverse rows.

LIFE COLOR AND PATTERN. Colored dark brown to light coffee above. Light stripe that often has brown border above, extends from posterior edge of nasal scales to front edge of orbit on each side of snout; dark supraocular scales; dark stripe, as long as eye diameter, extends from back edge of orbit; no pattern or seven to eight indistinct narrow dark transverse bands on dorsum; on upper surface of tail base indistinct dark spot can be present; 12 indistinct dark narrow transverse bands; these bands can run together on the middle of the tail forming "X" shaped pattern, which can partially or completely disappear on tail or back. Lower surfaces white.

SEXUAL DIMORPHISM AND AGE VARIATION. In comparison with the males (and with females of other species in this genus), females of the armored pygmy gecko {*A. loricatus* – MG} have poorly differentiated preanal pores, which can disappear almost completely with age. Besides that, the maximum female dimensions exceed those of the males. Significant differences ($t = 3.72$) between the mean values of male and female dimensions were found in the western part of the range. Age variation becomes apparent in some body ratios, such as a shorter tail in juveniles. We used the data only from adult specimens, because the values of this character become more uniform in pubescent geckos.

DISTRIBUTION. According to contemporary data, the range of the armored pygmy gecko is broken into two parts no less than 600 km apart. In the west, it inhabits Turkmenia along the left bank of Amu-Darja and the Fergan Valley in the east (Fig. 30).

{p. 67} LEGEND to Fig. 30.
A. l. szczerbaki (1 - 7): 1 - Kunya-Urgench (ZIK); 2 - 27 km along Tashauz - Kunya-Urgench road: Guldumsaz fortress [41]; 3 - Tahta Dist.: Izmykshir fortress [41]; 4 - Bederkand fortress (ZIK); 23 km SE of Tashauz [129]; 5 - Tahta settlement (ZIK); 6 - Hiva town [41]; 7 - Darghanata settlement (ZIK).
A. l. loricatus (8 - 16 - MG}: 8 - watershed of Angren and Syr-Darja Rivers (ZIL); 9 - Leninabad {Hodjent - MG} (ZIK); 10 - Kok-kurak (ZIL); 11 - Kanibadam (ZIK); 12 - Kirovo [160]; 13 - Naukat [160]; 14 - Mogoltau (ZIL); 15 - Mirzorobad {Murzarabat - MG}; 16 - Kurgancha Kishlak [147].

SYSTEMATICS AND GEOGRAPHIC VARIATION. According to the data of M. Golubev and T. Sattarov [42], from comparisons of western and eastern populations it becomes clear that despite partial or complete overlap in the extreme values of almost all the characters, their average values decrease (scales around dorsal tubercle; DTL; GVA; supralabials) or, vice versa, increase (L/LCD; preanal pores; subdigital lamellae) from west to east. A number of characters (L of males and females; EED; infralabials) are stable, and these characters do not exhibit geographic variation. The change in dimensions of scales on the upper surfaces is defined by a decrease from east to west in the size of loreal and postsupranasal scales, all scales of the snout, the increase in the number of scales across the head, decreasing size of dorsal tubercles, and, as a consequence, a decreasing number of scales around these tubercles. However, the appearance of a second additional nasal scale in the individuals from the western part of the range forms a hiatus between the western and eastern populations. This provided the foundation for separating the Turkmen specimens into a different form.

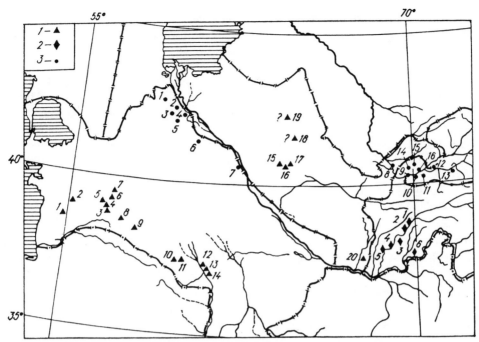

FIG. 30. Distribution of *Alsophylax laevis* (1), *A. tadjikiensis* (2), *A. loricatus* (3)

Key to the Subspecies of Alsophylax loricatus

1(2) Nostril between rostral, first supralabial, large supranasal scales and one
(100%) small nasal scale . . . . *A. l. loricatus* Str.

2(1) Nostril between rostral, first supralabial, large supranasal scales and two (94%;
very rarely, one - 6%) small nasal scales . . . . *A. l. szczerbaki* Golubev et Sattorov

ARMORED PYGMY GECKO — *ALSOPHYLAX LORICATUS LORICATUS* STR.

1979 - *Alsophylax loricatus loricatus* Golubev, Sattorov, Vestn. Zool., Kiev, no. 5, p. 9 (Species' type locality: Murzarabat, Mogoltau [northern Tadjikistan: Fergan Valley]).

DIAGNOSIS. Nostril between rostral, first supralabial, large supranasal scales, and one small additional nasal scale; distinct postsupranasals always contact each other; two to four large loreal scales equal in size to supranasals.

DEFINITION (from 125 specimens from Fergan Valley). The same type specimen as in species. L adult males (n = 25) 22.0-30.08 (27.28 ± 0.52) mm; L adult females (n = 42) 22.5-32.8 (28.87 ± 0.34) mm; L/LCD (n = 27) 0.64-0.82 (0.718 ± 0.008); EED (n = 34) 13-33 (22.94 ± 0.82); scales around dorsal tubercle (n = 122) 10-15 (13.22 ± 0.09); DTL (n = 90) 19-28 (23.18 ± 0.19); GVA (n = 87) 84-101 (92.05 ± 0.41); scales across head (n = 123) 8-13 (10.44 ± 0.09); supralabials (n = 250) 6-9 (7.23 ± 0.04); infralabials (n = 250) 3-6 (4.98 ± 0.03); preanal pores (n = 123) 8-11 (9.29 ± 0.06); subdigital lamellae (n = 243) 16-22 (18.15 ± 0.08). Besides the large supranasal scale, the only additional small nasal scale contacts nostril (98.4%); supranasals in contact with each other (99.2%), are followed by two postsupranasals also in contact with each other (94.8%); two to four large loreal scales, equal or slightly smaller than supranasals, between first supralabial and front edge of orbit.

DISTRIBUTION. Fergan Valley.

{p. 68} SZCZERBAK'S PYGMY GECKO — *ALSOPHYLAX LORICATUS SZCZERBAKI* GOLUBEV ET SATTOROV, 1979

Type locality: Eastern Turkmenia: Kunya-Urgench.

1979 - *Alsophylax loricatus szczerbaki* Golubev, Sattorov, Vestn. Zool., no. 9, p. 22.

DIAGNOSIS. Nostril between rostral, first supralabial, large supranasal scales, and two additional small nasal scales; two postsupranasal scales, variously defined, separated from each other by one (89%) or two (11%) scales; three to six small loreal scales distinctly smaller than supranasals.

HOLOTYPE. No. 13 ZIK, adult male. Turkmenia: Tashauz Dist.: Kunya-Urgench Town, June 26, 1976, collected by N. Szczerbak and M. Golubev. L = 25.0 mm; LCD = 32.0 mm; L/LCD = 0.78; Besides large supranasal scale, two small additional scales; supranasal scales, in contact with each other, followed by two postsupranasal scales, feebly defined and separated by one scale; 13 scales across head; 7 supralabials; 6 infralabials; two pairs of postmentals; 11 scales around dorsal tubercle; 89 GVA; nine preanal pores; 18 subdigital lamellae.

DEFINITION (from 104 specimens from eastern Turkmenia). L adult males (n = 40) 22.0-29.3 (25.89 ± 0.49) mm; L adult females (n = 37) 22.3-32.4 (27.89 ± 0.31) mm; L/LCD (n = 32) 0.70-0.95 (0.801 ± 0.009) (difference from eastern subspecies not found, t = 2.50); EED (n = 30) 17-33 (23.21 ± 0.43; t = 0.18); scales around dorsal tubercle (n = 98) 10-13 (26.00 {in error, error in original; probably ~12 ± – [Eds.]} ± 0.075 ; t = 10.6); DTL (n = 50) 14-30 (19.68 ± 0.34; t = 8.99); GVA (n = 78) 82-100 (89.52 ± 0.42; t = 3.83); scales across head (n = 99) 10-15 (12.14 ± 0.12; t = 7.38); supralabials (n = 208) 6-8 (6.78 ± 0.04; t = 4.84); infralabials (n = 208) 3-6 (4.95 ± 0.03; t = 0.20); preanal pores (n = 101) 8-12 (9.96 ± 0.09; t = 4.20); subdigital lamellae (n = 203) 16-21 (18.67 ± 0.08; t = 1.45). Besides large supranasal scale, two (94.2%) or rarely one (5.8%) additional small nasal scales contact nostril; supranasal scales contact each other (99.5%); followed by two distinctly (52%) or not distinctly (42.2%) defined postsupranasal scales, or no scales (5.8%); three to six small loreal scales, distinctly smaller than supranasals, between first supralabial and front edge of orbit.

DISTRIBUTION. Left bank of Amu-Darja River in northeastern Turkmenia {also found on the right bank of this river: Golubev, Streltsov, 1989; Golubev, 1990 – MG}.

HABITATS AND QUANTITATIVE DATA (Fig. 31). In Fergan Valley, this lizard has been caught in the cracks and ruins of old buildings [105, 108]. In eastern Turkmenia, it has been found on a small clay-saline section on the bank of an old aryk {a sort of irrigation canal in

FIG. 31. Biotope of *Alsophylax loricatus* and *Tenuidactylus caspius* (ruins of fortress rampart, Kunya-Urghench, Turkmen SSR).

Middle Asia – MG} between cotton fields [129]. All the known places of habitation of this species are connected with human activity and are located in direct proximity to clay-saline takyrs with the tamarisk {*Tamarix* sp. – MG}, camel spine {*Alhagi pseudalhagi* – MG}, *Lycium* sp. {MG}, and *Salsola* sp. {MG} growths. Numerous searches for this species that we conducted in the natural biotopes were fruitless. Probably, they have disappeared from there due to long term cultivation {*fide* Szczerbak; this lizard seems to be exclusively a synathropic species – MG}. All the known armored pygmy gecko finds are from ancient oases. Their range is within 200-400 m elevation. According to our observations, the pygmy geckos of this species are common and even numerous in the corresponding biotopes. On 26-27 June 1976, 70 individuals were caught at the base of an old fortress wall 80 m long, with western exposure (the greatest numbers; absent on the northern side), built from unfired loess brick. Individual geckos were found here at a distance of 30-40 cm from one another. Also here, on 6 June 1979, 52 individuals were noted. Judging by the abundance of sounds, heard on 18-20 June 1979 in the vicinity of Kanibadam, this lizard is also quite numerous here. S. Shammakov [129] caught 34 pygmy geckos on a 150 m long aryk bank near Dargan-Ata village.

{p. 69} The earlier, widely accepted opinion [23, 112, 121, 122] that the armored pygmy gecko is rare should be attributed to its secretive way of life and small, restricted range.

RESPONSE TO TEMPERATURE. According to S. Shammakov's observations [129, 130], in May the pygmy geckos were active with air temperatures of 21-24°.{p. 70} In Kunya-Urgench, at the end of June 1976 and the beginning of June 1979, active lizards of this species were observed at 2300 hrs with air temperatures of 21-26° [and] soil temperatures of 23-29°. The reduction of air temperature to 17° (Kanibadam, end of June 1978) did not affect the lizards' level of activity. According to T. Sattarov's observations (verbal report), the first armored pygmy geckos emerge after hibernation in Fergan Valley when the air temperature is 18-20°. In the summer, they are active when air temperatures are 30-35°, and take cover when the temperature exceeds 37°.

DAILY ACTIVITY CYCLE. In Kunya-Urgench, on 26 June 1976 and 6 June 1979, we noted the first individuals at 2100 hrs. Here they were active through the night until dawn. The greatest numbers were observed around 2300 hrs. According to T. Sattarov's data (written report), in northern Tadjikistan in May-June and September-October, these lizards enter a two-peak activity cycle; they are seen in the first half of the day until 1200-1400 hrs and in the second one until 1800-2100 hrs. These observations were made in residential buildings of the town Kanibadam, which has an abundance of secluded shaded areas, where, during the day, the conditions can resemble twilight.

SEASONAL ACTIVITY CYCLE. T. Sattarov (written report) noted the emergence of pygmy geckos from their winter shelters in Kanibadam on 10-11 March 1977, and a mass emergence on 7-12 April 1978. From our observations, the main emergence of these lizards in the vicinity of Leninabad occurred on 25 March 1979. During this time, young individuals predominate. In Fergan Valley, armored pygmy geckos disappear for the winter at the end of October.

SHEDDING. From the 49 pygmy geckos caught in eastern Turkmenia at the end of 1972, two individuals shed [129]. T. Sattarov encountered shedding geckos {p. 71} in northern Tadjikistan in May and August of 1977-1978. Shedding in individuals from Kunya-Urgench was noted on 3 May and 27 May 1983.

FEEDING. According to S. Shammakov's data [129, 130], the armored pygmy geckos feed on small spiders (encounter rate 58.3%) and ants (33.3%). Occasionally, click beetles, blister beetles, darkling beetles, leaf beetles, flies, and snout beetles are found (53 stomachs have been analyzed). According to our data from Kunya-Urgench (40 stomachs of pygmy geckos caught on 7 June 1979 have been dissected), in the first place are ants (56.7% encounter rate), small solpugids (30%), snout beetles (23.3%), hymenopterans (20.0%), butterflies (16.7%); spiders, orthopterans, cicadas, [and] heteropterans are encountered less frequently (Table 1). The analysis of Fergan Valley pygmy gecko stomachs (Sattorov, written report) showed that, here also, ants (76%), oniscomorphans (12%), and aphids (12%)

TABLE 1. Stomach contents of armored pygmy geckos
(Kunya-Urgench, Western Turkmenia, 7 April 1979, n = 40)

Food object	Stage	Absolute number of occurrences	% number of occurrences	Total number of specimens
Isopoda		2	6.7	4
Arachnida				
Pseudoscorpionidae	Imago	3	10.0	4
Fam. gen. sp.	"	1	3.3	1
Phalangidae	"	9	30.0	12
Arachnidae	"	1	3.3	2
Ixodidae	"	1	3.3	3
Insecta				
Orthoptera	"	3	10.0	7
Copeognatha	"	1	3.3	1
Homoptera				
Cicadellidae	"	3	10.0	6
Psyllidae	"	2	6.7	6
Homoptera	"	2	6.7	2
Hemiptera	"	4	13.3	6
Coleoptera				
Histeridae	"	1	3.3	1
Elatheridae	"	1	3.3	1
Coccinellidae	"	1	3.3	1
Chrysomelidae	"	1	3.3	1
Curculionidae	"	1	23.3	51
Coleoptera	Imago	3	10.0	3
Coleoptera	Larvae	1	3.3	1
Lepidoptera	Imago	5	16.7	7
Lepidoptera	Larvae	1	3.3	1
Hymenoptera				
Formicidae	Imago	17	56.7	103
Hymenoptera	"	6	20.0	6
Diptera	"	1	3.3	2
Reptilia, shed skin, *A. loricatus*		1	3.3	1
Totals		78		233

predominate; however, the diet is less varied (Table 2). In terraria, pygmy geckos readily ate aphids.

REPRODUCTION. From the eastern Turkmenia data [130], the female to male ratio is 1.0:1.4. Sexual maturity occurs at a minimum body length of 22 mm. According to T. Sattarov's observations, in northern Tadjikistan, pygmy geckos begin reproduction in April or at the beginning of May. At the end of June, in July, and in the first half of August they lay one or two clutches with one to two eggs with the dimensions of 8.6-8.8 × 5.5-6.1 mm (n = 5). In the middle of August, the hatching of the young begins. From our observation, incubation lasts 42 days. The length of newly hatched pygmy geckos is 15-17 mm.

The armored pygmy geckos (n = 20), observed in Kunya-Urgench on 14 April 1983, had not yet mated. In three females, caught here at the end of June 1979, we found a single mature egg. Three years later, a female caught here laid two eggs on 28 June.

GROWTH RATE. There are no data in the literature. As indicated above, hatchling pygmy geckos are 15-17 mm long. T. Sattarov reported to us that in October, that year's hatchlings are 18-19 mm long, individuals after hibernation (in April) are 20-21 mm, in May and June they are 23-25 mm, and in July and August, 27-29 mm. It is suspected that the individuals

TABLE 2. Stomach contents of armored pygmy geckos
(Tadjikistan SSR: near Kanibadam, coll. T. Sattorov, n = 25)

Food object	Stage	Absolute number of occurrences	% number of occurrences	Total number of specimens
Crustacea				
Isopoda	Imago	3	12	10
Arachnoidea				
Aranei	"	2	8	5
Insecta				
Orthoptera	"	1	4	1
Acrididae	"	1	4	1
Aphidodea	"	3	12	10
Coleoptera				
Elatheridae	"	2	8	6
Carabidae	"	1	4	3
Curculionidae	"	1	4	3
Neuroptera				
Chrysopidae	"	1	4	3
Hymenoptera				
Formicidae	"	19	76	152
Hymenoptera	Imago	2	8	4
Hymenoptera	Larvae	1	4	3
Psocoptera	Imago	1	4	4
Gastroliths		2	8	2
Totals		40		207

that hatched early participate in reproduction by the following spring. There are exceptions to that: in the sample from the vicinity of Leninabad, collected at the end of March {p. 72} 1979 (n = 42), there were young individuals 15-19 mm in size, probably from the last clutch.

ENEMIES. In Fergan Valley, these lizards are eaten by *Elaphe dione* {MG}, magpies {*Pica pica* – MG}, and domestic cats (Sattarov, written report). The lizards of this species were not found in the stomachs of two *Coluber karelini* {MG} and six owls {*Athene noctua* – MG} that were caught in the areas inhabited by the pygmy gecko in Kunya-Urgench.

The sharp decrease in the numbers of the armored pygmy gecko is due to the conversion of the land in their biotopes to human uses. Thus, in the places where S. Shammakov [129] reported large numbers of pygmy geckos, we were not able to find them in 1976, because the banks of the aryks had been dug recently and lightly covered with soil from the aryk bottom.

BEHAVIORAL ATTRIBUTES. This lizard moves in short runs. It prefers horizontal surfaces but can also move along vertical walls. On brick walls, it climbs to a height of 0.5 m. In its search for food it can climb onto grassy plants (observed in the terrarium). Spaces under mounds of clay or crevices in loess brick walls serve as summer shelter (from observations in Kunya-Urgench) at the depth of 5-20 cm. In northern Tadjikistan, these geckos have been found in cracks in the soil at the base of duvals {a kind of clay-brick wall around a Middle Asian house – MG} and other old earthen buildings (in these conditions old egg shells were also found), and in piles of trash and old cotton stems. Up to three individuals have been found in a single shelter. Winter shelters are in the same places but deeper (20-70 cm).

In the spring (the mating season), the peculiar voice of this lizard can be heard particularly often. It can be expressed as "chek..neeg." An agitated pygmy gecko can also emit a different sound, like "tsok...tsok...tsok," which resembles the clicking of metal sticks.

PRACTICAL SIGNIFICANCE AND PROTECTION. The continued existence of the armored pygmy gecko species depends entirely on humans. This lizard is of great interest to science as a unique component and a relic of the clay-saline biotopes, which have now disappeared. The already small range of this species (it is known from only a few localities) is constantly shrinking as the old buildings of historical interest in the areas inhabited by the pygmy gecko are restored and the old earthen domestic buildings are replaced with ones of brick and concrete. The species is included in the second edition of the *U.S.S.R. Red Book* and, on that basis, it should be declared a protected species, at least in one small area of its range of up to 1 hectare (create a micro-preserve) in Kunya-Urgench and around Leninabad or Kanibadam, because the armored pygmy gecko is not found in any of the existing preserves.

SMOOTH PYGMY GECKO — *ALSOPHYLAX LAEVIS* NIK., 1907

Type locality: Karry-Bent kishlak (southern Turkmenia).

Karyotype: 2n = 36; 28 acrocentrics; 4 subtelocentrics {error in original – MG}; 2 metacentrics; 2 submetacentrics; NF = 44.

1907 (1905) - *Alsophylax laevis* Nikolsky, Ann. Zool. Mus., vol. 10, no. 3-4, p. 333.
1934 - *Alsophylax microtis*, Chernov, Tr. soveta poizuch. proizvodit. sil. Leningrad, no. 6, p. 257 (part., err. syn.).
1949 - *Alsophylax pipiens*, Terentjev, Chernov, Opredelitel presmikayuschihsya i zemnovodnih. Moscow, Sov. Nauka, p. 130 (part).
1964 - *Alsophylax kashkarovi* Andrushko, Mikkau, Vestn. Leningrad Univ., no. 9, ser. biol., no. 2, p.14.

DIAGNOSIS. No dorsal tubercles; nine to 11 preanal pores; only large supranasal scale present, no additional small nasal scales; four to seven wide bands along body; tail white below.

HOLOTYPE lost; type locality is flooded now by Tedjen reservoir.

NEOTYPE. SR no. 314 ZIK, adult female. Turkmenia: Tedjen Dist.: Takyr Station, 10 May 1976, M. Golubev and N. Szczerbak coll. L = 33.6 mm; LCD = 35.0 mm; 8/8 supralabials; 5/6 infralabials; 14 scales across he- {p. 73} -ad; 93 GVA; 9 preanal pores; 16/15 subdigital lamellae; supranasals separated from each other by one scale; no postsupranasals; no postmentals, trapezoid mental scale bordered behind by five gular scales, those in corner slightly larger than others.

Neotype comes from a site close to the type locality.

DEFINITION (Fig. 29) (from 290 specimens in ZIK, ZIL, and IZT). L adult males (n = 98) 25.0-35.4 (31.07 ± 0.24) mm; L adult females (n = 114) 25.0-37.8 (33.12 ± 0.33) mm; L/LCD (n = 66) 0.82-1.00 (0.900 ± 0.001); HHW (n = 231) 58-72 (62.73 ± 0.22); EED (n = 169) 13-26 (19.54 ± 0.23); supralabials (n = 578) 5-9 (7.05 ± 0.02); infralabials (n = 579) 4-7 (5.76 ± 0.03); scales across head (n = 290) 12-19 (15.08 ± 0.08); GVA (n = 90) 87-105 (95.07 ± 0.42); preanal pores (n = 158) 8-13 (10.37 ± 0.08); subdigital lamellae (n = 556) 12-19 (15.46 ± 0.05). Nostril between rostral, first supralabial, and large supranasal scales (up to 7% of specimens may have one additional small nasal scale); supranasals contact each other or can be separated by one to three scales; behind supranasals, postsupranasal scales may be present, variously defined; mental scale pentagonal or triangular to trapezoid; none to two pairs of postmentals; no dorsal tubercles, however, some specimens from different samples have feebly defined ones in the sacral area that sometimes extend [forward] on the lower back.

LIFE COLOR AND PATTERN. Sandy ochre above. On each side of head, two to three scales-wide dark brown stripe runs from first labial scale through eye and high above ear opening. These stripes can run together on neck, forming horseshoe-shaped pattern. Between nostril and eye the stripes have lighter edging, and between these light edges an indistinct dark brown pattern is present on upper surface of snout from upper part of rostral scale to the level of the front edge of the orbit. Four to seven dark {p. 74} brown stripes of various widths (as a rule, wider than interspaces) run from neck to [dorso-] lumbar [region]. These stripes in the middle of the dorsum may be interrupted and shifted relative to one another along the longitudinal axis. Up to 11 wide transverse bands of the same color on supracaudal surface. Upper surfaces of limbs also covered with indistinct transverse bands. Body white below.

SEXUAL DIMORPHISM AND AGE VARIATION. According to our data, sexual dimorphism in this species appears as a greater maximum and average size of females and the weaker expression than in males of preanal pores, which may disappear with age.

Some proportions are variable with age. As one may see in Fig. 33, juveniles have shorter tails than adults (hatchling lizards have tails longer than the body). Besides, juveniles have a larger eye diameter in comparison to body length. These parameters become more uniform in adult geckos. In this case, we considered only data from adult specimens.

DISTRIBUTION (Fig. 30). Western and southern Turkmenia, central and southern Uzbekistan. Probably also found in northern Iran and northern Afghanistan {*i.e.,* within the limits of South Tadjik or North Afghan Depression – MG}.

LEGEND to Fig. 30 (according to our data, [36]).
{*Turkmenia* – MG} 1 - Meshkhed-i-Misrian; 2 - takyr between Kyuren-Dagh ridges and Malij Balkhan; 3 - four km SW of Kyzyl-Arvat; 4 - four km NW of Kyzyl-Arvat; 5 - 10 km NW of Kyzyl-Arvat; 6 - 30 km N of Kyzyl-Arvat; 7 - 40 km N of Kyzyl-Arvat; 8 - 8 km NW of Bamy Station; 9 - vicinity of Bakharden Station; 10 - Takyr Station; 11 - 33 km W of Tedjen {village-MG}; 12 - 24 km from Tedjen {village – MG} to Serakhs; 13 - Tedjen River Valley; 14 - 22 km from Serakhs to Tedjen {village-MG}; {Uzbekistan – MG} 15 - Shuruk settlement on the southern train of Kuldjuktau Mauntains; 16 - 30 km W of Ayakagytma; 17 - Ayakagytma; 18 - Karakatta Hollow; 19 - Mynbulak Hollow; 20 - Karasu around Termez.

SYSTEMATICS AND GEOGRAPHIC VARIATION. For a long time, the smooth pygmy gecko was considered as related to the squeaky pygmy gecko [98; 112; 121, *et al.*]. Since the validity of this species was confirmed [20], an independent species, *A. kashkarowi* [6, 9], was recognized. S. Shammakov and Ch. Atayev [131] denied its distinctiveness, however, not supporting their viewpoint by any arguments. Having studied around 300 individuals of the smooth pygmy gecko from different parts of the range, we [36] concluded that this form is not valid. As it became clear, some of the meristic and variable characters of this species are subject to clinal variation. Thus, from southeast to northwest, supranasals and postsupranasals, as well as the scales of the snout, become larger and coarser, the frequency of contacts of these scales increases (the line between Meshkhed and Takyr Station is an isophene of cline of the last character); some absolute body dimensions and the number of preanal pores increase. In the same direction, per every 100 km, about 15% of the geckos lose the trapezoid shape of the mental scale and the well developed pair of postmentals.

Despite some peculiarities of the Kyzylkum specimens, in particular, the lower average value of subdigital lamellae number and the ochre brown (instead of dark brown) body, tail, and limb bands, it is not advisable to recognize them as a different subspecies.

HABITAT AND QUANTITATIVE DATA (Fig. 32). The smooth pygmy gecko is a typical denizen of takyrs, that is of the naked, flat clay surfaces almost free of vegetation in the sand desert zones. In Turkmenia, we would catch lizards of this species on a smooth takyr with multiple cracks and thin grassy vegetation made up of ephemeral grasses and dry *Salsola* {MG}, and less frequently, on takyr-like saline areas among mounds from the dead saxaul {*Haloxylon* sp. – MG} and *Salsola* sp. {MG} bushes on the boundary between the clay and sand deserts (on Messerian Plateau, in the vicinity of Malij Balkhan Crest, near the town Kyzyl-Arvat and Bami and Takyr stations). In Uzbekistan, we observed the pygmy geckos of this species on a saline takyr with occasional small bushes of *Salsola* sp. {MG}, in sections with and without vegetation (the {**p. 75**} vicinity of Ayakagytma settlement, to the southwest of Kuldjuktau Crest). The takyr salinity levels in Kyzylkum do not affect the numbers of these lizards. It is a different case in western Turkmenia. On the Messerian Plateau, on 5 May 1984, we dug up the pygmy geckos shortly after a rain. The saline takyr took on a dark color and the purely clay areas that do not absorb moisture were lighter. The lighter spots were distributed over the saline takyr without order. The results of our work quickly suggested to us that the digging should be done only in the clay sections, because the lizards were absent from the nearby saline areas. This can probably be explained by the fact that the western Turkmen takyrs are frequently flooded in the spring with rain water that will not penetrate a layer of clay even 5 cm thick. The saline soils are easily penetrated by water and

FIG. 32. Biotope of *Alsophylax laevis* (takyr on the Kesserian Plateau, Turkmenia)

hold it for a long time. The areas inhabited by the smooth pygmy gecko are located at elevations of 200-250 m.

According to the census data that were collected using truck headlights on the Messerian Plateau on 2 May 1975, 13 individuals were counted along 16 km of the route; the highest count of up to three individuals was noted in a 10 m^2 area. In the same region, on 12 May 1975, 14 individuals were caught along 12 km of the route, and on 7 May 1983, 25 individuals were observed along 17 km of the route with an observation strip 5 m wide. On the same night, the highest count of 5 individuals was discovered in an area 100 × 25 m; that is 1 individual per 500 m^2. In the vicinity of the Meshkhed-i-Messerian ruins, on 5 May 1984, we encountered areas where lizards were found 5-6 m from each other. It is here that the highest numbers of the smooth pygmy gecko were registered in Turkmenia. In other areas, we only came across single individuals, although according to the literature [25, 131], the lizards of this species were quite numerous in the foothills of Kopetdagh and Malij Balkhan, in the Tedjen Valley and near Takyr Station. Evidence of our persistent search for the smooth pygmy gecko is our thorough examination of 930,000 m^2 of this lizard's biotopes

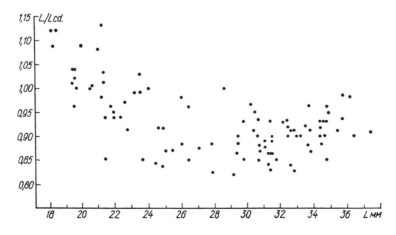

FIG. 33. Growth change index *L/LCD* for *A. laevis*

that we conducted in recent years during 12 field trips. They were not found on the takyrs of Tedjen River Valley and the vicinity of Khauskhan, of the town Tedjen, between the rivers Meana and Chaacha, Takyr Station, between Kompressornaya and Uch-Depe, between Archman and Bakharden, 25 km east of Kazandjik, between Kyzyl-Arvat and Kazandjik, near Butentau (Zaunguz Karakums), near the Bami Station (the last pygmy gecko caught on 8 May 1975), in many areas in southern and central Karakums.

{p. 76} The highest numbers of five individuals for 3 km of the route, *i.e.,* one individual per 600 m, was noted in Uzbekistan, in the vicinity of Ayakagytma settlement, on 14 June 1976. A large population density of this species was found in the vicinity of Kuldjuktau Crest on 25-29 June 1979; lizards were noted every 5-8 m in an area 500 × 200 m, *i.e.,* there were no fewer than 1600 individuals living here. On 18 June 1976, we found no smooth pygmy geckos on the takyrs of Mynbulak Depression.

The aforementioned demonstrates that currently this species is experiencing a progressive decline in numbers and a shrinking distribution.

RESPONSE TO TEMPERATURE. In western Turkmenia, in May (1975, 1983, 1984), active pygmy geckos were observed on the surface when the air temperature was 21°, the soil temperature was 18°. They were absent when the air temparature was 18°. On 26 June 1979, in the vicinity of the Kuldjuktau crest, the air temperature fell to 15° by 2200-2300 hrs and the pygmy gecko activity sharply decreased, although occasional lizard squeaks could be heard until 0200-0300 hrs.

DAILY ACTIVITY CYCLE. According to our observations in western Turkmenia on 2 May 1975, the pygmy geckos appeared around 2200 hrs, were encountered in greatest numbers at 2300 hrs, and disappeared at 2400 hrs. At the same location, on 12 May 1975, the first lizard was encountered at 2330 hrs; individuals were active until 0230 hrs. On 7 May 1983 they were discovered at 2230 hrs, the greatest numbers were noted at 0145 hrs, and active pygmy geckos (the field trip ended at 0240 hrs), judging by their sounds, remained until dawn.

In Kyzylkum on 14 June 1976, the first pygmy gecko was caught at 2100 hrs. On 26 June 1979, they were first seen soon after sunset (the first ones were observed without a flashlight) at 2030 hrs, and the main emergence occurred within a half hour. The activity peak was observed within 2-3 hours after the appearance and it levelled off by 2300-0100 hrs. Under favorable conditions, inactive individuals remained on the surface until dawn — 0400-0430 hrs. In summation, it can be concluded from these observations that the pygmy geckos emerge from shelter at twilight from 2030 to 2200 hrs, the activity peak is noted 2 to 3 hrs after they appear on the surface, and they disappear en masse after 2400-0100 hrs, although some individuals may remain on the surface until dawn. In the springtime, these lizards' activity period is somewhat shortened and lasts 2-3 hrs.

SEASONAL ACTIVITY CYCLE. The first occurrence of a pygmy gecko with a full stomach is from 10 March 1968 [25]. As a rule, they {p. 77} appear later, at the beginning of April. S. Shammakov [130] encountered them for the first time on 6 and 7 April 1970, at which time in 1971 they were still hibernating. According to his data, the last active pygmy geckos were caught on 21 October 1967 and 20 October 1968, and their complete departure for hibernation is presumed to be at the beginning of November. According to our observations on the Messerian Plateau, on 1 October 1981 the pygmy geckos no longer appeared.

SHEDDING. S. Sammakov [130] noted shedding individuals in May, September, and October, and skins in the stomachs of pygmy geckos were discovered in May, July, and September. We have noted mass pygmy gecko shedding in southern Kyzylkum at the end of June 1979, and on 10 August 1983 in Turkmenia. A freshly shed skin of this species was found in a termite colony near Meshkhed-i-Messerian on 28 September 1981. Newborn pygmy geckos shed a few hours after hatching. Probably, they shed no fewer than four times in a season.

FEEDING. The most complete feeding data for this species can be found in S. Shammakov's work [130]. The author had analyzed 182 lizard stomachs that were procured between April and October during the years from 1967 to 1970. As it turned out, spiders are

TABLE 3. Stomach contents of smooth pygmy gecko
(Western Turkmenia: Messerian Plateau, May 1975; n = 30)

Food object	Stage	Absolute number of occurrences	% number of occurrences	Total Number of specimens
Crustacea				
Isopoda	Imago	8	22.2	12
Arachnida		—	—	—
Pseudoscorpionidae		—	—	—
Chelifer sp.	"	3	8.3	3
Arachnida	"	4	11.1	6
Insecta				
Mantoidea	"	1	2.8	1
Homoptera	"	1	2.8	1
Hemiptera	"	2	5.6	3
Coleoptera				
Staphilinidae	"	1	2.8	1
Elateridae	"	1	2.8	1
Scarabaeidae	"	1	2.8	1
Hymenoptera				
Formicidae	"	5	13.9	10
Lepidoptera	Larvae	1	2.8	1
Diptera	Imago	1	2.8	1
Insecta sp.		5	13.9	5
Totals		36		48

the most important [items] in the smooth pygmy geckos' diet (50.5% encounter rate), as well as termites (20.6%) and beetles (10.9%), while other groups of invertebrates are rarely encountered. In April and May pygmy geckos feed primarily on termites (40.2%) and small beetles (31.2%), and in the summer and fall — on spiders (up to 80% encounter rate). Our analysis of smooth pygmy gecko stomachs from western Turkmenia (n = 30) demonstrated that they most frequently eat *Oniscomorpha* (22.2% encounter rate), ants and flies (13.9% each), and spiders (11.1%) (Table 3). As a rule, these are small invertebrates with soft chitinous shells.

REPRODUCTION. The sex ratio in our collections is close to 1:1. Pygmy geckos become sexually mature as they reach body lengths of 25-29 mm. Mating occurs at the beginning of April, judging by the greatest testicle mass [130]. In eight females that we caught on the 16 May 1975 near the Meshkhed-i-Messerian ruins, we found one egg in the oviducts [of each individual]. These were nearly ready for laying (8-9 × 5 mm). In seven dissected lizards [in which we found two oviducal eggs in each], the larger egg was somewhat smaller (6-8 mm) [than that reported above], and the second egg [was even] less well developed (2-4 mm), which suggests staggered clutches. A pygmy gecko female caught in the same place and at the same time as the ones mentioned above, laid one egg and another female laid two on 22 May 1975. S. Shammakov [130], on 14 July 1972, discovered an egg ready for laying. According to our materials (n = 14), the egg dimensions are 8.5-10 × 5.9-7.2 (9.27 ± 0.13 × 6.51 ± 0.1) mm. A pygmy gecko with a 18.5 mm body length and 16.5 mm tail length hatched on 5 August 1979 from an egg that was found on 25 June 1979. A hatchling with the respective dimensions of 17.3 and 17.0 mm hatched on 10 August 1979 from an egg that was found in the same place. Judging by the data presented above, incubation lasted no fewer than 42-47 days. Later discoveries of hatchlings indicate a second clutch.

RATE OF GROWTH. According to S. Shammakov's materials [130], the body length of a pygmy gecko increases by 6-10 mm in the first 9-10 months. It is supposed that in June and

July some individuals enter the group of sexually mature animals and are included in the reproductive cycle. In subsequent years, the distinctions between two- and three-year-old individuals diminish.

ENEMIES. The smooth pygmy gecko has been found in the stomachs of a racerunner (*Eremias intermedia*), Karelin racer (*Coluber karelini*), and saw-scaled viper (*Echis carinatus*). It is possible that they are eaten by solipugids, which are numerous in the biotopes of this species. As indicated above, continuing decline in the numbers of the smooth pygmy gecko and shrinking of their range have been noted in recent years in Turkmenia. The main cause of this is the transformation of its biotopes due to human activities in the area. The Turkmen Canal, which cut through the takyrs near Kopetdagh, strongly stimulated their cultivation. Thus, in the vicinity of Takyr Station, where these lizards were once common, they disappeared after the takyr has been plowed. Right now the canal reaches Kizyl-Arvat and will continue further on the Messerian Plateau, which threatens the complete annihilation of this species.

{p. 78} These pygmy geckos have already disappeared in the Tedjen River Valley.

The Karrybent type locality was particularly rich with the pygmy geckos. Thus, O. P. Bogdanov [25] reported that here he dug out 62 individuals from their cover over the distance of 200 m. Also here, as S. Shammakov reported, he collected up to 10 lizards during a one-hour excursion in 1971. In recent years, persistent and numerous searches have been unsuccessful. After a flood in 1977 that was caused by a breach in a reservoir dam, smooth pygmy geckos competely disappeared from there. On 5 May 1984, on the Messerian Plateau, we would dig up pygmy geckos from the cracks and insect burrows, and they were so close to the surface (cloudy day) that even a simple drive over the takyr would undoubtedly kill some of them.

BEHAVIORAL ATTRIBUTES. This lizard moves in a manner typical of geckos of the plains – its body is arched and tail is lifted. When disturbed, the pygmy gecko runs quickly 2-5 m and then freezes. From observations in the terrarium, they can also move on vertical surfaces and climb up to 50 cm.

For summer shelter pygmy geckos use cracks in the soil, termite hill passages, old beetle burrows, hollows in the saxaul and *Salsola* sp. stumps, and other lizards' burrows. They have been found in burrows with entrances nearly sealed or washed shut by the rain. As we determined for moist substrates, pygmy geckos can dig their own burrows 0.5-0.6 cm in diameter. On cloudy days they stay at a depth of 3-5 cm and deeper on the hot ones. The pygmy geckos also lay their eggs here. Thus, in western Turkmenia, we found clutches in an old termite hill at the depth of 15-20 cm (six eggs – a colonial clutch), and under an old *Salsola* sp. bush at the depth of 5 cm. Fragments of 16 pygmy gecko eggs were found at the boundary between dry and moist substrates in a termite hill 2 m in diameter and 20 cm tall, and were concentrated in the northern section of the dome. S. Shammakov [130] found pygmy geckos in termite hills at a depth of 5-11 cm, in sunwatcher {*Phrynocephalus helioscopus* – MG} burrows (depth of 12 cm, burrow length 17-30 cm), in the roots and trunks of saxaul, [and] in long-tailed desert lizard {*Mesalina guttulata* – MG} burrows. Usually, one or two individuals hibernate together. Once, five lizards were found together.

Smooth pygmy geckos emit a single sharp sound that can be expressed as "ee-eek ... ee-eek." When they discover prey or when in danger, the call frequency increases. They can be heard especially often during the mating period.

PRACTICAL SIGNIFICANCE AND PROTECTION. As a species with declining numbers and range and at our urging, the smooth pygmy gecko was included in the second edition of the *U.S.S.R. Red Book*. Unfortunately, this species is not found in the existing preserves. For its protection it would be advisable to set aside a 6-10 km^2 section of a saline takyr to the north of Bami Station (as an extension of the Kopetdagh Preserve), where, aside from the smooth pygmy gecko, a unique spotted toad-headed lizard species (*Phrynocephalus maculatus*) is found.

TADJIK PYGMY GECKO — *ALSOPHYLAX TADJIKIENSIS* GOLUBEV, 1979

Type locality: Valley of Vakhsh River in southern Tadjikistan.
Karyotype: 2n = 36; 30 acrocentrics; 4 subtelocentrics; 2 submetacentrics, NF = 42.

1962 - *Alsophylax pipiens*, Said-Aliyev, Trans. IZIP Akad. Nauk Tadj.S.S.R., vol. 22, p. 103.
1977 - *Alsophylax laevis*, Muratov, N., and Muratov, P., in: Vopr. gerpetologii, avtoref. dokladov, Leningrad, Nauka, p. 150.
1979 - *Alsophylax laevis tadjikiensis* Golubev, Trans. ZIN Akad. Nauk S.S.S.R., vol. 89, Ecol. i sist. amph. i rept., Leningrad, izd. Akad. Nauk S.S.S.R., p. 62.
1984 - *Alsophylax tadjikiensis*, Golubev, Vestn. Zool., no. 2, p. 73.

DIAGNOSIS. No dorsal tubercles; five to seven preanal pores; besides large supranasal scale, one (around 93%), very rarely two (approx. 7%) additional small nasal scales present; no wide dorsal bands; live specimens have ochre orange tail.

{p. 79} HOLOTYPE. ZIL 18929, adult male. Tadjikistan: Kurghantyube Dist.: 7 km from Kyzylkala settlement down Vakhsh River, 25 April 1965, collector unknown. L = 28.2 mm; LCD = 34.5 mm; L/LCD = 0.82; large supranasal scale and one additional small nasal scale present; behind supranasals, which contact each other. There is a pair of weakly defined postsupranasal scales which are separated by one scale; 6/6 supra- and infralabials; 15 scales across head; six preanal pores; 98 GVA; 16/17 subdigital lamellae.

DEFINITION (Fig. 22) (from 20 specimens in ZIK and ZIL). L adult males (n = 8) 27.4-30.2 (28.30 ± 0.34); L adult females (n = 6) 25.0-31.0 (28.18 ± 1.05); L/LCD (n = 3) 0.80-0.88 (0.833 ± 0.024); HHW (n = 11) 60-70 (64.82 ± 0.91); EED (n = 16) 13-26 (19.50 ± 0.94); supralabials (n = 40) 5-7 (6.10 ± 0.08); infralabials (n = 40) 5-6 (5.38 ± 0.08); scales across head (n = 13) 12-15 (13.58 ± 0.23); GVA (n = 14) 89-102 (96.50 ± 1.08); preanal pores (n = 18) 5-7 (6.22 ± 0.15); subdigital lamellae (n = 34) 14-18 (15.88 ± 0.18). Nostril between rostral, first supralabial, large supranasal scales and one (around 93%) or, very rarely, two (around 7%) additional small nasal scales; supranasal scales in contact with each other (90%) or separated by a scale, are followed by postsupranasal scales (30%) variously defined and separated by one scale (90%) or in contact with each other. Mental scale trapezoid (60%), hexagonal (25%), triangular or pentagonal (15%); 70.5% of specimens have one or two (27.3%) pairs of postmentals, 2.2% lack these scales. No dorsal tubercles, but, as in the previous species, they may appear sometimes in sacral and lumbar areas.

LIFE COLOR AND PATTERN. Color of upper body surfaces ochre with olive (khaki) tint. Five or more transverse bands, lighter than main color, may be on body between fore- and hindlimbs. Brown transverse wavy stripe on body-tail boundary; tail unicolor bright ochre with orange tint; brown stripes (up to eight) or spots may be scattered without order. From the tip of snout through eye up to the stomach area and sides of neck is a dark brown stripe, which may continue as separate spots along dorsum-abdomen boundary to hindlimbs. Above this stripe runs a yellowish stripe of the same width from nostril to supraocular scales. Light supraocular stripes come together on tip of snout; the interspace between nostrils and eyes darker than main color. Regenerated tail colored as nonregenerated. Lower surfaces (belly, thighs, crus) light grayish; lower surface of tail ochre orange.

SEXUAL DIMORPHISM AND AGE VARIATION. Compared to the males, female Tadjik pygmy geckos have less defined preanal pores. Females are [also] slightly larger than males. Despite the relatively small number of specimens studied, a tendency toward the change of some body proportions, characteristic of all the species of this genus, was noted, such as the age-related increase of tail length and decrease of eye diameter relative to body size.

DISTRIBUTION (Fig. 30). Southern Tadjikistan.

LEGEND to Fig. 30.
1 - 7 km from Vakhsh settlement downwards along Vakhsh River [81]; 2 - Sumbulak settlement (ZIK); 3 - Kolkhozabad (ZIL); 4 - right bank of Vakhsh River in front of Djlikul near foothills of Aruktau Mountains [107]; 5 - 53-56 km on road to Shaartuz [81]; 6 - 18 km from crossing of Kyzylsu River in Karatau foothills [107] [this locality] needs confirmation).

GEOGRAPHICAL VARIATION was not found. This species is related to the smooth pygmy gecko and until recently was united with the latter.

HABITATS AND QUANTITATIVE DATA (Fig. 34). We caught these lizards in the foundations of the ruins of loess buildings and walls, but the main places inhabited by this species are the clay slopes with rocks scattered {p. 80} on the tops and with rodent burrows, occasional *Ephedra* sp. bushes and grasses (observed near Sumbulak village, 13 km south of Kyzyl-Kala settlement). They were also encountered in dry pebble channels that remained after spring floods and slides (clay washes out of the conglomerates, [leaving] in the stone piles many cracks and hollows). Areas of habitation of this species are located at an elevation of 450 m. An area 500 × 500 m, according to our count by their voices, is inhabited by about 20 individuals. In some cases individuals were encountered 25 m from one another.

FIG. 34. Biotope of *Alsophylax tadjikiensis* (stony semi-desert along the foothills of Aktau Ridge in the vicinity of the Sumbula Settlement, Tadjik SSR).

RESPONSE TO TEMPERATURE. Active lizards were seen on 27-28 April 1980 when the soil temperature was 20-27° and the air temperature was 17-18°. With an air temperature of 12°, on 22 April 1978, pygmy geckos did not appear.

DAILY ACTIVITY CYCLE. According to observations near Sumbulak village at the end of April, voices of the pygmy geckos could be heard even before darkness at 2045 hrs. The first lizard was caught at 2100 hrs. By 2230 hrs, the level of activity fell off sharply and vocalizations stopped almost completely. The last individual was caught around 0100 hrs.

SEASONAL ACTIVITY CYCLE. This species emerges from cover after hibernation in March [81]. In 1984, after a long and cold spring, lizards were still absent from the surface at the beginning of April. The time of disappearance for the winter has not been established.

SHEDDING was observed on 29 April 1975 [81].

FEEDING. According to our investigations (6 stomachs of the lizards caught in April were dissected), spiders predominate (37.5% encounter rate) and the other invertebrate groups are equally distributed (pseudoscorpions, cicadas, aphids, beetles, and butterflies are 12.5% each).

REPRODUCTION has not been studied. Digging in the pygmy gecko's biotopes has yielded egg shells, which allowed the determination of the following egg dimensions – 8.3-8.7 × 5.1-6.1 mm (n = 6).

GROWTH RATE has not been studied. Drawing an analogy with a close species (smooth pygmy gecko), sexual maturity is reached probably by the beginning of the following season.

ENEMIES have not been discovered. There may be opportunistic attacks on pygmy gecko by the steppe ribbon snakes {*Psammophis lineolatus* – MG}, and by solpugids and scorpions, which are all found in this lizard's habitat.

BEHAVIORAL ATTRIBUTES. Unlike the smooth pygmy gecko, the Tadjik pygmy gecko frequently moves around on vertical surfaces in nature. Several times we found them on rocks up to 0.5 m high. Once we found one sitting {**p. 81**} on the side of a large 1.5 m-tall boulder, on a vertical surface with a negative angle, its head pointing down.

For shelter this species uses cracks in rock piles, spaces under rocks, rodent burrows and cracks in loess ruins. They may also spend their winters there (at a depth of 40 cm). We found clutches of eggs when digging in the remains of loess buildings at their bases and at a depth of 5-20 cm. Judging by their numbers, it can be supposed that two or three females may lay their eggs in a single place. We also found the remnants of a clutch in a rodent's {*Meriones* sp. – MG} passage at the depth of 40 cm and 1 m from the burrow entrance.

In the spring, after nightfall the voice of the Tadjik pygmy gecko is heard frequently. It can be imitated by the word "petz" with a metallic ring to it. These sounds come in series of three to four and then pause to repeat 0.5-1 min or more later.

PRACTICAL SIGNIFICANCE AND PROTECTION. The Tadjik pygmy gecko is a rare species with a narrow range that inhabits the zone of active human cultivation. It is not found in the Tadjikistan preserves. This species should be included in the *Tadjik S.S.R. Red Book* and an area (13 km south of Kyzyl-Kala) 500 × 1000 m between the highway and Vakhsh River, with its high population density, should be declared a micropreserve.

Subgenus *Altiphylax* Jeriomtschenko et Szczerbak, 1984

Type species: *A. tokobajevi*
1984 - *Altiphylax* Jeriomtschenko et Szczerbak, Vestn. Zool., no. 2, p. 46.

DIAGNOSIS. Digits with slight angular bend; preanal pores only in males; length of forelimb approximately 30% of body length, when adpressed reaches tip of snout with fingers; adpressed hindlimb reaches middle of neck with toes; caudal segments feebly differentiated (up to 9-10 segments); caudodorsal tubercles are [also] on lateral surfaces to [level of] segments 6-8. Subgenus includes one species.

TJAN-SHAN PYGMY GECKO — *ALSOPHYLAX TOKOBAJEVI* JERIOMTSCHENKO ET SZCZERBAK, 1984

Type locality: Kirghizia: Naryn Dist.: vicinity of Baygonchek village.
Karyotype: 2n = 34; 26 acrocentrics, 4 subtelocentrics, 4 submetacentrics, NF = 44.
1984 - *Alsophylax tokobajevi* Jeriomtschenko, V. K., Szczerbak, N. N., Vestn. Zool., no. 2, p. 46-50.

DIAGNOSIS. Small roundish flat dorsal tubercles more or less without order; six to seven preanal pores; large supranasal scale and two additional small nasal scales present 8-13 supralabials; 4-7 wide crossbands on body; tail ochre.

HOLOTYPE. Re. No. 16 ZIK, adult male. Kirghiz S.S.R.: Naryn Dist.: spurs of Baybichee-Too Ridge: near Baygonchek village, 1800 m elevation, 11 May 1983, coll. V. K. Jeriomtschenko. L = 34.2 mm; LCD (regenerated) = 30 mm; 18 scales across head; 10/12 supralabials; 7 infralabials; one to three pairs of postmentals; 100 GVA; 6 preanal pores; 26 subdigital lamellae.

DEFINITION (Fig. 35) (from 61 specimens in ZIK). L adult males (n = 22) 33.6-45.4 (40.58 ± 0.60) mm; L adult females (n = 27) 30.8-49.5 (40.80 ± 1.08) mm; L/LCD (n = 18) 0.68-0.87 (0.77 ± 0.01); HHW (n = 51) 47-64 (55.33 ± 0.53); EED (n = 33) 18-28 (23.64 ± 0.44); scales around dorsal tubercle (n = 61) 7-10 (8.34 ± 0.09); DTL (n = 49) 7-12 (9.57 ±

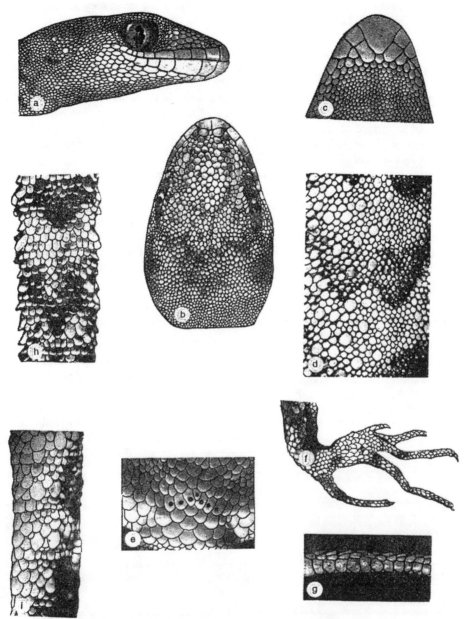

FIG. 35. *Alsophylax tokobajevi* {error in original – MG}:
a, head from the side; b, head from above; c, chin; d, dorsum; e, preanal pores;
f, hind foot from above; g, toe from below; h, tail from above; i, tail from below
(ZIK RF 16/3, Baygonchek, Kirghizia).

0.16); supralabials (n = 119) 8-13 (10.44 ± 0.09); infralabials (n = 122) 5-8 (6.72 ± 0.07); scales across head (n = 60) 13-21 (16.37 ± 0.21); GVA (n = 56) 91-109 (102.04 ± 0.53); preanal pores (n = 21) 5-7 (6.33 ± 0.16); subdigital lamellae (n = 121) 20-27 (23.5 ± 0.12).

{p. 82} Besides large supranasal scales, nostril contacts two (100%) additional small nasal scales; supranasals in contact (19.7%) or separated by one (75.4%) to two (4.9%) small {error in original – MG} scales; postsupranasal scales defined very weakly (14.7%) or completely not defined (85.3%); mental scale hexagonal (42.6%) or trapezoid (57.4%); one (4.2%), two (59.7%) or three (36.1%) postmentals; these scales in first pair separated by one (42.6%), two (54.1%) or three 3.3%) gular scales; small roundish flat dorsal tubercles arranged in not completely distinct longitudinal and transverse rows; they increase noticeably in size from mid-dorsum to flanks.

LIFE COLOR AND PATTERN. Main color of upper body surface grayish with shade of violet, it can become dark, up to brown, tail grayish ochre. From rostral scale to upper part of orbit runs a pair of ochre brown stripes. The same kind of stripes run from nostril through eye and temporal area toward the neck where they merge or are separated by main [body] color. Indistinct brownish pattern on head. Six to seven angularly bent brownish transverse bands with black distal edges from neck to base of tail. Bands of the same type also run through upper surface of tail (up to 13), but black lines are transformed into small black triangular spots. Upper surfaces of limbs also have a transverse brown pattern but with color less bright and without black fringes. Lower surfaces white.

SEXUAL DIMORPHISM AND AGE VARIATION. As in other species of this genus, the females are larger than males. Preanal pores in females are not defined (but scales perforated by pores are found in 7.4% of females examined). Body ratios change with age as in other species of the genus (Fig. 23). The tail is orange-yellow in juvenile specimens.

DISTRIBUTION. Kirghizia, central Tjan-Shan, Naryn Dist., adyrs {low foothills or isolated hills within mountain valleys – MG} along right bank of Alabuga River from Ugyut village in the northeast, to Kosh-Debe village in the southwest, 1,800 to 2,500 m elevation (Fig. 89).

LEGEND to Fig. 89 [56, 57].
1 - Baygonchek village; 2 - Kochnok village; 3 - Dzergetal village; 4 - Kosh-Debe village.

HABITATS AND QUANTITATIVE DATA (Fig. 36). The typical areas of habitation for this species are denuded clay adyrs that are nearly free of vegetation or covered with occasional *Salsola* sp. bushes. Pygmy geckos are found here under slabs of loose schist and in washed out depressions on slopes. The steepness of the slopes reaches 45°. The lizards have been found on the southern, western, and eastern slopes. The relative height of the hills is 250 m and the pygmy geckos can be encountered from the base to the crest of the hills. In the vicinity of Kongorchok village we have also dis- {p. 83} -covered the Tjan-Shan pygmy gecko on loess cliffs of seasonal and permanent streams that feed into Alabuga River and

FIG. 36. Biotope of *A*[*lsophylax*] *tokobajevi* (ruins of a fortress and adyrs on the floodplain of Alabuga River, Central Tian Shan, Kirghizia).

connect the adyrs with the river's flood-lands. Woody vegetation composed of willows and bushes (sweetbrier, hawthorn) can be encountered in these ravines. Two individuals were seen on cliffs with a southern exposure (500 m in length) and up to 2 m in height. Using the ravine cliffs, Tjan-Shan pygmy geckos can also take up residence in human structures, especially the ruins of medieval fortresses. Thus, a considerable density of pygmy geckos has been noted on the remains of loess-brick walls of the Cholok-Korgon fortress (elev. 1950 m) in an area 30 × 50 m, where mounds [of debris] from buildings still remain. Taking into account the wall perimeter and structure remnants, which combined make a 170 m transect, we found 30 pygmy geckos here — an average of an individual per 5.6 m. In this case the lizards were encountered mainly on the walls with a southern exposure, and more rarely on the ones with western and eastern exposures. There were no lizards on the northern walls. Here, the greatest density of the pygmy geckos (three individuals) was found on a 2 × 2 m pile of loess bricks, up to 40 cm in height, with rodent burrows. Nearby, a 1 km distance from the fortress, we caught five individuals under rocks on the slopes of clay-gravel hills during a two-hour search.

To the west of this place, in the vicinity of Kosh-Debe village, the number of schist fragments and the gravel content of the slopes increase, and so, Tjan-Shan pygmy geckos are encountered more frequently. Thus, on 6 June 1984, in one hour (a 200 × 20 m area was explored) six individuals were found (the least distance between them – 10 m). Taking into account the absence here of the gray rock gecko {*Tenuidactylus russowi* – MG} and the Nikolsky racerunner, the Tjan-Shan pygmy gecko should be treated as the common species of the desert adyrs, where the fauna and flora are [otherwise] extremely scant.

RESPONSE TO TEMPERATURE. The area inhabited by the Tjan-Shan pygmy gecko is characterized by its continental climate, steep temperature gradients {it is not clear from the original text whether this refers to seasonal or daily temperature extremes – MG}, and summer precipitation. The average temperature at which they emerge from shelter was established to be 18° by measurements on 4 June 1984 of substrate temperatures in locations where active pygmy geckos were found on the surface at night (n = 10). The minimum activity temperature was noted on 5 June 1984 in the vicinity of Kongorchok village at 14.5°. The data on 182 temperature records were collected during laboratory research on 3 July 1984 with five in- {p. 84} -dividuals in a thermal gradient apparatus: the lizards chose selected areas with the temperature from 6° to 40° ($\overline{X} \pm$ se = 25.95° ± 0.84°). This species' preferred temperature is significantly lower than that of the other gecko species of the U.S.S.R. fauna (for *T. russowi*, 33.3° ± 0.83°; for *T. caspius*, 32.7° ± 1°).

DAILY ACTIVITY CYCLE. Pygmy geckos eat the insects that enter their shelters in the daytime. On 4 June 1984, their appearance on the surface was noted at 2220 hrs local time, right at twilight. The greatest numbers were seen between 2300 and 2400 hrs. A sudden temperature drop after 2400 hrs prevented them from staying above ground. It can be assumed that their feeding and mating is assured by sufficiently high daytime temperatures (to 30°). They appear on the surface at twilight and at the beginning of night.

SEASONAL ACTIVITY cycle has not been studied. Judging by the climate of the area under study and by the condition of the gonads, it can be supposed that the Tjan-Shan pygmy geckos are active from the beginning of May to the middle of October.

SHEDDING was observed by us on 5 June 1984. On 8 June, it was observed in the majority of the individuals captured for laboratory research. A pair of shedding lizards was seen on 16 and 24 June 1984. One more individual was shedding on 15 July 1984, and two more on 30 July 1984.

FEEDING. We analyzed the contents of 25 stomachs of the pygmy geckos collected in the vicinity of Kongorchok village on 4 and 5 June 1984 (Table 4) and 45 fecal samples from newly captured lizards. The following predominate in the Tjan-Shan pygmy gecko diet, the total numbers by percentage: snout beetles (44 and 26.3% respectively), cicadas (24 and 16.3%), butterfly larvae (20 and 7.8%), as well as spiders (40 and 18.4%). These data correspond, primarily, to the materials obtained from the fecal analysis.

TABLE 4. Stomach contents of *A[lsophylax] tokobajevi*
(n = 25, 16 males and 9 females) captured 4 June 1984 near Kongorchok village

Food items	Cases		Total number	
	Specimens	*%*	*Specimens*	*%*
Arthropoda				
Arachnoidea				
Araneida	10	40	14	18.4
Acarina	1	4	1	1.3
Trombicidae	2	8	2	2.6
Chilopoda	1	4	1	1.3
Isopoda	1	4	1	1.3
Insecta				
Homoptera	1	4	1	1.3
Cicadidae	6	24	12	16.3
Coleoptera	2	8	2	2.6
Carabidae	1	4	1	1.3
Cuculionidae	11	44	20	26.3
Elateridae	1	4	1	1.3
Scarabidae	1	4	1	1.3
Chrysomelidae	2	8	2	2.6
Hemiptera	4	16	4	5.2
Pentatomidae				
(*Eurigaster*)	1	4	1	1.3
Tysanura	2	8	2	2.6
Hymenoptera				
Formicidae	3	12	3	3.9
Lepidoptera (l[arvae])	5	20	6	7.8
Lepidoptera (i[mago])	1	4	1	1.3

The remains of 44 insects were found in the 45 fecal samples (sand and indeterminate chitinous fragments were found in one of them). The frequency of the invertebrates {p. 85} found here is distributed as follows: snout beetles – 22.7% (10 specimens), heteropterans – 20.4% (9 spec.), homopterans – 18.2% (6 spec.), butterfly larvae – 13.8% (6 spec.), ants and beetles – 6.8% each (3 spec. each), flies – 4.5% (2 spec.), spiders – 2.3% (1 spec.). Thus, the stomach content and fecal analyses indicate that this pygmy gecko prefers to feed on snout beetles, hemipterans, cicadas, and butterfly larvae. Comparing these data with those on the diet analysis of the gray rock gecko from Kirghizia [162] reveals the clear differences in the diets of these species; the gray thin-toed gecko prefers orthopterans (84.2%), hemipterans (36.8%), and beetles (15.7%).

REPRODUCTION. Tjan-Shan pygmy geckos reach sexual maturity at body lengths of 38 mm for females and 37 mm for males. Mating probably takes place shortly after emergence from the winter {p. 86} shelters, at the beginning of May, because at the beginning of June the majority of adult females were encountered with clear signs of being gravid (the follicles in an individual captured on 4 June 1984 were 8.2 × 6 mm). On 5 June 1984, a freshly laid clutch of two eggs was discovered in a crack in the wall of the medieval fortress Cholok-Korgon. The pygmy geckos captured here laid their eggs on 13, 20, 27 June and 2 July 1984 (three clutches on this day). As a rule, the adult females would lay two eggs and the young ones – just one (the latter are encountered half as often). The eggs are (n = 28) 10.5-12.5 × 6.8-9 mm (correspondingly $\overline{X} \pm se = 11.56 \pm 0.11 \times 8.48 \pm 0.09$). The eggs are laid at night (from the laboratory observations). The incubation lasts 94 days. The eggs, obtained from the females fertilized in nature, were incubated in the laboratory in an incubator at 26-30° and hatched on 3 October 1984 (Fig. 37). The newly hatched pygmy geckos had 21.4 mm bodies and 19.2 mm tails.

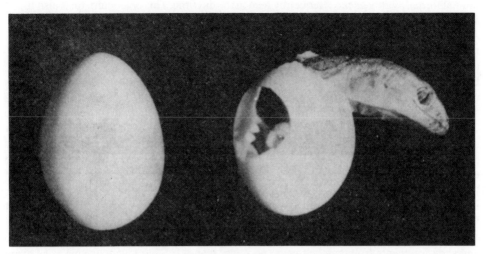

FIG. 37. A[lsophylax] *tokobajevi*, hatching with simultaneous shedding.

GROWTH RATE. The smallest specimens with juvenile characters (brightly colored underside of the tail), caught at the beginning of May, had body lengths of 20-22 mm; 27-34 mm for those from the beginning of June. In comparison with the sizes of the newly hatched pygmy geckos, it can be seen that in the period from the hibernation up to June (three months of activity) the body length increased, on average, by 10 mm. Taking into account the subsequent slowing of the rate of growth, it can be supposed that pygmy geckos reach sexual maturity in the third calendar year after birth and maximum size is reached in the fourth or fifth year (males – 45.5 mm, females – 47.0 mm).

ENEMIES. The data on the numbers of the pygmy geckos in various biotopes may be an indication of possible competition with the gray thin-toed gecko. We have established that in areas where both are found, *A. tokobajevi* is rare, whereas in places where there are no gray rock geckos, it is quite numerous. We have found Dione ratsnakes (*Elaphe dione*) in the Tjan-Shan pygmy gecko biotope. A young specimen of this snake that was found under a rock on a slope, where the pygmy geckos were also found, regurgitated a young Nikolsky racerunner. Given an opportunity, the snake, undoubtedly, also attacks the pygmy geckos. As it turned out, the majority of *A. tokobajevi* individuals have regenerated tails (59.5%). From terrarium observations, they can lose the tail in fights with individuals of their own {p. 87} species. *Falco tinnunculus*, a potential enemy of this lizard, was noted several times in areas of their habitation.

BEHAVIORAL ATTRIBUTES. As evident from the above data, of all the U.S.S.R. gecko fauna, the Tjan-Shan pygmy gecko is found at the highest elevations and is the most resistant to cold. Life in such harsh mountain desert conditions led to a number of behavioral features. It can freely move on vertical surfaces, which allows it to occupy an ecological niche characteristic of the thin-toed geckos.* As a rule, the pygmy geckos live at the bottom of walls, but are found on them up to 1 m high, and indoors have been encountered at a height of up to 2 m. In the genus *Alsophylax*, the ability [to climb] is also evident from the completely atypical structure of the hindlimbs, which are very much like those in the genus *Tenuidactylus*. When hunting insects, the Tjan-Shan pygmy geckos can jump further than their body length, although they are not as fast as the gray thin-toed geckos.

Judging from the insect remains in the stomachs and feces, among which there are many

---

* {The use of vertical surfaces is characteristic of all thin-toed geckos. Even the mostly plains-dwelling *Tenuidactylus caspius*, often searches for food on horizontal surfaces, but when inactive or in danger they take up a vertical position, and when in shallow-sloped mammal burrows, they usually concentrate on the ceiling. In contrast, *Alsophylax* prefer horizontal surfaces, although they can easily move on vertical ones. Because *A. tokobajevi* is no exception, the view stated here, *i.e.*, that this species occupies the ecological niche of thin-toed geckos, is not convincing.– MG}

diurnal forms, pygmy geckos frequently feed in the daytime. This characteristic is also found among other mountain-dwelling nocturnal animals, particularly of the diploid [sic] toads. At night, the temperature here falls sharply, which limits the possibilities for poikilotherms.

For cover, Tjan-Shan pygmy geckos use the spaces under rock fragments (as a rule, these are schists, but they are not found under gypsum crusts or pebble conglomerate fragments), cracks in loess cliffs, gullies in clay slopes, the spaces between unfired bricks of ancient ruins, and, sometimes, rodent burrows. They also lay their eggs here, but because there are few convenient and sufficiently heated places for incubation, more frequently the eggs are laid collectively by several females in one place. We have found clutches under schist slabs (six eggs from previous years near Kongorchok village), in cracks in loess cliffs (one egg from the previous year at a height of 1.6 m) and in hollows in an old wall from unfired loess bricks. Among the ruins of Cholok-Korgon fortress, we discovered a sort of "incubator," which the pygmy geckos used for colonial egg laying over many years. Thus, in a wall projection running north to south, at its southern end (it was heated by the sun for the entire day from the east, south, and west), 1.2 m above the ground, and at the depth of 10 cm (the area 50 × 50 cm), old dried up eggs were found, as well as a current year's clutch and shell fragments from hatched eggs (no fewer than 60). There were groups of as many as ten eggs. They were glued to the walls of the cracks or were lying on the horizontal {**p. 88**} surfaces (Fig. 37). Also here, two gravid females were caught, and in one place, large accumulations of droppings suggested frequent visits by the pygmy geckos. Also, the number of winter shelters – deep gullies and rodent burrows – is limited here.

The individual territory of a pygmy gecko is limited to its shelter and the immediate surroundings. Overall, they are quite tolerant towards one another, which is suggested by the discovery of two individuals of the same sex under one rock, colonial egg laying, and observations in the terrarium. The occurrence of the Tjan-Shan pygmy geckos in ruins located up to 1 km from their natural surroundings, [that are] separated by loess ravines, which are not typical localities, suggest the ability to migrate. Probably, these migrations are most frequent to places for egg laying and winter shelters.

Judging by the marks on the bodies of females, during mating the males hold them by biting down on the sides of the abdomen above the hips.

In the terrarium, the pygmy geckos adjust to the new conditions well (cover has to be provided), are not picky about food (they will eat the small larvae of mealworms, aphids, and other small insects caught with a net), will mate in captivity, and lay eggs (a gravid female accidentally died on 16 February 1984, its two eggs in the membrane had reached the dimensions of 7.5-7.8 × 6.2-6.5 mm).

PRACTICAL SIGNIFICANCE AND PROTECTION. This is a beneficial species. As a rare species with a narrow range, endemic to the Soviet Tjan-Shan, it should be protected and is included in the *Kirg*[*izistan*]*S.S.R. Red Book*. Considering the facts that the Tjan-Shan pygmy gecko inhabits areas not subject to active human cultivation and that its numbers are stable, according to the 1983-1984 observations, special protection measures are not necessary.

Genus Tuberculated Geckos — *Bunopus* Blanf., 1874

Type species: *Bunopus tuberculatus* Blanf.
1874 - *Bunopus* Blanford, Ann. Mag. Nat. Hist., London, vol. 13, p. 454.
1885 - *Alsophylax*, Boulenger, Cat. Liz. Brit. Mus., vol. 1, p. 20 (partim).

DIAGNOSIS. Digits slightly angularly bent, clawed, cylindrical, not widened and not compressed, covered below by one row of transversely enlarged subdigital lamellae, which are tuberculated or spiny on their free edges. Lateral digital scales with more or less distinct spines; no fringes (as it is in *Crossobamon* or in part of *Stenodactylus*). Pupil vertical with serrated edges. Enlarged dorsal tubercles, arranged in more or less transverse and longitudinal rows, always present. Preanal pores (3-16) distinctly marked only in males. Two to five granular nasal scales almost equal in size to each other. Tail segmented. First segment of proximal part of tail covered by two to three whorls of scales and by one whorl of caudal tubercles. They can be separated into semicircles by one to two scales in the center of the segment. Tubercles in each semicircle are in close contact with each other by the entire lateral edge. These segments covered below by three whorls of scales, which can sometimes be altered by a large scale in the middle of each whorl. Height of the first supralabial scale from nostril to the edge of mouth is greater than, equal to or insignificantly smaller than its width along the edge of mouth. Each supralabial scale is in the same plane and drops straight to the edge of mouth. Ciliary growths present in the upper part of orbit.

ADDITIONAL CHARACTERS. Nostril surrounded by quadrangular rostral, first supralabial and three to five small convex nasal scales almost equal each other in size; first nasal scales separated from each other by one to three scales; no postsupranasal scales; scales of snout flat or slightly keeled, granular, {p. 89} heterogeneous, irregularly polygonal, becoming smaller toward the neck, where they mix with tubercles; ear opening small. Gular scales conical, with a pore on the top, or flat. Back covered by smooth or keeled flat pentagonal scales, which become gradually enlarged toward the flanks with no border between them; abdominal scales rhomboid or subtriangular, imbricate, flat or keeled; preanal pores in almost straight line on scales larger than ventrals; several large scales, similar to ventrals, gradually decreasing [in size] toward vent, between the preanal pore row and vent. Adpressed forelimb reaches tip of snout with fingertips, adpressed hindlimb reaches axilla by toes; both covered above and below with flat, feebly keeled or smooth scales, more or less homogeneous and are similar to body scales; tubercles, similar to dorsal ones, can be scattered among these scales on upper surfaces of arms. Digits covered above with two, and from sides with two scale rows; base of claw covered with small upper and large lower lamellae, which have many longitudinal keels.

Members of the genus are distributed on the Arabian Peninsula and in Southwest Asia to the Indus River. The northern border of range runs along about 35°N. The genus includes three species, previously united with species of the genus *Alsophylax* by mistake.

Key to Species of the Genus Bunopus

1 (4) Postmental scales present; three to six preanal pores in males 4
2 (3) Enlarged subcaudal scales present; GVA less than 130; gular scales flat
. . . . *B. crassicauda* Nik.
3 (2) No enlarged subcaudal scales; GVA more than 130; gular scales semiconical
. . . . *B. spatalurus* J. Anderson
4 (1) No postmental scales; 5-16 preanal pores in males . . . . . *B. tuberculatus* Blanf.

SOUTHERN TUBERCULATED GECKO — *BUNOPUS TUBERCULATUS* BLANF., 1874

Type locality: "Gedrosia in Persique meridionali" (terra typica restricta: Smith, 1935 {p. 36 – MG} Baluchistan).
Karyotype: 2n = 42 (no more detailed data) [312].
1874 – *Bunopus tuberculatus* Blanford, Ann. Mag. Nat. Hist. London, vol. 13, p. 454.
1885 – *Alsophylax tuberculatus*, Boulenger, Cat. Liz. Brit. Mus., vol. 1, p. 20.
1887 – *Bunopus blanfordii* Strauch, Mem. Acad. Sci. St. Petersbourg, 35 (7), 2, p. 61.
1938 – *Bunopus biporus* Werner, Zool. Anz., Leipzig, vol. 121, p. 267.
1967 – *Bunopus abudhabi* Leviton et Anderson, Proc. Calif. Acad. Sci., ser. 4, vol. 35, {no.} 9, p. 164.

DIAGNOSIS. No postmental scales; 5-16 preanal pores in males.
LECTOTYPE. BMNH 74.11.23.85 RR 1946.8.22.84, "near Bampur, Baluchistan, coll. W. T. Blanford." Adult female. L = 50.7 mm; tail regenerated. Nostril between rostral, first supralabial and three small convex nasal scales, the middle one slightly larger then others. 14 scales across head. 13/14 supralabial scales, 10 infralabials. Mental scale trapezoidal, with irregular rear edge and with four adjoining small gular scales. 153 GVA, about 30 ventral scales across midbody, 18 subdigital lamellae.
DEFINITION (from 132 specimens in ZIK, ZIL, AMNH, BMNH, CAS, FMNH, MCZ, MNHP, MZSF, MNMB, NMW, SAM, UMMZ, USNM) (Fig. 38). L adult males (n = 54)

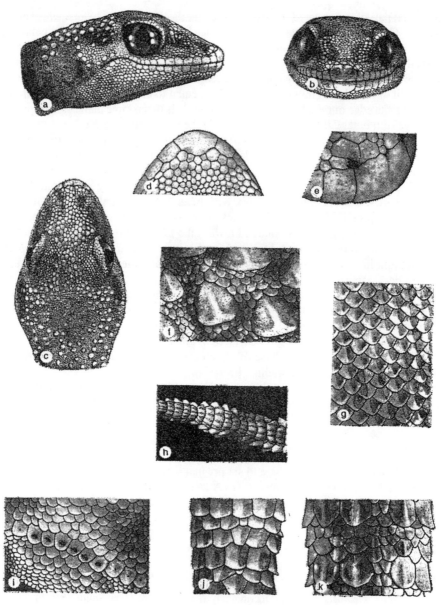

FIG. 38. *Bunopus tuberculatus*:
a, head from the side; b, snout in front; c, head from above; d, chin; e, nasal scales; f, dorsum; g, belly; h, toe from below; i, preanal pores; j, tail from below; k, tail from above (a-c, CAS 97847, Abu Dhabi; others, ZIL 11488, Abineft, Mesopotamia).

24.6-53.8 mm (41.98 ± 0.72); L adult females (n = 59) 31.9-55.7 mm (43.58 ± 0.78); L/LCD (females and males) (n = 54) 0.69-1.09 (0.83 ± 0.030); HHW (n = 69) 50-68 (60.78 ± 0.41); EED (n = 75) 37-61 (48.20 ± 0.74); supralabials (n = 220) 10-15 (12.42 ± 0.09); infralabials (n = 220) 7-14 (9.97 ± 0.12); across head (n = 115) 10-19 (14.72 ± 0.23) scales; DTL (n = 96) 11-39 [probably in error in original; range likely 11-19 – Eds.] (11.88 ± 6.12); scales around dorsal tubercle (n = 103) 11-18 (12.82 ± 0.20); GVA (n = 105) 122-170 (145.31 ± 1.28); ventral scales across midbody {p. 90} (n = 76) 25-34 (30.17 ± 0.26); preanal pores, males (n = 50) 6-16 (8.26 ± 0.36), females (n = 6) 6-13 (9.33 ± 1.05); subdigital lamellae (n = 236) 16-24 (20.15 ± 0.10). First nasal scale separated from nostril by two (21.01%) or three (78.99%) scales; these scales contact each other (0.8%) or are separated by one (69.9%), two (15.5%), or three (13.8%) scales. Ear opening oval. Mental scale trapezoid (100%) with irregular rear edge and followed by three (6.67%), four (24.4%), five (50.54%), six (11.01%), or seven (7.34%) gular scales. Gular scales subconical with a pore on the tip. Dorsum covered by smooth or slightly keeled scales and 12-16 longitudinal rows of large keeled trihedral tubercles. Abdominal scales smooth or keeled.

COLOR on the upper side is in light-brown hues. The head is covered with a poorly defined dark brown pattern. {p. 91} A brown stripe runs from the eyes along the sides of the head and neck, continues onto the body as separate spots to the area of the forelimbs. Four to five wide stripes of the same color, with smooth edges run across the body. On the tail, there are 8-11 of them. Vague, thin, transverse brown stripes are found on the limbs. The pattern on the upper side of the body may partially disappear. The lower surface is light.

SEXUAL DIMORPHISM AND AGE VARIATION. Males differ from females by the presence of preanal pores; sometimes females have only vestiges of these. No size difference between sexes, unlike the *Alsophylax* species. Nature of age variability of L/LCD index is given in Fig. 41.

DISTRIBUTION. From extreme south of USSR (Badkhyz), Afghanistan, and eastern Pakistan through Iran, Iraq, and Syria up to Arabian Peninsula inclusive (Fig. 39).

LEGEND to Fig. 39.
U.S.S.R., *Turkmenia*: 1 - Badkhyz: Er-Oylon-Duz (ZIK).
Syria: 1 - Damesin's Camp (FMNH {error in original – MG}); 2 - Deir-ez-Zor (MNHP).
Lebanon: 1 - Beirut (AMNH).
Jordan: 1 - Jerusalem (ZIN) {placed here in error – MG}; 2 - Akaba (MCZ).
Israel: 1 - Ein-Hussub (NMW); 2 - Negev: Ein Yahav SAM); 3 - S of Wadi Arava: Yotvata (MZSF).
Iraq: 1 - Al-Hadhr [269]; 2 - Abu-Naft (ZIL); 3 - K 3 Station on pipeline near Haditha (MCZ); 4 - Ghazilla River (ZIL); 5 - Mandaly (ZIL); 6 - Wassit Prov.: 10 km N Al Kut [269]; 7 - Es-Samava (NHMB); 8 - between Zubaideh and Ash-Shobaikah [269].

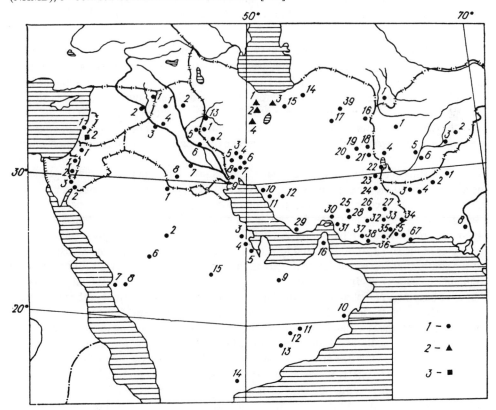

FIG. 39. Distribution of *Bunopus tuberculatus* (1), *B. crassicauda* (2), *Tenuidactylus amictopholis* (3).

Arabian Peninsula: 1 - Jumaimah, E of Rafha (MCZ); 2 - 8 km W of Unaizah (CAS); 3 - Dhahran (FMNH); 4 - Abqaiq (CAS); 5 - Al Uqair [228]; 6 - Nabhahiyah (CAS); 7 - Jiddah, 21°20'N, 39°10'E (CAS); 8 - Burayman (CAS); 9 - Abu Dhabi: within the circle of about 15 km radius centered 23°45'N, 53°33'E (CAS); 10 - Oman: Khalil {error in original – MG} [177]; 11 - Burza (19°25'N, 54°25'E) [177 {error in original – MG}]; 12 - Bin Khautar [177 {error in original – MG}]; 13 - Shisr (18°15'N, 53°40'E) [177 {error in original – MG}]; 14 - Hadramaut (MNHP); 15 - Mishash al Kharrarah (CAS); 16 - Ed-Daud (ZFMK).

{p. 92} *Iran*: 1 - Ghenghir River (ZIL); 2 - Dehluran (UMMZ); 3 - Dizful (ZIL); 4 - Shuster (ZIL); 6 km W of Gotiwand (MCZ); 5 - Mahor Birinji (MCZ); 6 - Gholghir (Gulsh, or Ghurjir) (ZIL); 7 - Ahvaz - Haft Kel Road (CAS); 8 - Ahvaz (Nasrieh-e-Ahvazeh) (ZIL); 9 - Kharun River at locality Kharma, Ziarat Makhram-Amin-el-Mumunad (ZIL); 10 - 52-56 km NW Ganaveh on Jari-Behbehan Road (CAS); 11 - 10-13 km S of Borazdan on Road from Bushrie (CAS); 12 - Ahram, Fars Prov. (CAS); 13 - Khasre-Shirin (ZIL); 14 - Dhamghan (ZIL); 15 - Semnan, 65 km E of Tehran (MNHP); 16 - Niyazabad in Haf (ZIL); 17 - Halwan (NMW); 18 - Chah-i-Bena Well (ZIL); 19 - Sarr Chah [59]; 20 - Between Faizabad and Beshiran [92]; 21 - Chah-i-Guisheh Well (ZIL); Between Khadji-i-du-Chaghi and Guisheh [59]; 22 - Neizar in Seistan (ZIL); 10 km SW Helmand, 31°03'N, 61°38'E (CAS); Khouzdar; Chah-i-Novar in Seistan; Ghumbez-i-novar in Seistan (ZIL); 23 - Hormak (ZIL); 24 - 52 km S of Zahedan Mirjaveh Rd, 29°08'N, 61°20'E (CAS); 25 - Sabzawaran [315]; Sarghad Dist.: locality Ando (ZIL); 26 - Locality Duz-ab at Taghab lowland between Shurab and Pensareh [60]; 27 - Kalghan (ZIL); 28 - Khanu, 100 km S of Sabzawaran (NMW); 29 - Bandar-Langeh [223]; 30 - 13-40 km N of Bandar-Abbas (NMW); 31 - N of Minab (CAS); 32 - 32 km W of Bampur (27°10'N, 60°10'E) (CAS); near Bampur (BMNH); Iranshahr (ZIL); 33 - Shurab in Megas (ZIL), Khash [303]; 34 - Dehek (ZIL); Rigan Mountain [172]; 35 - Picrus [172]; 36 - Bahu Kalat [172]; 37 - Ziarat (NMW); Makran: Bagu (ZIL); 38 - Chahbahar [223]; 39 - Doruneh [303].

Afghanistan: 1 - Adraskan (Herat) (NMW); 2 - 30 km S of Mukur [209]; 3 - Kandahar [172]; 4 - 35 km S of Farakh (CAS); 5 - Chah-i-Angir (CAS); 6 - 10 km S of Darweshan (CAS).

Pakistan: 1 - Sibi Dist.; Kach [263]; 2 - near Nushki [263]; 3 - Dalbandin [258]; 4 - Kharan [258]; 5 - Mand [172]; 6 - Dasht River [172]; 7 - SE of Pasni [258]; 8 - Hyderabad [263].

Not found on maps: (a) Ziarat Seid-Khanat, Mesopotamia (ZIL); (b) Iran: Isfandak, Tumb Island; Pakistan: Baluchistan: Saman [172]; (c) Leb-e-Kal [310].

SYSTEMATICS AND GEOGRAPHIC VARIATION. K. Schmidt [284 {error in original – MG}] expressed his doubts that the tuberculated geckos, which inhabit such an extensive territory, do not show geographical variation. Yu. Gorelov and his co-authors [48] had supposed that in the east of its range this species is represented by a different subspecies, that is smaller and has a smaller number of preanal pores and scales behind the mental scale. At the same time, E. Arnold [177] united this species with two others, *B. blanfordi* Str. and *B. abudhabi* Leviton and Anderson, on the basis of the absence of clear distinguishing characters among these forms.

The results of our studies of tuberculated geckos from different parts of the range incline us to subscribe to the latter point of view. However, weak geographical variation in several characters can be recognized. Thus, the number of preanal pores gradually increases from Badkhyz and Baluchistan, in the eastern part of range, to Mesopotamia and then decreases again toward the southern part of the Arabian Peninsula. The level of expression of keeling on scales of the snout, abdomen, and lower tail surface increases from east to west and then to the south. In the extreme southern part of the range (from United Arab Emirates to the Sultanate of Oman), keeling of abdominal scales decreases again, the keels on the subdigital lamellae become smaller, and a row of plates appears on the posterior part of the tail's lower surface. However, the latter characters are characteristic, in varying degrees, of all populations that inhabit the Arabian Peninsula, which was the basis for uniting the aforementioned species [177]. Thus, the weak differentiation of characters and the presence of transitions do not yet allow us to define any taxonomically significant forms within this species.

HABITAT AND QUANTITATIVE DATA. N. Zarudny [60] had reported that this species is widely distributed in eastern Iran, where it is quite common in many depressions and extends into the mountains on similar soils. The author notes that "it avoids, if possible, the rocky soils and does not particularly like loose and deep sands, and appears most readily on clays, loam and sandy loam, partly uncovered and partly covered with a thin layer of sand and occasionally saline" (pp. 8-9). In the United Arab Emirates, the southern tuberculated gecko is found on takyrs, cemented sands, and loose, wind-blown sands [177]. S. Minton [263] believes that in the vicinity of the town Nushki (Pakistan), this lizard is common only in sandy areas. However, in the Baluchistan mountains the southern tuberculated gecko was caught at an elevation of 700 m in the flood-plain of a small mountain river among boulders and in short veg- {p. 93} -etation. It reaches an elevation of 1425 m along the left Helmand tributary – Tornak River (Afghanistan), and near Gaomi-Faringhi it occurs up to 2100 m [172, 245].

According to S. Minton's written report, in Palestine the southern tuberculated gecko is found in fairly arid areas on sandy soils with thin shrub vegetation. He observed three lizards of this species near an abandoned building. Two individuals of the southern tuberculated gecko were caught, for the first time in the U.S.S.R., in April of 1972 in the Er-Oilon-Duz Depression (Badkhyz, elevation 315-460 m) (Fig. 40). The lizards were found under a rock on effusive outlier and near the entrance to a *Meriones* sp. {MG} burrow. Also here, on 15 May 1976 we caught a southern tuberculated gecko at the bottom of a short clay hill, entirely devoid of vegetation, in immediate proximity to a moist sor. On 10 April 1977 we caught one more specimen of southern tuberculated gecko 2-3 m from a sor on a fairly steep clay slope with multiple cracks and without vegetation, and 10 m higher began the outlier's foundation with *Salsola* sp. {MG} bushes and gravel slopes. In subsequent years (May of 1982, April 1984), thorough searching for this lizard in the Er-Oilon-Duz Depression was unsuccessful, which causes us to consider it one of the rarest species in the fauna of the U.S.S.R.

RESPONSE TO TEMPERATURE. We caught an active southern tuberculated gecko on the surface with a soil temperature of 17°. According to published data [172], active lizards of this species with a body temperature of 26-34° have been encountered, and in individuals dug out of burrows, the body temperature exceeded 33.4-37°.

DAILY ACTIVITY CYCLE. N. Zarudny [60] considered the southern tuberculated gecko to be a "true nocturnal lizard," and other authors [288] confirm this or attribute this species to the twilight animals [313]. We caught our first specimen at 2100 hrs and a second one at 2130 hrs. There are still no further detailed data.

FIG. 40. Biotopes of *Bunopus tuberculatus* and *Tenuidactylus caspius* (gypsum-saline depression of Er-Oylon-Duz, Badghyz, Turkmenia).

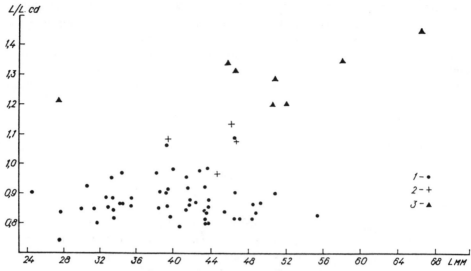

FIG. 41. Age variation of L/LCD index of *Bunopus tuberculatus* (1), *B. crassicauda* (2), *B. spatalurus* (3).

SEASONAL ACTIVITY CYCLE. In southern Turkmenia we caught the first active individual on 10 April 1977. That year had a cool spring, thus, in this case, the discovery in the U.S.S.R. can be considered to be the first one. The material collected in Iraq was obtained from January through October. From N. Zarudny's collection in Iran, the spring collections are from 13 February 1901; 10 January, 18 February and 25 February 1904; the last lizard in the fall was found on 8 October 1903 (old [Russian] calendar) (to estimate current calendar date, add 12 days to the old date – MG).

{p. 94} SHEDDING. No data available.

FEEDING. In captivity, southern tuberculated geckos ate termites [48]. In two stomachs from the lizards we dissected, the remnants of a spider (1), orthopteran (1), an ant (1), and hymenopteran (not determined more precisely, 1) were found.

REPRODUCTION. The females, caught on 14 and 21 May in northern Baluchistan, had one to two eggs nearly ready for laying [263]. There are no data on the reproduction of this species in the U.S.S.R.

ENEMIES. A. Leviton [242] found the southern tuberculated gecko in the stomach of an Afghan awl-headed snake (*Lytorhynchus ridgewayi*).

BEHAVIORAL ATTRIBUTES. It was N. Zarudny [60] who noted that this lizard "unlike the majority of . . . geckos, is not a wall animal but a ground one . . . because it is rarely seen on the vertical surfaces" (pp. 8-9). This species, which runs on the ground and only occasionally climbs on vertical surfaces, was also noted by other authors [313]. For cover, the southern tuberculated gecko most likely uses rodent burrows, possibly the salt crust along the sor shores, [and] spaces under rocks. At least in Er-Oilon-Duz, the southern tuberculated gecko was once caught near a *Meriones* sp. {MG} burrow [48]. On the Arabian Peninsula they have been caught near burrow entrances, in rodent burrows, and under pieces of wood [177].

PRACTICAL SIGNIFICANCE AND PROTECTION. Being extremely rare, this species has been included in the second edition of the *U.S.S.R. Red Book* (1984). It is found in the Badkhyz Preserve and currently does not require additional protection measures.

ARABIAN TUBERCULATED GECKO — *BUNOPUS SPATALURUS* J. ANDERSON, 1901

Type locality: Wadi Jimil near Aden, Yemen.
Karyotype: not studied.

1901 – *Bunopus spatalurus* J. Anderson, Proc. Zool. Soc. London, no. 3/4, p. 137.
1959 – *Trachydactylus jolensis* Haas and Battersby, Copeia, no. 3, p. 200.

1967 – *Trachydactylus spatalurus*, Leviton and Anderson, Proc. Calif. Acad. Sci., ser. 4 , vol. 35, no. 9, p. 166.

DIAGNOSIS. Postmental scales present; males with three to six preanal pores; gular scales conical; more than 130 scales along lower surface of body; no subcaudal plates.

HOLOTYPE. BMNH 99.12.13.40 RR 1946.8.23.59, "Jimil Valley, Southern Arabia, coll. A. B. Percival." Juvenile male. L = 34.6 mm, tail regenerated with slightly swollen tip. Nostril between quadrangular rost- {p. 95} -ral, first supralabial, and three convex small nasal scales, more or less equal to each other. First nasal scales separated from each other by one scale. Fourteen scales across head. Fifteen supralabials, 11/10 infralabials; four preanal pores; 16 subdigital lamellae.

DEFINITION (from 16 specimens in BMNH and MNHP, Fig. 42). L adult (n = 12) 42.6-66.5 mm (50.88 ± 1.79); no sexual dimorphism. L/LCD general (n = 8) 1.20-1.45 (1.29 ± 0.03); HHW (n = (n = 12) {p. 96} 57-67 (63.08 ± 0.84); EED (n = 11) 38-53 (45.50 ± 1.49); supralabials (n = 32) 13-17 (15.10 ± 0.20); infralabials (n = 32) 10-13 (11.50 ± 0.16); scales across head (n = 16) 12-16 (13.81 ± 0.32); DTL (n = 16) 7-10 (8.13 ± 0.18); GVA (n = 15) 133-160 (143.4 ± 1.96); ventral scales across midbody (n = 12) 30-50 (36.3 ± 1.54); preanal pores (n = 2) 4-5; subdigital lamellae (n = 32) 14-17 (16.09 ± 0.16).

Three to five nasal scales; first nasal scales separated from each other by one (14.2%), two (73.7%), or three (7.1%) scales; height of the first supralabial scale from nostril to the edge of mouth greater than its width along the edge of mouth; 7-7.5 supralabials reach front of orbit. Snout scales smooth or slightly keeled. Mental scale pentagonal (50%), hexagonal (35.8%), septangular or trapezoid with uneven rear edge (7.1% each); one pair of postmental scales, sometimes almost undefined (about 30% of specimens examined have very weakly defined second pair), the scales have common suture (46.5%), or separated by one (40%) or two (13.5%) scales; gular scales conical. Abdominal scales smooth or keeled. Smooth or keeled dorsal tubercles, variously defined. Limbs covered by uniform scales, among which individual large scales, similar to dorsal tubercles, are distinguished only on upper surface of thighs. Subdigital lamellae with rows of several tubercles, ending with a small spine. Each tubercle surrounded at the rear by a semicircle of much smaller spiny tubercles. Preanal pores developed much less clearly than in other representatives of this genus.

COLOR AND PATTERN (as fixed in alcohol). The main background of the upper surface is brownish. There is no pattern on the head. Four to five dark chocolate-brown broad transverse stripes occur from the neck to the lumbar area. Each stripe has a slightly curved front edge and a straight rear edge. There are 9-10 wide, straight stripes of the same color on the tail. The limbs are covered with several thin transverse bands of the same color. The lower surfaces are white.

{p. 97} DISTRIBUTION. Southern and southeastern coast of Arabian Peninsula (Fig. 43).

LEGEND to Fig. 43.
1 - Wadi Jimil (BMNH) [175]; 2 - Zamakh, Hadramaut (BMNH); 3 - Oman: 16 km S of Thamarit [177]; 4 - Mashirah Island [175]; 5 - Sultanate of Oman: Awabi [175]; 6 - Wadi Sahtan [175]; 7 - Rostaq [175]; 8 - Wadi Qid, Jebel Aswad [175]; 9 - Maskate (MNHP); 10 - S. of Aqqah [175]; 11 - Idhn [175]; 12 - Manama [175]; 13 - Siji [175]; 14 - Wadi Asimah [175]; 15 - Masafi [177]; 16 - Tayyibah Plain [175]; 17 - Jebel Fayah [175]; 18 - Khawr Fakkan area [175]; 19 - Ras Dibba [175]; 20 - Dhayah, N. of Ras al Khaymah [175]; 21 - S of Ash Sha'am [175]. {localities 4 through 14 and 16 through 21 are mistakenly identified in the original text, as published in [177] – MG}

SEXUAL DIMORPHISM AND AGE VARIATION. The only distinction between the sexes in this species is in the presence of preanal pores in males. Also, preanal pores in male Arabian tuberculated geckos are much less well-developed than in the males of other species in the genus. We could not study age variation because of the insufficient numbers of juvenile specimens in our sample. One assumes that the L/LCD index in this species increases with age (contrary to *Alsophylax* species).

SYSTEMATICS AND GEOGRAPHIC VARIATION. The Arabian tuberculated gecko was returned to the genus *Bunopus* by E. Arnold [175]. At present, it is represented by the

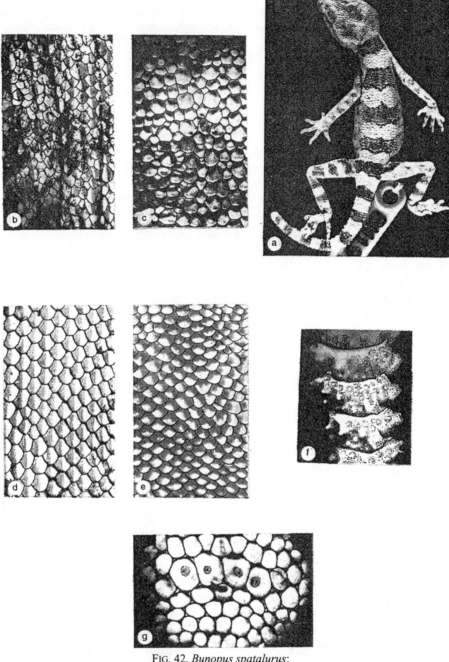

FIG. 42. *Bunopus spatalurus*:
a, general view from above; b-c, dorsum; d-e, belly; f, toe from below;
g, preanal pores (a, c, e, BMNH 1978.1095, Dhofar, Oman – *B. s. spatalurus*;
others, BMNH 1971.1207, Masifi, Oman [error in original – MG} – *B. s. hajarensis*)

northern (Persian Gulf coast) and southern (Aden Gulf coast) groups of populations, which are assigned the status of subspecies [177].

Key to the Subspecies of B. spatalurus

1 (2). Dorsal scales smooth . . . . *B. s. spatalurus* J. Anderson
2 (1). Dorsal scales keeled . . . . *B. s. hajarensis* Arnold

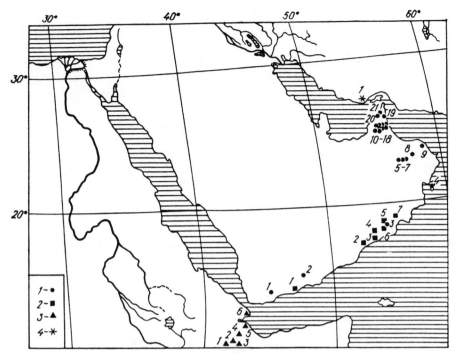

FIG. 43. Distribution of *Bunopus spatalurus* (1), *Tropiocolotes scortecci* (2), *T. tripolitanus* (3), *T. steudneri* (4).

ARABIAN TUBERCULATED GECKO — *BUNOPUS SPATALURUS SPATALURUS* J. ANDERSON

1980 – *Bunopus spatalurus spatalurus*, Arnold, J. Oman Stud. Spec. Rep. no. 2, p. 273-332.

DIAGNOSIS. Scales of head and dorsum (including weakly defined tubercles) smooth; three (rarely four) nasal scales; L max. = 50-55 mm.

Description of the subspecies, based on type specimens of species and of *Trachydactylus jolensis* (BMNH 1953.1.6.96), as well as on a specimen from Oman (BMNH 1978.1095) does not have significant differences from species description in other characters.

Inhabits southern Arabian Peninsula from Aden to Dhofar [177].

HAJAR TUBERCULATED GECKO — *BUNOPUS SPATALURUS HAJARENSIS* ARNOLD, 1980

Type locality: 25°18.5′N, 56°10′E {error in original – MG}, United Arab Emirates.
1980 – *Bunopus spatalurus hajarensis* Arnold, J. Oman Stud. Spec. Rep., no. 2, p. 277.

DIAGNOSIS. Scales of head and dorsum (including large but very flattened dorsal tubercles) distinctly keeled; four (rarely, five) nasal scales; L max. = 65-70 mm.

HOLOTYPE [177]. Adult female. Wadi Ham Masafi {error in original – MG} (25°18.5′N, 56°10′E) United Arab Emirates: BM 1973.1801, (E. Arnold, 27.5.1973).

L – is not given. Internasal scales separated by two scales; nostril between rostral, first supralabial, supranasal and three small nasal scales; 15/14 supralabial scales (12/11 to middle of eye); 11 infralabials (9 to middle of eye). One pair of enlarged scales (weakly defined postmental scales) behind mental, separated from each other by two gular scales; 18 subdigital lamellae.

DESCRIPTION of the subspecies is based on 10 paratypes from the British Museum and three juvenile specimens from the Paris Museum; in the other charac- {**p. 98**} -ters, except those included in the diagnosis, there are no significant differences from the species description. Inhabits eastern Arab Emirates, northern Oman south to Mashirah Island [177].

ECOLOGY almost unknown. Some data on this subject appear only in the work of E. Arnold and M. Gallagher [178]. They report that *Bunopus spatalurus* is a common nocturnal terrestrial gecko, which is widely distributed in various parts of the rocky plains, as well as in dry river and stream beds and stone hills adjacent to rocks. It is in places like these that all 59 specimens described by the authors were collected. All areas of occurrence are at elevations of 270-500 m. The sex ratios are: males – 41.6%, females – 50.0%, juveniles – 8.4%.

Judging from the samples, the lizards were captured from February through December.

Active animals frequently rise up on straight limbs in a manner typical of many open-ground-dwelling desert geckos. They move slowly and often in search of food. The body temperature of two feeding individuals was 35°[C] immediately after capture.

Small stones are often the daytime cover for these lizards. In sandy areas along the edges of ridges they are replaced by *B. tuberculatus* and by species of the genus *Stenodactylus*, specifically *S. leptocosymbotus* Leviton and S. Anderson.

THICK-TAILED TUBERCULATED GECKO — *BUNOPUS CRASSICAUDA* NIK., 1907

Type locality: Kum, Malyat-Abad and Khara-Magommed-Abad in Iraq-Adjemi Prov. (northwestern Iran).

Karyotype not studied.

1907 – *Bunopus crassicauda* Nikolsky, Ann. Zool. Mus., v. 10, nos. 3-4, p. 261.
1963 – *Alsophylax crassicauda*, S. Anderson, Proc. Calif. Acad. Sci., ser. 4 , vol. 31, p. 474.

DIAGNOSIS. Postmental scales present; three to six preanal pores in males; gular scales flat, smooth; 130 GVA; subcaudal plates present.

LECTOTYPE. Because this species was described from a series of syntypes, we selected the best preserved specimen as the lectotype, ZIL 10233, adult male, Khara-Magommed-Abad, 10 May 1904, coll. N. Zarudny.

L = 40 mm; tail regenerated. Nostril between quadrangular rostral, first supralabial, and three convex small nasal scales, the middle one slightly larger than others; first nasal scales separated from each other by one scale. Scales across head 14. 9/10 supralabials, 9 infralabials on left (right row damaged). About 120 GVA. Five preanal pores, in an almost straight line; 19/20 subdigital lamellae.

DEFINITION (from 21 specimens in ZIL, AMNH, CAS, SAM, and MNHP) (Fig. 44). L adult male (n = 12) 39.9-45.3 mm (44.35 ± 0.82); L adult female (n = 9) 37.3-54.5 mm (46.30 ± 1.70); L/LCD (females and males) (n = 4) 0.97-1.13 (1.06 ± 0.03); HHW (n = 18) 57-71 (61.61 ± 0.83); EED (n = 18) 27-46 (38.0 ± 1.17); supralabials (n = 41) 9-12 (10.40 ± 0.12); infralabials (n = 41) 8-11 (9.07 ± 0.12); scales across head (n = 21) 13-17 (15.10 ± 0.24); DTL (n = 20) 11-16 (13.2 ± 0.31); scales across dorsal tubercle (n = 20) 8-10 (9.40 ± 0.15); GVA (n = 21) 101-127 (112.3 ± 1.46); ventral scales across midbody (n = 19) 26-31 (27.60 ± 0.34); preanal pores (n = 12) 3-6 (5.16 ± 0.27); subdigital lamellae (n = 40) 16-20 (17.45 ± 0.19).

Three convex nasal scales, the middle one frequently slightly larger than others or equal to them; first nasal scales separated by one (81%) or two scales; height of the first supralabial scale from nostril to the edge of mouth slightly smaller than its width along the edge of mouth; 4.5 to 6 supralabials reach the orbit; mental scale pentagonal; two (16%), three {p. 99} (63%), or four (21%) pairs of postmental scales, the first much larger than others, the scales contact each other across wide sulcus. Gulars large and flat. Intermixed with smooth dorsal scales are 9-12 longitudinal rows of roundish at the base, subconical, smooth tubercles with weak edge at the end. Abdominal scales large, flat, smooth, roundish-hexagonal, much larger than dorsal scales. Tail scales flat above, {p. 100} tail tubercles with keels; lower surface of several proximal tail segments with three scale rows [per segment], each central scale becomes larger from first to last whorl [within each segment; sometimes central scale is divided] (1 > 1 > 1; sometimes 1 > 1 > 2 or 1 > 2 > 2 {error in original; should read, anterior to posterior, 1< 1< 1, sometimes 1 < 1 < 2 ⟨2 refers to a divided scale⟩ or 1 < 2 < 2

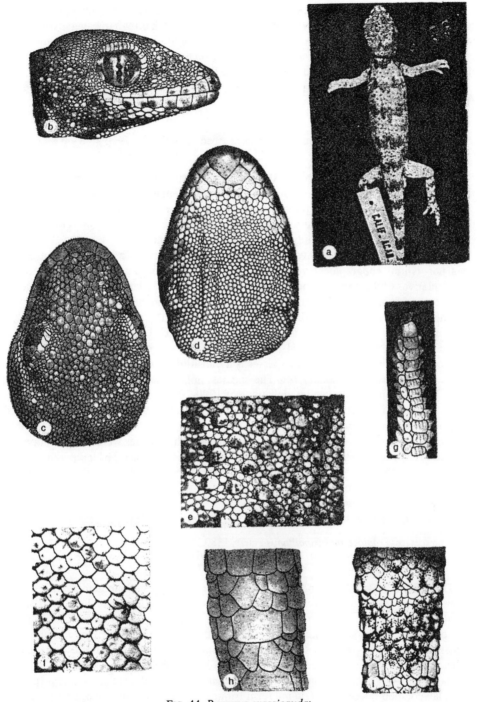

FIG. 44. *Bunopus crassicauda*:
a, general view from above; b, head from the side; c, head from above; d, head from below {error in original – MG}; e, dorsum; f, belly; g, toe from below; h, tail from below; i, tail from above. (a, e, h, i – CAS 140599, Ghazwin, Iran; others – AMNH 99663, Tehran Prov., Iran).

– MG}; after two to three segments, subcaudal plates appear; about ¼ of tail length, there are three plates on the first segment (1 = 1 = 1), then two plates on each following segment (1 = 1). Unlike other species of the genus, scales of lateral digital surfaces are flat with very weak keels.

COLOR AND PATTERN (as fixed in alcohol). The main background of the upper surfaces is from light chocolate to sandy yellow. Four to five wide dark brown transverse bands with straight edges run from the neck to the lumbar area. Eight to ten stripes of the same color are found on the tail. There is a vague, dark pattern on the head. The limbs are covered with thin, dark brown transverse bands. The lower surfaces are white.

DISTRIBUTION. This species possesses a very small range to the south and west of Tehran (Iran) (Fig. 39).

LEGEND to Fig. 39.
Iran: 1 - Kazwin (CAS); 2 - Khara-Magommed-Abad; Chaldjatabad [92]; 3 - 8 km S of Tehran (SAM); 4 - Kum [92].

SEXUAL DIMORPHISM AND AGE VARIATION. In this species, the males are distinguished from females by the presence of preanal pores. The females occasionally exhibit only vestiges of these pores. Among the specimens we studied, the females are somewhat larger than the males, however, this variation is not reliable. The lack of young individuals in our sample made it impossible to study age variation in sufficient detail.

SYSTEMATICS AND GEOGRAPHIC VARIATION. This species has a small distribution and does not exhibit geographical variation but is of interest because it differs from other representatives of the genus in some characters – keeling of the lateral digital scales, dorsal tubercles, gular scales, and caudal scales partially or completely absent; the gular and abdominal scales are large and flat, which results in a smaller number of scales along the longitudinal body axis; there are fewer supra- and infralabial scales; there are very large subcaudal scales ("*B. t. abudhabi*" also has these, but they are much less distinct). F. Werner [309] noted that, in his opinion, this species is a transitional one between the species of the genera *Bunopus* and *Gymnodactylus*. The species, unlike the other representatives of the genus, lives in a milder climate (at elevations of 1000-2000 m), which was the likely cause for the reduction in keeling of the scales and plates, as well as other differences connected, probably, with evaporation.

ECOLOGY has not been studied. Two individuals of this species were caught at noon on 5 August 1965 [265] on the edge of a cultivated alluvial plain [265]. The temperature in the shade was 53°. One lizard was found under a clump of earth and another, probably startled out of its shelter, was moving among short bushes.

Genus Dwarf Geckos — *Tropiocolotes* Peters, 1880

Type species: *Tropiocolotes tripolitanus* Peters, 1880.
1880 – *Tropiocolotes* Peters, Monatsber. Königl. Preuss. Acad. Wiss. Berlin {error in original – MG}, p. 306.
1885 – *Stenodactylus*, Boulenger, Cat. Liz. Brit. Mus., vol. 1, p. 16 (part.).

DIAGNOSIS. Nasal shield and scales flat; free (upper) edges of supralabial scales curve up onto the upper surface of snout so that each one is not in a single plane; height of the first supralabial scale from nostril to the edge of mouth distinctly less than its width along the edge of mouth; dorsal scales uniform, flat or slightly convex, keeled or smooth; no dorsal or tail tubercles; tail segments not distinguishable from above; tail covered below by uniform scales or by one row of enlarged plates (two per segment); none or no more than seven preanal pores (only in males); digits slightly angularly bent, not dilated or compres- {p. 101} -sed, clawed, their lateral scales do not form fringes (small spines may be present), covered below by one row of smooth or keeled subdigital lamellae. Pupil vertical, with serrated edges. Ciliary growths in the upper part of orbit present.

Distributed in North Africa (north of 10°N), Arabian Peninsula, and in Southwest Asia to Indus River. Northern border of range runs near 35°N. The genus includes eight species, which we divided into three subgenera.

Key to the Subgenera of the Genus Tropiocolotes

1 (2). Subdigital lamellae with keels . . . . *Tropiocolotes* Peters

2 (1). Subdigital lamellae smooth 3
3 (4). Three nasal scales contact nostril, the first one (supranasal) much larger
than others . . . . *Microgecko* Nik.
4 (3). Two to three nasal scales, almost equal in size, contact nostril . . . .
. . . . *Asiocolotes* Golubev et Szczerbak {error in original – MG}

Subgenus *Tropiocolotes* Peters, 1880

Type species: *T. tripolitanus* (s. str.).

DIAGNOSIS. Subdigital lamellae keeled; a pair of flat supranasal scales not touching nostrils; no postsupranasal scales; besides rostral and first supralabial scales, nostril is in contact with two scales, that do not significantly differ in size from supranasal or other scales of snout (except *T. scortecci*, whose supranasal scales are significantly larger than other scales, yet not in contact with nostrils); upper edges of supralabial scales curve up onto the upper surface of snout so that each one is not in a single plane; body scales flat, rhomboid, keeled or smooth; dorsal ones can sometimes be a little smaller than abdominal scales (usually larger or equal); no subcaudal scales; as a rule, no preanal pores (two poorly developed spots in place of preanal pores are sometimes present in some males of *T. steudneri* and *T. scortecci*).

ADDITIONAL CHARACTERS. Snout scales flat, irregularly polygonal, become slightly smaller toward neck. Ear opening small, roundish. Mental scale triangular or pentagonal, sometimes with blunt rear edge. Postmental scales may be absent, but usually first pair is present, the scales contact each other or are separated by gular scales; often it is followed, along infralabial scales, by one to two pairs of scales that gradually decrease in size. Number and position of postmental scales vary in different species. Gular scales flat, irregularly polygonal; several gular scales are larger than others. Dorsal scales uniform, rhomboid, imbricate; abdomen also covered by flat, uniform, rhomboid scales. Tail cylindrical, slightly thicker at the base, thining in the last quarter, covered above and below by flat imbricate scales. Adpressed forelimb reaches the mid-point between nostril and front edge of orbit or the tip of snout; adpressed hindlimb almost reaches shoulder. Extremities covered above and below by scales similar to body scales. Digits covered above by one row, on sides by two scale rows.

DISTRIBUTION. North Africa (north of 10°N), Sinai and Arabian Peninsula; a single record from southern Iran. The subgenus includes three species.

Key to the Subgenus Tropiocolotes

1 (4). All scales of snout, including scales which contact nostril, almost equal
each other in size 4
2 (3). Dorsal scales distinctly larger than abdominal ones; all body scales with
distinct longitudinal keels . . . . *T. tripolitanus* Peters
3 (2). Dorsal scales equal to or insignificantly larger than abdominal ones; body
scales smooth . . . . *T. steudneri* (Peters)
4 (1). Internasal scales distinctly larger than other snout scales but do not contact
nostrils . . . . *T. scortecci* Cherchi et Spano

{p. 102} TRIPOLI DWARF GECKO — *TROPIOCOLOTES TRIPOLITANUS* PETERS, 1880

Type locality: "Wadi M'belem" (Tripoli, Libya).
Karyotype: not studied.

1880 – *Tropiocolotes tripolitanus* Peters, Mon. Berlin. Acad., p. 306.
1942 – *Tropiocolotes somalicus* Parker, Bull. Mus. Comp. Zool., vol. 91, no. 1, p. 46.
1942 – *Tropiocolotes occidentalis* Parker, Bull. Mus. Comp. Zool., vol. 91, no. 1, p. 47.

DIAGNOSIS. Body scales heavily keeled; dorsal scales distinctly larger than ventrals; supranasal shields no larger than other scales, they do not touch nostrils.

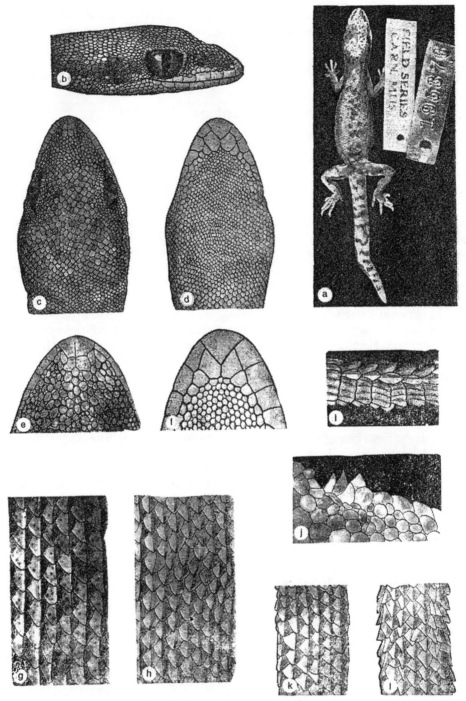

FIG. 45. *Tropiocolotes tripolitanus*:
a, general view from above; b, head from the side; c, head from above; d, head from below; e, snout from above; f, chin; g, dorsum; h, belly; i, toe from below; j, upper edge of orbit; k, tail from above; l, tail from below. (USNM 195376, Tunisia, Gabes).

DEFINITION (from 106 specimens in AMNH, BMNH, USNM, CAS, MNHP, MZSF) (Fig. 45). Type specimen is lost (ZMB 9668, Dr. G. Peters, pers. commun.). L total = 11.7-39.8 (26.31 ± 0.59) mm; HHW 56-67 (62.47 ± 0.46); scales across head 8-13 (10.52 ± 0.10); supralabials 7-10 (8.81 ± 0.05); infralabials 5-8 (7.05 ± 0.04); GVA 77-115 (100.43

± 1.20); ventral scales across midbody 34-48 (43.48 ± 0.28); subdigital lamellae 11-17 (14.10 ± 0.11); 2.7-3 supralabials reach orbit. As a rule, postmental scales well developed, these scales in the first pair contact each other by a narrow suture. Keels on all scales combined in characteristic longitudinal rows.

COLOR AND PATTERN (as fixed in alcohol). The main body background is light yellowish. Wide dark brown stripe that disappears before reaching the shoulder runs on each side of the head from the nostril, through the eye and to the ear (in the individuals from Somalia, these stripes merge on the neck in a horseshoe pattern). Usually, the head and the upper body surface are densely covered with dark brown dots and small spots. Sometimes, there are up to seven vague transverse bands on the body and the same stripes (up to 10) may appear on the tail. The lower surfaces are light.

DISTRIBUTION. North Africa from 10°N in the east to the lower Nile and Gulf of Aden (Djibuti and Somalia) (Figs. 43, 46).

LEGEND to Fig. 43.
Somalia: 1 - 10°20'N, 42°50'E [272]; 2 - near Borama: 10° 45'N, 43°E [272]; 3 - Biji [272]; 4 - 11°25'N, 43°15'E [272]; 5 - Djibuti (MNHP); 6 - Obock (MNHP).

LEGEND to Fig. 46.
Egypt: 1 - Gebel Mokattam above Cairo [249]; 2 - Gyza Prov.: Abu Rawash (USNM); 3 - Gyza [249]; 4 - Faiyoum Prov.: near Qarun Lake (USNM).
Libya: 5 - Gialo [249]; 6 - El Ageila (MCZN); 7 - Sirtica [249]; 8 - Jaafer [249];
Tunisia: 9 - Foum Tatahouine [249]; 10 - Gabes [249]; 11 - Gafsa [249]; 12 - Oum Ali near Gafsa [249]; 9 km NE Gafsa (CAS); 13 - Tunis (ZFMK); 14 - Taferma between Gafsa and Tozeur [249].
Algeria: 15 - Bon (SMF); 16 - Biskra [249]; 17 - Ghardaja (MZFS); 18 - Djebel Djara, Ain Sefra (MNHP); 19 - Figuig [249]; 20 - Beni Ouinf (MNHP); 21 - Benni Abbes (MNHP); 22 - Colomb Beshar (CAS); 23 - Kenadsa [249].
Morocco: 24 - Bou Haiara [198]; 25 - Bou Denib, Bou Anane [198]; 26 - hamada du Guir [198]; 27 - Meski (Tafilalit) [198]; 28 - 40 km NE Tindouf (CAS); 29 - 7 km N Taidalte (RMNH); 30 - 10 km S Taidalt (RMNH); 31 - Goulimime (MNHP); 32 - el-Hassane [198]; 33 - Torkoz (CAS); 34 - 15 km S Torkoz (CAS); 35 - 10 km S Torkoz (CAS); 36 - Assa [198]; 37 - Aouinet-Ait-Oussa [198]; 38 - Taskala [198]; 39 - Bou Guejouf [198]; 40 - Foum el Guid (oued Dra) (AMNH).
Western Sahara: 41 - Seguet-el-Hamra: Amuiserat [198]; 42 - 10 km N Semara (CAS); 43 - 20 km W Aaiun (CAS); 44 - Rio-de-Oro [249].
Mauritania: 45 - Port Etienne [249]; 46 - 15 km N of Nouakchott Airport (CAS).
Mali: 47 - Sangha [271]; 48 - Bandiagara [271]; 49 - Hombori [271].
Niger: 50 - 18 km N Niamey (CAS); 51 - Azzel [271]; 52 - Dabaga (Air) [271]; 15 km S Arlit (ZFMK); 8 to 10 km N Agades (ZFMK); 53 - Teouar [271]; 54 - 80 km S Aguelhok (CAS); 55 - Tessalit (CAS).

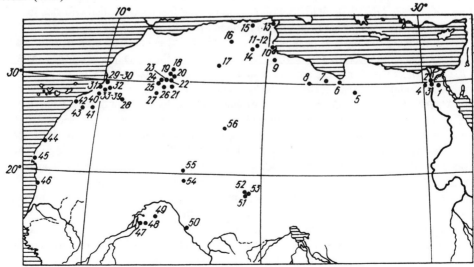

FIG. 46. Distribution of *Tropiocolotes tripolitanus*.

Algeria: 56 - Ahaggar [249]; Qued Amra (ZFMK); Tin-sig (ZFMK).

Not located on maps: Wadi M'belem, Tripoli [249]; Tunisia: Djebel Bu Hedma [249]; Governm. de Gabes: Gabes, 21 km W, 19 km N Busack (USNM); Gol. de Tafana (MNHP); Algerie: Carie (MNHP); Sahara et Souss. Jusqu'a [274]; Mauritania N'Keder (MNHP); Rio-de-Oro: Audebdinet [271]; Morocco: zw. Tazzarine und Alnif: O El Arba, Trummerfeld (ZFMK); 16 km W Bouanane sandiges Trocken oued (ZFMK); 5 km SW Bou-Rbia Fuss des I. Bani (ZFMK); Touantia (ZFMK).

{p. 103} SEXUAL DIMORPHISM AND AGE VARIATION. Males of this species are distinguished from females only by insignificant swellings at the base of tail. Females are slightly larger than males but this difference was not established with certainty because of inadequate samples available to us. Age variation of body proportions is similar to that in species of the genus *Alsophylax* (Fig. 47).

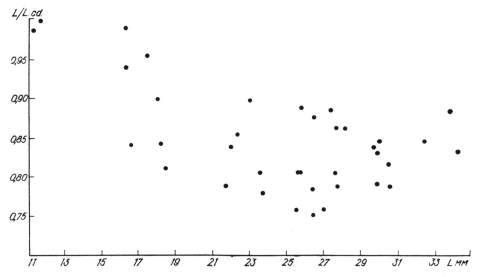

FIG. 47. Age variation of L/LCD index in *Tropiocolotes tripolitanus*.

{p. 104} SYSTEMATICS AND GEOGRAPHIC VARIATION. H. W. Parker [272] described two new species that were closely related to the Tripoli dwarf gecko. According to his viewpoint, *T. somalicus* differs from the latter [the Tripoli dwarf gecko] in having [both] posterior postmentals and small scales in the upper ocular region. On the other hand, in *T. occidentalis* the characteristics of the posterior postmentals and enlarged scales in the upper ocular region are very unclear and difficult to discern. Also, both new species were distinguished from the type species by 35-41 scale rows around midbody (44-48 in *T. tripolitanus*).

In revising the African geckos, A. Loveridge [249] considered both of the Parker's species only as subspecific forms of *T. tripolitanus*. In addition, he described the subspecies *T. t. algericus* (its characters are: short suture between first pair of postmental scales and second infralabials; small scales in upper ocular region; 44 scale rows around body; range in Algerian Sahara) and designated the nominate subspecies (wide suture between first pair of postmental scales and second infralabials; no contact between the second pair of postmentals and first infralabial scales; 42-48 scale rows around mid- {p. 105} -body; distributed from Egypt through Libya to Tunisia). He specified that the second pair of postmentals in *T. t. somalicus* has a wide suture with the first infralabials, but the postmental scales in *T. t. occidentalis* are poorly defined and their first pair cannot contact the infralabials at all. A. Loveridge did not mention the number of specimens examined but indirect data allow the conclusion that he studied isolated specimens.

From several localities in Mali, T. Papenfuss [271] described the subspecies *T. t. apoklomax* in which he found 46-54 scale rows around the midbody and no contact between the second postmentals and the first infralabials. But he mentioned variations in the first character: Kombori specimens have 46-52 scales around midbody, but [specimens from] Sanghi and Bandiagara (120 miles southwest) have – 52-54.

Thus, at present, five subspecies of this species are known, which differ from each other, in general, in the number of scale rows around the midbody, different degree of definition of infralabials, and the presence of their contact with infralabial scales.

The comparison of small samples and single specimens from Morocco, Spanish Sahara, Rio-de-Oro, Mauritania, Mali (to north of *T. t. apoklomax* range), Niger, Algeria, Tunisia, Egypt, and Somalia demonstrated that the characters used for the diagnoses of these forms probably revealed clinal variation. Thus, the size of the first and second pairs of postmentals increases from southwest to northeast, and the third and fourth pairs of these scales appear, so the width of sutures between the scales in first pair of postmentals, as well as between these and the second infralabials, distinctly increases. Additionally, along the line connecting the western and northern shores of Africa (Algiers), a relatively stable (isophene) number of scales is observed around the midbody and, from below, along the longitudinal body axis. From this line, toward the center of the continent, the numerical value of these characters increases. The only exception is the sample from Somalia. Considering the fact that the distribution of the Tripoli dwarf geckos remains totally unexplored over such a wide territory, we believe that separating this species into subspecies is premature. It can be supposed that, in reality, there exist only the Somalian form, which has discrete characters and is separated by the Abyssinian Plateau and, probably, the Mali form. However, the final resolution of this issue should be based on more representative collections.

ECOLOGY unknown. A. Loveridge [249] reports, based on his observations, that the nominative form in Algiers and Egypt (near the Giza pyramids) lives under rocks, on pebble surfaces, and on loose sands (100-300 m elevation). It feeds on bloodsucking insects, small spiders, very small beetle larvae (mealworms), and crickets.

According to his data, the distribution of the Somalian subspecies is limited by the arid shore zone. Seven of eight specimens, which constitute the type series, were caught on the ground in a sandy area covered with stones, nearly devoid of vegetation, at about 960 m elevation.

The Algerian subspecies from the eastern mountain slopes in the vicinity of Ferme Dufour Farm was found under rocks (its elevational range – 100 to 500 m). The specimens were collected from February to December. It is also known [219] that it feeds on ants, termites, and, possibly small moths. It is considered a nocturnal species (from the observations in captivity). Twice, lizards were caught in the daytime deep in a burrow and under a rock. In the spring, summer, and fall, the dwarf gecko emerges for foraging at sunset for the whole night. It is probably also active in the winter, especially on the jebels, where the temperature is always higher. According to W. Klingelhöffer's data [236], *T. tripolitanus* is encountered under stones in rocky areas that migrate with strips of sand. It is active in the evening. In the terrarium, it feeds on small spiders, fruit flies, and moths.

{p. 106} STEUDNER'S DWARF GECKO — *TROPIOCOLOTES STEUDNERI* (PETERS, 1869)

Type locality: Sennar, Sudan.
Karyotype: $2n = 38$ (not described in detail [312]).
1869 – *Gymnodactylus steudneri* Peters, Monats. Acad. Wiss., Berlin, p. 788.
1891 – *Stenodactylus petersii* Boulenger, Transact. Zool. Soc. London, vol. 13, p. 108.
1900 – *Tropiocolotes nattereri* Steindachner, Denkschr. Acad. Wiss. math.-naturw. Kl., Bd 69, no. 1, p. 326.
1900 – *Tropiocolotes steudneri* Steindachner, Denkschr. Acad. Wiss. math.-naturw. Kl., Bd 69, no. 1, p. 326.

DIAGNOSIS. Body scales smooth or with weak keels; dorsal scales equal to or slightly larger than abdominal ones; supranasal scales not larger than others, do not touch nostrils.

DEFINITION (Fig. 48) (from 38 specimens in ZIK, FMNH, USNM, CAS, MNHP, MZSF). Type specimen lost (Dr. M. Häupl, NMW, pers. commun.). L general = 15.3-34.8 (25.24 ± 0.87); HHW 55-68 (60.17 ± 0.82); scales across head 12-17 (14.10 ± 0.21); supralabials 8-10 (9.13 ± 0.08); infralabials 6-8 (6.87 ± 0.07); GVA 95-118 (107.90 ± 0.98);

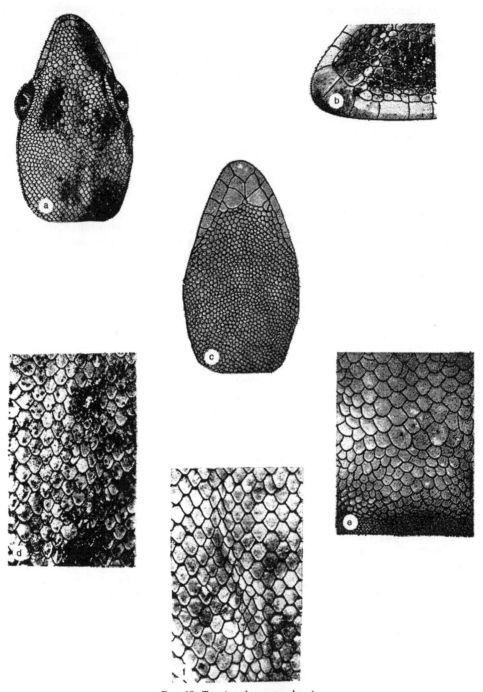

FIG. 48. *Tropiocolotes steudneri*:
a, head from above; b, snout from the side; c, head from below; d, dorsum; e, preanal pores; f, belly. (d-f – USNM 136355, near Cairo, Egypt; others – FMNH 78393, Red Sea coast, Egypt).

ventral scales across midbody 46-55 (50.68 ± 0.45); subdigital lamellae 15-18 (16.70 ± 0.12); 3.5-3.8 supralabials reach orbit. Two (28.6%), three (31.4%), or four (40%) pairs of postmentals; the first pair is the largest, scales in it joined by wide suture. Sometimes males have a pair of feebly defined preanal pores, frequently separated from each other by one to two scales.

COLOR AND PATTERN (as fixed in alcohol). The main body background is sandy yellow.

On each side of the head, a wide dark brown stripe runs from the nostril, through the eye and toward the ear without reaching the shoulder blades. The pattern on the back varies. The individuals from Egypt are characterized by a large number of dark spots and dots or by vague wide transverse bands. Unique to the individuals from Palestine are four or five thin, wavy, dark transverse bands that are edged on the rear with a light stripe. There are up to 13 dark transverse bands, with varying degrees of prominence, on the tail.

DISTRIBUTION. Northeastern Africa from Algeria in the west to northern Sudan in the south and the Jordan River in the east; and an isolated locality in southeastern Iran (Figs. 43, 49).

LEGEND to Figs. 43, 49.
Iran: 1 - Bandar-Langeh [223].
Algeria: 1 - Oued Igharghar, C Sahara [249]; 2 - Amguid (MNHP); 3 - 5 km NE Tamanrasset (CAS); 4- 3 km E Tamanrasset (CAS); 5 - Tassili N'Ajjer (MNHP).
Libya: 6 - 25°30′N, 16°E [273]; 7 - Cufra Oasis, Et Tag [249]; 8 - Cufra Oasis, El Giof [249]; 9 - Cufra Oasis, el Telib {error in original – MG} (MCZN); 10 - Cufra Oasis, Es Zurgh [306].
Egypt: 11 - el Giza [249]; 12 - Harraniya (USNM); 13 - Aburoash [249]; 14 - Embaba Dist. {error in original – MG}; 15 - 24 km W Cairo [251]; 16 - Wadi Ghuweibba [251]; 17 - 60 km from Cairo to Suez (ZIK); 18 - Gebel Mokattam, E Cairo [249]; 19 - Gebel el Anquabiya [249]; 20 - Wadi Gindali [251]; 21 - Maadi [251]; 22 - Helwan [249{error in original – MG}]; 23 - Wadi Hof [249]; 24 - Wadi Asyuti [251]; 25 - Ras Gharib [251]; 26 - Kalamshah, Fayum [249]; 27 - Hurghada [251{error in original – MG}]; 28 - Luksor [249]; 29 - el Karnak (ZFMK); 30 - Isna [249]; 31 - Wadi Abbad [249]; 32 - Philae [249]; 33 - Bir Abraq (USNM).
Sudan: 34 - Halaib [249] (is mentioned on the maps as Egyptian territory); 35 - Nubia [249]; 36 - Kosheh [249]; 37 - Sennar [249].
Sinai: {error in original – MG} 38 - Sinai: Nuveiba Coast of Acaba Gulf [295]; 39 - Sinai: Bir al Mashiya [295].
Israel: 40 - Wadi Roman, Negev [249]; 41 - Ein Gedi (SAM).
Not located on maps: Algeria: 15 km Terhenanet Raud lines Wadi m. Felsen; Libya: El Airenat [249]; Israel: Negev: Hamakhtesh Haqatan (MZSF).

SEXUAL DIMORPHISM AND AGE VARIATION. Males of this species often have a couple of weakly defined preanal pores. In addition, there are weakly developed swellings beneath the base of the tail. The females are probably slightly larger than the males, but inadequate samples in our hands do not allow [us] to establish this with certainty. We were not able to study age variation for the same reason.

{p. 107} SYSTEMATICS AND GEOGRAPHIC VARIATION. While studying the systematic position of some species in this genus, S. Minton with coauthors [263] noticed differences

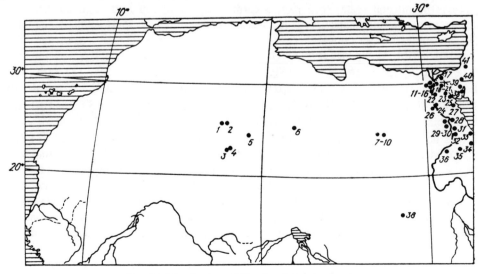

FIG. 49. Distribution of *Tropiocolotes steudneri*.

between specimens from Egypt and Palestine. They noted that specimens from the west have a somewhat higher number of supra- and infralabial scales (6-10 and 6-8 as opposed to 7-8 and 5-7), scales across midbody (38-50 as opposed to 41-43), and subdigital lamellae (14-19 as opposed to 14-16). Differences in pattern were also found.

{p. 108} We made an attempt to compare small samples from Algeria, Egypt, and Palestine and the only specimen known from Iran. This comparison demonstrated that, in the northern parts of the range, the decrease in the number of scales along and across the body indeed occurs from west to east. However, we did not note any variation in the number of supra- and infralabial scales and subdigital lamellae.

Considering the facts that the distribution of this species within the range is not clear, comparative material from the type locality (northern Sudan) is absent, and [available] samples in collections are not large enough, the issue of intraspecific systematics of *T. steudneri* cannot yet be resolved.

HABITAT. This species is encountered in rocky and sand deserts, under rocks, and, in the Sahara, it lives in tree trunks. All the areas of habitation are on vegetation-free, rocky hills that are dry, windy, and severely heated by the sun. The elevation is 100-300 m [249].

BEHAVIORAL ATTRIBUTES. In the terrarium, these lizards do not dig burrows but spend all their time on top of large stones. They are very cautious and agile, however, [they] quickly become tame. The collections were obtained from February through October [249].

FEEDING. Steudner's dwarf geckos have twice been seen eating *Lepisma sakharina*. In captivity they daily ate various flying insects [249].

SCORTECCI'S DWARF GECKO — *TROPIOCOLOTES SCORTECCI* CHERCHI ET SPANO, 1963

Type locality: El-Safa, Hadramaut, Yemen.
Karyotype not studied.
1963 – *Tropiocolotes scortecci* Cherchi et Spano, Boll. Mus. Ist. Biol. Univ. Genova, vol. 32, no.188, p. 29.

DIAGNOSIS. Body scales slightly keeled; dorsal scales slightly larger than ventrals; supranasal scales distinctly larger than other scales on snout (do not contact nostrils).

HOLOTYPE. MIZG-1. "El Safa, Hadramaut 2 May 1962, G. Scortecci coll." Adult male. L = 21.5 mm; LCD = 20.2 mm; L/LCD = 1.06. Nostril between quadrangular rostral, first supralabial, and two small flat nasal scales (the first somewhat larger than second), not distinguishable from other scales of snout; internasal scales larger than surrounding scales, do not touch nostrils. Eleven scales across head. {p. 109} Nine supralabials (10, according to description); seven infralabials (8, according to description). Mental scale hexagonal; two pairs of postmental scales (one pair, according to description), scales of the first pair separated by one gular scale; GVA 106; 50 scales across midbody. 15-16 subdigital lamellae (17, according to description).

DEFINITION. Composed from all currently known specimens (three type specimens in MIZG and 11 specimens from BMNH), Fig. 50. L general = 17.2-29.8 (23.99 ± 1.07); L/LCD gen. (n = 6) 0.87-1.31 (1.14 ± 0.06); HHW (n = 10) 58-70 (62.20 ± 1.07); scales {p. 110} across head (n = 14) 10-16 (12.36 ± 0.45); supralabials (n = 28) 9-11 (9.61 ± 0.13); infralabials (n = 28) 6-9 (7.36 ± 0.14); GVA (n = 13) 85-107 (94.54 ± 2.50); [ventral] scales across body (n = 13) 36-50 (43.54 ± 1.43); subdigital lamellae (n = 28) 12-19 (14.75 ± 0.81). Mental scale triangular or pentagonal (78.7%), hexagonal (14.2%), or trapezoid (7.1%); one (28.6%), two (57.1%), or three (14.3%) pairs of postmentals, scales in the first pair are in contact (78.7%), or separated by one (14.2%) or two (7.1%) scales. Sometimes, males have a couple of poorly developed preanal pores, often separated by one to two scales. Along row of infralabials, one (34.6%), two (23.1%), or three (7.7%) pairs of increasingly smaller scales can follow the second pair of postmentals; 34.6 % of our specimens do not have postmental scales.

COLOR AND PATTERN. The main body background is light yellow (according to other data [178], the color in life is pinkish yellow). On each side, a wide dark brown stripe runs

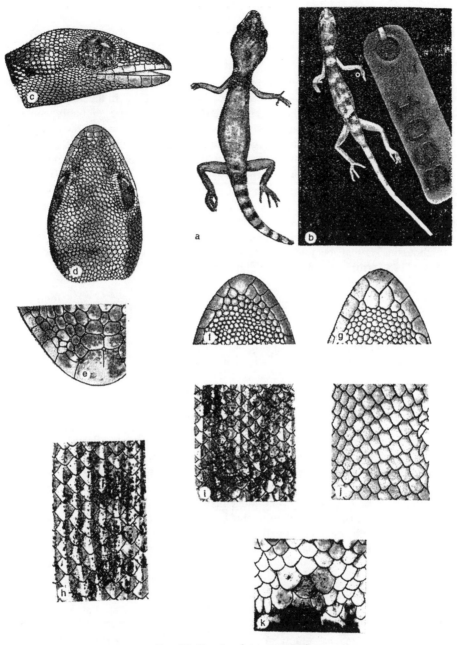

FIG. 50. *Tropiocolotes scortecci*:
a-b, general view from above; c, head from the side; d, head from above; e, snout from above; f-g, chin; h-i, dorsum; j, belly; k, preanal pores (c, d, f – MIZG 1, Hadramaut, Yemen; a – MIZG 2, same locality; others – BMNH 1978.1098 Dhofar, Oman).

from the nostril through the eye and toward the ear and stops beyond the shoulder blade. In addition, the stripes branch on each side above the shoulder blade. These branches come together on the neck, forming a collar-like pattern. There are no stripes on the back. Each dorsal scale is darker in the center than the main body color. On the tail, there are 10-12 dark brown broad stripes that run from the base to the tip. The upper surfaces of the limbs are the same color as the back. The lower surfaces are whitish.

DISTRIBUTION. Southern part of Arabian Peninsula (Fig. 43).

LEGEND to Fig. 43.
1 - El-Safa, Hadramaut, S Yemen (MIZG); 2 - Wadi Raykhut; 3 - Wadi Ghayz; 4 - Ayun; 5 - Thamarit; 6 - 16 km S Thamarit; 7 - Birba Shu'aythan {error in original; localities 2-7 from [177] – MG}.

SEXUAL DIMORPHISM AND AGE VARIATION. As in the other species of the subgenus, the sexual dimorphism is not pronounced in Scortecci's dwarf geckos. In the males, a swelling is sometimes noticeable beneath the base of the tail. The females are, probably, somewhat larger than the males, but the inadequate size of the sample available to us made it impossible to confirm the validity of this variation. Age variation remains unstudied for the same reason.

SYSTEMATICS AND GEOGRAPHIC VARIATION. The fourth specimen, captured after the three type specimens, already showed variation in some scalation characters. First of all, this was evident in the contact between the postmental scales of the first pair (in the type specimens these scales are separated by one or two small scales). In comparison with the types, this specimen has more scales between the orbits, more supralabial scales and subdigital lamellae, and fewer scales around midbody and along the longitudinal body axis on the underside. If this specimen's capture site, 500-600 km east of the type locality, is considered [175 {error in original – MG}], it may be assumed that the indicated characters demonstrate geographic variation.

ECOLOGY has not been studied. At the type locality, these geckos were found under rocks in an area with plentiful bushy vegetation, as well as in sandy gravel and gravel areas. One of the individuals was seen during the day. In an attempt to hide it slipped under a rock, where there was another gecko – *Pristurus karterii* [*sic*; *P. carteri*][208]. The areas of this species' habitation (type locality) are located at an elevation of 400 m.

Subgenus *Microgecko* Nik., 1907

Type species *Microgecko helenae* Nikolsky, 1907.

DIAGNOSIS. Smooth subdigital lamellae; supranasal scales significantly larger than other nasal scales and always contact nostrils; one to two pairs postsupranasal scales well developed; all body scales (including subdigital lamellae) smooth; dorsal scales distinctly smaller {p. 111} than ventrals; no enlarged subcaudal plates nor preanal pores. Distributed in southern part of Southwest Asia. The subgenus includes three species.

ADDITIONAL CHARACTERS. Snout covered by convex, heterogeneous, roundish-polygonal, flat scales, become distinctly smaller toward neck; ear opening very small, roundish. Gular scales roundish-polygonal, convex, flat, homogeneous. Dorsum covered by small, convex, smooth, oval scales, gradually increasing toward the flanks, so that the border with abdominal scales cannot be discerned. Ventral scales roundish-rhomboid, large, flat, smooth, imbricate, distinctly larger than dorsal scales. Tail cylindrical, somewhat thicker at base, becoming sharply thinner in the last third; covered by roundish, flat, smooth scales; no subcaudal plates. Adpressed forelimb reaches front edge of orbit, or a little farther, by finger tips; hindlimb reaches elbow or almost to axilla. Limbs covered above by scales similar to dorsal ones, below – similar to ventrals. Digits covered above by one [and] on the sides by two scale rows; claw base contacts upper and lower smooth lamellae; these lamellae followed by one [scale] above and two scales on the sides.

Key to the Species of the Subgenus Microgecko

1 (4). Postmental scales present.
2 (3). No more than seven supralabial scales . . . . *T. helenae* (Nik.)
3 (2). No less {error in original – MG} than eight supralabial scales . . *T. persicus* (Nik.)
4 (1). No postmental scales . . . . *T. latifi* Leviton et Anderson

KHUZESTAN DWARF GECKO — *TROPIOCOLOTES HELENAE* (NIK., 1907)

Type locality: Alkhorshir in Khuzestan (Zagros Ridge, western Iran).
Karyotype: not studied.
1907[1] – *Microgecko helenae* Nikolsky, Ann. Zool. Mus. Imp. Acad. Sci., 1905, vol. 10, no. 3-4, p. 265.
1970 – *Tropiocolotes helenae*, Minton, S. Anderson and Jer. Anderson, Proc. Calif. Acad. Sci., ser. 4, vol. 37, no. 9, p. 345.

DIAGNOSIS. Six to seven supralabial scales; postmental scales present (usually do not contact each other in the first pair).

LECTOTYPE. ZIL 10242. Adult female. Persia, Alkhorshir Village, 30 Sept. 1903, N. Zarudny. L = 27.5 mm; LCD = 30.5 mm; L/LCD = 0.90. Nostril between pentagonal mental, first supralabial, large flat supranasal scales and two small nasal scales; supranasal scales separated from each other by one scale; a pair of large postsupranasal scales, which are separated from each other by one row of two scales, follows supranasals. 19 scales across head. Six supralabial scales on both sides; 5-6 infralabials; postmental scales in the first pair separated from each other by two gular scales; GVA 121; 68 scales across midbody. 14-14 subdigital lamellae.

DEFINITION (from 60 specimens in ZIK, ZIL, USNM, and MCZ), Fig. 51. L gen. = 15.5-28.5 (22.03 ± 0.46); HHW 56-66 (60.41 ± 0.48); scales across head 15-23 (18.57 ± 0.23); supralabials 6-7 (6.68 ± 0.040); infralabials 5-7 (5.66 ± 0.05); GVA 101-122 (113.08 ± 0.71), scales across midbody 60-75 (66.41 ± 0.51); subdigital lamellae 11-15 (12.90 ± 0.07). Internasal scales contact each other (55%) or separated by a row of granules (45%); large postsupranasals in single pair in contact (12%) or separated from each other by one (78%) or two (10%) scales; 2.3-3 supralabials reach front edge of orbit. Mental scale trapezoid (87%) or sometimes pentagonal (8%) or triangular (5%); one (95.8%) or very rarely two (4.2%) pairs of postmental scales (the second pair very small); scales of first pair in contact (11.7%) or separated by one (25%), two (55%), or three (8.3%) gular scales.

{p. 112} COLOR AND PATTERN. The main background of the upper body surface is yellowish gray. A dark brown stripe runs from the nostril through the eye and above the ear opening. It ends at the neck or shoulder blade level. Thin transverse bands of the same color are found on the back (5-7) and the tail (up to 10). All the bands are edged with light color in the rear. The bands on the back are straight or zig-zagging. Sometimes they can be interrupted {p. 113} in the middle and shifted along the longitudinal body axis in such a way that halves of the same band do not touch. The regenerated tail is dark brown, almost black. The lower surfaces are light. There is no pattern on the limbs. The pattern on the back can partially or completely disappear in some individuals [286].

DISTRIBUTION. Found in numerous localities along the Zagros Ridge (western Iran); probably also inhabits northern Iraq (Fig. 52).

LEGEND to Fig. 52.
Iran: 1 - Dize, 125 km W Kermanshah on the road to Baghdad (ZSM); 2 - 25 km W Khorremabad, Lurestan, on the road to Ahvaz (ZSM); 3 - Masheid-Soleyman (ZIK); 4 - Khuzestan Prov.: 16 km S Mashejd-Soleyman (USNM); 5 - Khuzestan Prov.: Bidezar (ZIL); 6 - Agulyashker (ZIL); 7 - Alkhorshir (ZIL); 8 - Esfahan [92]; Khuzestan Prov.: 9 - 35 km E Gach Saran on road to Khazerun (USNM); 10 - 80 km W Shiraz (ZSM); 11 - Fars Prov.: Mehkuh {error in original – MG}, 80 km W Shiraz to Miankotal (ZSM).

SEXUAL DIMORPHISM AND AGE VARIATION. Sexual dimorphism not present. Age variation is seen in the larger eye diameter to body length ratio in juveniles in comparison with adults.

SYSTEMATICS AND GEOGRAPHIC VARIATION. Geckos from the western part of the range (Lurestan and Kermanshah Provinces) were recently separated into the subspecies *T. h. fasciatus* [286]. The only difference from the nominate form (according to the authors' viewpoint, type locality is "Bidezar," Bid Zard) was in the straight bands across the body.

---

[1] Ann. Zool. Mus. Imp. Acad. Sci., vol. 10 (for 1905) was published in 1907. Thus, the date of publication is 1907 rather than 1905.

However, the newly rediscovered syntypes lack pattern (designated lectotype came from Alkhorshir rather than from Bid Zard; this was required because of the the poor condition of other types). Having examined collections of this species from various parts of the range, as well as the literature data, we noticed that, in A. M. Nikolsky's [92] and R. Tuck's [302]

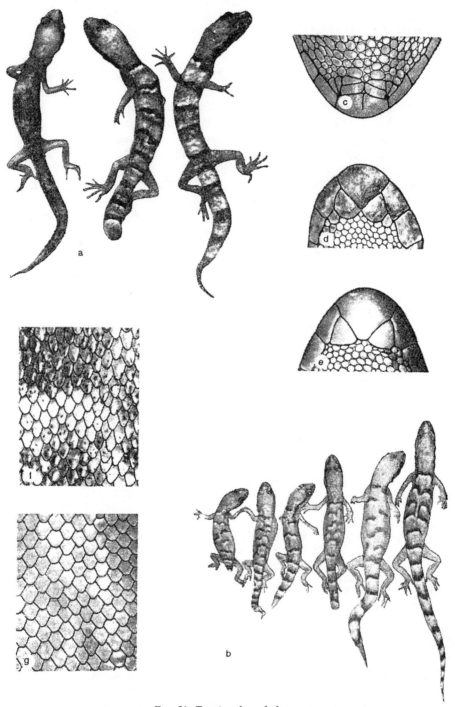

FIG. 51. *Tropiocolotes helenae*:
a-b, general view from above; c, snout from above; d-e, chin; f, dorsum; g, belly
(c, e – ZIL 10242, Alchorshir, Iran; b – ZSM 500/68, Mekuh, Iran;
others – ZIK SR 569, Masheid-Soleyman {error in original – MG}.

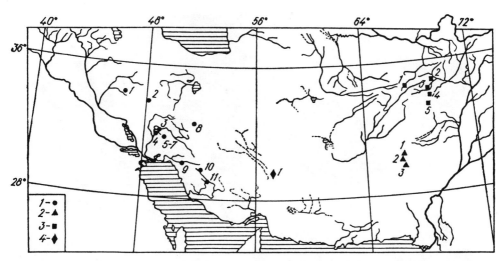

FIG. 52. Distribution of *Tropiocolotes helenae* (1), *T. depressus* (2), *T. levitoni* (3), *T. latifi* (4).

works, the dorsal bands in the pictured lizards are slightly sinuous. The specimens from Mashejd-Soleyman, available to us, have the same bands (Fig. 51). However, in the geckos from the southeastern part of the range (Fars Prov.), the dorsal bands are sharply twisted and often torn in the middle (Fig. 51). Thus, clinal variation in this character is well developed, and so, the subspecific separation of *T. helenae* does not appear advisable. Our study of scalation variation supports this conclusion.

ECOLOGY. A small amount of biotope data for the Khuzestan dwarf gecko is found in an article by J. and F. Schmidtler [286], which states that individuals of this species were caught near the town Mekukh on mountain slopes covered with a thin growth of pistachios and with occasional single oaks, where outcrops of bedrock are sometimes found. The geckos would hide under small – the size {p. 114} of a human hand – flat rocks in shady areas at an elevation of about 1200 m.

R. Tuck [302] reports that several geckos of this species were caught near the Gach Saran - Kazerun road in a dry channel on the western slope of the Zagros crest on 8 February 1964. This area is a mountain valley with steep gypsum and limestone slopes, where the soil is littered with small flat stones. The vegetation is thin, {p. 115} made up of occasional grasses, lichens, thorny bushes, euphorbias, and relict oaks. The daytime temperature is 19-42°, relative humidity 89%; there is rain and high clouds. The geckos were extracted from under rocks in the middle of the dry steam bed. To the south of Mesdjede-Soleiman, also on the western slopes of the Zagros crest, a young dwarf gecko was caught on 26 February 1964. The area is crisscrossed by dry channels, covered with thin vegetation, and has some limestone outcrops. The air temperature was 28-44°. The author noted that, among the eight females, not a single one was gravid.

I. S. Darevsky reports that he caught these geckos in the vicinity of a settlement in the cracks of loess brick duvals {Middle Asian loess-brick walls – MG} and ruins.

The specimens known to us were obtained from January through December.

PERSIAN DWARF GECKO — *TROPIOCOLOTES PERSICUS* (NIK., 1903)

Type locality: "Vikus Degak" Dehek (Dyzek), Iranian Baluchistan.
Karyotype not studied.

1903 – *Alsophylax persicus* Nikolsky, Ann. Zool. Mus. Imp. Acad. Sci., vol. 8, p. 95.
1970 – *Tropiocolotes persicus*, Minton, S. Anderson, and Jer. Anderson, Proc. Calif. Acad. Sci., ser. 4, vol. 37, no. 9, p. 348.

DIAGNOSIS. Eight to 10 supralabial scales; postmental scales present (usually first pair in contact).

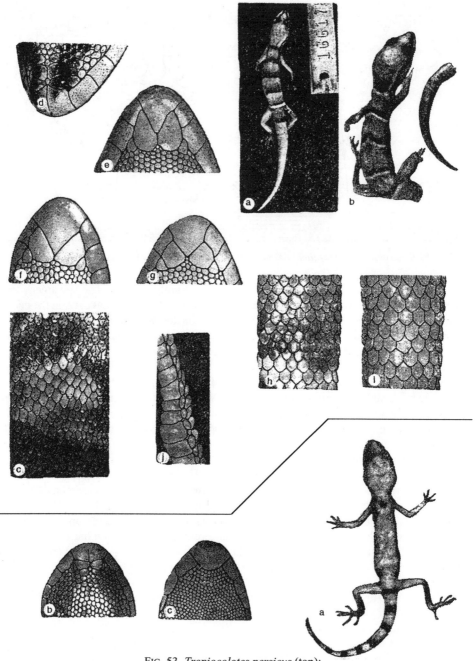

FIG. 53. *Tropiocolotes persicus* (top):
a-b, general view from above; c, dorsum; d, snout from above; e-g, chin; h, tail from above; i, tail from below; j, toe from below (b, d, g – MNHN 1966.17, *T. p. persicus*, Iranshehr, Iran; b, f – CAS 86408, *T. p. bakhtiari*, Masheid-Soleyman to Sari-Gach, Iran; others – USNM 166178, *T. p. euphorbiacola*, Hub Chawki, Pakistan).
Tropiocolotes latifi (bottom):
a, general view from above; b, snout from above; c, chin (CAS 134365, Kerman, Iran).

HOLOTYPE. ZIL 10005, "Vikus Degak in terra Dizak, Persia Orientalis, 9 Feb. 1901, N. Zarudny coll." L = 25.3 mm; tail regenerated; nostril between pentagonal rostral, first supralabial, large, flat supranasal scale and two small additional nasal scales; supranasal scales in contact; followed by two pairs of postsupranasal scales that contact each other {in the pairs – MG}. Nineteen scales across head. Nine supralabials on each side, 7 infralabials; two pairs of postmental scales, the first pair is the largest, the scales in it separated by one gular scale.

DEFINITION (from 61 specimens in ZIL, ZIK, AMNH, USNM, UMMZ, CAS, and MNHP [= MNHN]), Fig. 53. L general = 17.8-35.9 (26.18 ± 0.52) mm; HHW 56-70 (53.31 ± 0.75); scales across head 15-20 (17.64 ± 0.18); supralabials 8-10 (8.81 ± 0.06); infralabials 6-9 (7.18 ± 0.04); GVA 113-130 (121.03 ± 0.64); [ventral] scales across midbody 66-78 (71.21 ± 0.42); subdigital lamellae 11-15 (13.50 ± 0.08).

{p. 116} ADDITIONAL CHARACTERS. Internasal scales in contact (47.5%) or separated by a scale (52.5%); followed by one to two postsupranasal scales that contact each other in the first pair (95%); 3.4-3.8 supralabial scales reach front edge of orbit. Mental scale triangular or pentagonal with pointed (very rarely blunt) rear edge; between three and one poorly developed pairs of postmentals; as a rule, scales of first pair in contact or, very rarely, separated by a gular scale.

COLOR AND PATTERN (as fixed in alcohol). The main background of the upper body surfaces is light yellowish. A wide dark or chocolate-brown stripe runs from the nostril through the eye and ear opening to the shoulder blade (sometimes, along the sides of the body and to the base of the hindlimbs). Five to seven such stripes, sometimes edged with white at the rear, run from the neck to the sacrum. In some individuals, the intensity of color grows toward the distal part of each stripe (this character is subject to geographical variation); the ratio of stripe width to width of the interspace is highly significant in determining the subspecies. There are 6-10 stripes on the tail. The head and limbs have no pattern. The lower surfaces are white. The regenerated tail is white.

DISTRIBUTION. Occurs in southwestern and southern Iran and southeastern Pakistan (Fig. 54).

LEGEND to Fig. 54.

T. p. bakhtiari: 1 - Iran: between Mashijd-i-Suleiman and Sar-i-Gach 31°57'N, 45°21'E (CAS).

T. p. persicus: 1 - Chagai Dist.: Mirjawe (in Iran; Chagai Dist. in Pakistan) (SAM); 2 - 100 km N Iranshar (MNHP[= MNHN]); 3 - 20 km SW Pip [256]; 4 - Degak in terra Dizak (ZIL).

T. p. euphorbiacola: Pakistan: 1 - 36 km N Bela (AMNH); 2 - Upper Hub River, Las Bela Prov. (MNHP[= MNHN]); 3 - 10 km N Diwana (AMNH); 4 - 10 km N Uthal (AMNH); 5 - Las Bela Dist.: Pub Hills (CAS); 6 - 10 km N Karachi (UMMZ); 7 - 8 km E Landhi (AMNH); 8 - 35 km NW Bund Murad Khan [263]; 9 - 9 km S Sehwan (AMNH); 10 - 8 km E Thana Bula Khan [263]; 11 - Tatta Dist.: Haledji Lake (AMNH); 12 - Nabisar (AMNH).

SEXUAL DIMORPHISM AND AGE VARIATION. Females are somewhat larger than males.

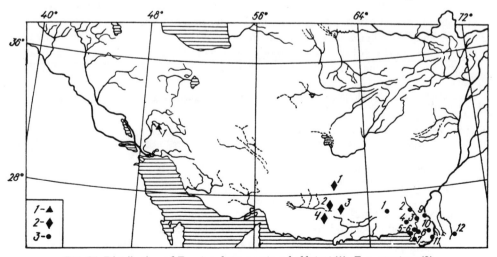

FIG. 54. Distribution of *Tropiocolotes persicus bakhtiari* (1), *T. p. persicus* (2), *T. p. euphorbiacola* (3).

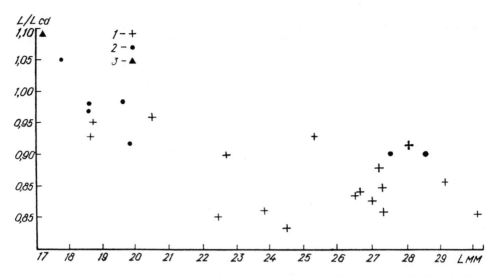

FIG. 55. Age variation of L/LCD index in *Tropiocolotes helenae* (1), *T. persicus* (2), *T. latifi* (3).

Age variation is defined by a larger eye diameter to body length ratio in juveniles in comparison to adults. Also, in comparison to adults, juveniles have shorter tail (Fig. 55).

SYSTEMATICS AND GEOGRAPHIC VARIATION. For a long time, this species, which was described as an *Alsophylax* and closely related {p. 117} to *T. helenae*, was not recognized by specialists, which led to a long list of synonyms. Having looked into its confused systematics, S. Minton and co-authors [265] proved that this species belongs in the genus *Tropiocolotes* and described three subspecies. The distinctions are in the different ratios between the width of dorsal bands and the interspaces between them.

A study of small samples and single specimens of the Persian dwarf gecko showed that, within the limits of the eastern subspecies' range (which is the best represented in our samples), some characters vary within fairly wide limits, but the diagnostic ones mainly form a hiatus between the subspecies. However, it is posssible that transitional forms between the nominal and eastern subspecies will be found. At present, three subspecies of Persian dwarf gecko can be distingiushed.

Key to the Subspecies of Tropiocolotes persicus

1 (2). Shields in first pair of postmentals separated by one gular scale or in point contact
. . . . *T. p. persicus* (Nik.)
2 (1). Shields in first pair of postmentals contact each other by wide suture.
3 (4). Width of dorsal crossbars wider than interspaces . . . *T. p. bakhtiari* Minton, et al.
4 (3). Width of dorsal crossbars equal to or slightly narrower than interspaces (but always more than half the width of these interspaces) . . *T. p. euphorbiacola* Minton, et al.

PERSIAN DWARF GECKO — *TROPIOCOLOTES PERSICUS PERSICUS* (NIK., 1903)

Type locality: of species.
1903 – *Alsophylax persicus* Nikolsky, Ann. Zool. Mus. Imp. Acad. Sci., vol. 8, p. 95.
1956 – *Tropiocolotes helenae*, Mertens, Jh.Ver. vaterl. Naturk. Württemberg, vol. 3, p. 92 {error in original – MG}.
1963 – *Bunopus persicus*, S. Anderson, Proc. Calif. Acad. Sci., ser. 4 , vol. 31, p. 474.
1966 – *Microgecko helenae*, Guibé, Bull. Mus. Nat. Hist. Natur., ser. 2, vol. 38, p. 98.
1970 – *Tropiocolotes persicus persicus* Minton, S. Anderson, and Jer. Anderson, Proc. Calif. Acad. Sci., ser. 4, vol. 37, no. 9, p. 348.

DIAGNOSIS. Width of dorsal crossbars narrower than half width of interspaces; every such crossbar has distinct light posterior edge; width of tail bands narrower than interspaces; scales in first pair of postmentals separated from each other by one gular scale or have point contact.

DEFINITION (from 3 specimens in ZIL, MNHP [= MNHN] and SAM). L gen = 25.3-

31.8 mm; 16-19 scales across head; 9-10 supralabials; 7-9 infralabials; 130 GVA; 78 [ventral] scales across midbody; 13-14 subcaudal plates. Supranasal scales contact each other (one case) or separated by a scale (two cases); postsupranasal scales contact each other; mental scale pentagonal or triangular with blunt posterior edge (two cases) or trapezoid (one case); two pairs of postmental scales; scales in the larger, first pair separated by a scale (two cases) or have point contact (one case).

DISTRIBUTION. Baluchistan along the Iran-Pakistan border.

BAKHTIARI PERSIAN DWARF GECKO —*TROPIOCOLOTES PERSICUS BAKHTIARI* MINTON ET AL., 1970

Type locality: Mashijd-i-Suleiman, Khuzestan, Iran.

1961 – *Tropiocolotes helenae* (not *Microgecko helenae* Nik., 1907), S. Anderson, Wasmann. {error in original – MG} Jour. Biol., vol. 19, p. 287.
1963 – *Microgecko helenae* (not Nikolsky, 1907), S. Anderson, Proc. Calif. Acad. Sci., ser. 4, vol. 31, p. 440.
1970 – *Tropiocolotes persicus bakhtiari* Minton, S. Anderson, and Jer. Anderson, Proc. Calif. Acad. Sci., ser. 4, vol. 37, no. 9, p. 351.

{p. 118} DIAGNOSIS. Width of dorsal and caudal crossbars more than the width of interspaces; no white posterior edge on these crossbars; scales in first pair of postmentals contact each other by wide suture.

DEFINITION. Composed from holotype CAS 86408. Adult male (?) "between Masjid-i-Suleiman (Masjed Soleyman) and Sar-i-Gach, Khuzistan Prov., Iran, 13 May 1958, S. C. Anderson." L = 20.5 mm; LCD = 29.8 mm; L/LCD = 0.96; 18 scales across head (22, according to description); 9/10 supralabial scales; 7/7 infralabial scales; 113 GVA; 74 [ventral] scales across midbody; 12 subdigital lamellae. Supranasal scales separated from each other by one scale, postsupranasals - by two. Mental scale is pentagonal, followed by two pairs of postmental scales (the first significantly larger than second), scales in the first one contact each other by wide suture. It is mentioned in the description that the subdigital lamellae in this specimen are feebly keeled. Having examined the holotype, we concluded that the keel-like structures on the subdigital lamellae could have appeared as a result of desiccation of this specimen. They are probably smooth in nature.

DISTRIBUTION. It is known only from type locality.

SPURGE PERSIAN DWARF GECKO —*TROPIOCOLOTES PERSICUS EUPHORBIACOLA* MINTON ET AL., 1970

Type locality: Las Bela Prov., near Karachi, Pakistan.

1962 – *Tropiocolotes helenae* (not Nikolsky, 1907), Minton, [Bull.] Amer. Mus. Nat. Hist., vol. 134, p. 81.
1970 – *Tropiocolotes persicus euphorbiacola* Minton, S. Anderson, and Jer. Anderson, Proc. Calif. Acad. Sci., ser. 4, vol. 37, no. 9, p. 354.

DIAGNOSIS. Width of dorsal crossbars equal to or slightly narrower than interspaces (but always more than half the width of these interspaces); regenerated tail always yellow without dark crossbars.

HOLOTYPE. CAS 93939. "Lower Pub Hills, Hab Chowki, Las Bela Dist., W. Pakistan," about 150 m elevation, 27 Dec. 1962, Jer. A. Anderson coll. Adult female {according to description, male – MG}. L = 28.5 mm; LCD = 32.5 mm; L/LCD = 0.87; supranasal scales in contact, followed by two large postsupranasal scales that also contact each other; 8/9 supralabial scales; 8 infralabials; two pairs of postmentals, scales in first pair significantly larger than in second and contact each other by wide suture. 14 subdigital lamellae.

DEFINITION (from 58 specimens in ZIK, CAS, SAM, AMNH, MNHP[= MNHN], SMF, and USNM). L gen = 17.8-35.9 (26.01 ± 0.53) mm; scales across head 15-20 (17.67 ± 0.19); supralabials 8-10 (8.78 ± 0.06); infralabials 6-8 (7.18 ± 0.04); GVA 111-130 (120.89 ± 0.60); [ventral] scales across midbody 66-77 (71.00 ± 0.41); subdigital lamellae 11-15 (13.55 ± 0.09). Supranasal scales contact each other (45.6%) or separated by one scale (54.4%); postsupranasal scales contact each other (96.5%) or very rarely separated by one scale (3.5%). Mental scale pentagonal or triangular, followed by one (18.4%), two (65.8%), or three (14.0%) pairs of postmentals, scales in the first pair significantly larger than others (in only one specimen, which is 1.8% of all examined, there is one pair of poorly developed postmental scales); these scales in first pair contact each other by wide suture (only in the aforementioned specimen separated by a gular scale).

DISTRIBUTION. This subspecies occupies the eastern part of the species' range, southernPakistan to Indus River on the east.

ECOLOGY. Some data on the ecology are contained only in the works by S. Minton [263] and S. Minton *et al.* [265]. The observations are from Pakistan and have to do with the subspecies *T. p. euphorbiacola*.

HABITAT AND QUANTITATIVE DATA. This dwarf gecko is widely distributed from the vicinity of Bela to Haleji (near town of Tatta). Its density here {p. 119} is quite high. The geckos stay in the areas where there are accumulations of the dead parts of *Euphorbia caudiciphola*, and find cover in the hollow trunks of these plants. It is probable that the geckos also live among the living plants; however, searching for the geckos is difficult here. Local populations have been encountered by the authors in analogous conditions in the vicinity of Lakhi (S of Sehvan), Diwana (along the upper Hab River valley) and Nabisar (in the Thar Desert east of Indus River). In the vicinity of Lakhi, the Persian dwarf gecko has been caught around the elevation of 225 m. For the nominate form, the elevation was 1880 m [223].

DAILY AND SEASONAL ACTIVITY CYCLE. This species is numerous from the middle of November until the end of February. It can be quite numerous following the winter rains and fog. They are observed on the surface on the humid warm nights in September and October running among the rocks. According to the collection materials, the specimens of this species were caught from January to December.

FEEDING. In the stomach of one gecko, caught on 28 July 1958 in southern Pakistan, we found the remains of a heteropteran and an ant. Persian dwarf geckos drank water in the terrarium by licking it off of a watch lens. The lizards were not discriminating eaters and ate various invertebrates of acceptable size, young spiders, and especially termites. One gecko was counted eating more than 20 termites at one time. Just captured geckos have a thicker tail than those [held] a long time in captivity and poorly fed.

REPRODUCTION. Adult individuals usually live in pairs. The female lays a single 8 × 5.7 mm egg in the dead, hollow roots of *Euphorbia caudiciphola* {MG}, in crevices, or under small roots of other plants. The egg-laying period lasts from April until August. Immediately after being laid, the egg is soft, but in a few hours the shell hardens and becomes brittle. It is elongated at the ends, glossy-white, with longitudinal stripes that disappear after three or four weeks. During this time the egg enlarges slightly. Incubation lasts five to eight weeks (35-60 days). Hatchling geckos are 13-15 mm long, and the young are colored like the adults.

BEHAVIORAL ATTRIBUTES. The lizards attempt to take cover when disturbed. They move in a snake-like fashion by bending the body and tail, almost without using the limbs. In addition, they can escape by jumping dexterously from branch to branch on small bushes. Researchers have noted that in the terrarium the Persian dwarf geckos willingly climb up the tree trunks; however, in nature, they have never been seen in the trees.

LATIFI'S DWARF GECKO — *TROPIOCOLOTES LATIFI* LEVITON ET ANDERSON, 1972

Type locality: Kirman, Iran.
Karyotype: not studied.
1972 – *Tropiocolotes latifi* Leviton and S. Anderson, Occas. Papers Calif. Acad. Sci., no. 96, p. 37.

DIAGNOSIS. No postmentals.

DEFINITION (Fig. 53) (from holotype, the only known specimen) CAS 134365. "Kirman, Kirman Prov., Iran, 1965." Juvenile male (?). L = 17.1 mm; LCD = 15.7 mm; L/LCD = 1.09; EED 0.38. Nostril between pentagonal rostral, first supralabial, large and flat supranasal scales and one (on left) or two (on right) additional nasal scales (according to description, one such scale is present on both sides); supranasal scales in contact with each other and followed by a pair of large postsupranasal scales separated from each other by one scale; 16 scales across head; 6/7 supralabials (7/7, according to the description); 6 infralabials (5 according to the description); 120 GVA; 72 [ventral] scales across midbody; 14/13 subdigital lamellae {p. 120} (14/14, according to description). Mental scale trapezoid with irregular posterior edge, bordered by eight gular scales.

COLOR AND PATTERN (as fixed in alcohol). The main background of the upper surface is yellowish white. A dark stripe runs on each side from the rear edge of the eye, through the top of the head and above the shoulder blade, and ends at the level of the elbow joint of the forelimb, when adpressed rearward. Between the shoulder blades, in the center, there is a small dark spot. Two spots of the same color and size are found on the back, in front of the hindlimbs.

There are nine narrow (narrower than the interspaces), straight dark transverse bands noticable on the tail. There is no pattern on the head and limbs. The ventral surfaces are white.

DISTRIBUTION. Known only from type locality in central Iran, on the boundary of the southern extreme of the Zagros Ridge and a large dessert massif.*

LEGEND to Fig. 52.
Iran: 1 - Kirman (CAS).

SYSTEMATICS. Absence of postmental scales in this specimens allowed A. Leviton and S. Anderson to separate this species. However, in the most closely related species, *T. persicus* (nominate subspecies) and *T. helenae*, the frequency of appearance of the character "absence of contact between scales in first pair of postmentals" is quite high (especially in *T. helenae*), which indicates a tendency towards the disappearance of postmental scales. In addition, variation of this character in other species (*T. tripolitanus, T. depressus, T. scortecci, A. laevis,* etc.) shows that the advisability of using this character for species differentiation is doubtful. However, because the knowledge of this group of species is poor, we conditionally consider *T. latifi* as a separate species.

ECOLOGY not studied. Caught at the elevation of 1710 m.

Subgenus *Asiocolotes* Golubev, 1984

Type species: *Tropiocolotes levitoni* Golubev et Szczerbak.

DIAGNOSIS. Subdigital lamellae smooth; a pair of flat supranasal scales do not contact nostrils; postsupranasal scales not defined; besides rostral and first supralabial scales, nostril surrounded by two (very rarely three) scales that do not differ significantly in size from supranasals nor from other scales of snout; body scales flat, dorsal scales small, distinctly smaller than ventrals; at least last third of lower surface of unregenerated tail covered by various combinations {in number and size – MG} of enlarged scales; males always with two to eight well defined preanal pores. Distributed in eastern part of Southwest Asia (Pakistan and Afghanistan). The genus includes two species.

ADDITIONAL CHARACTERS. Snout covered by slightly convex smooth polygonal scales, becoming distinctly smaller toward the neck. Gular scales flat, smooth, roundish polygonal; upper gulars slightly smaller than others. Dorsum covered by flat, smooth, roundish quadrangular small scales, gradually increasing in size toward the flanks in such a way that border between these scales and abdominal ones not discernable. Ventral scales flat, smooth, roundish pentagonal, imbricate, distinctly larger than dorsal scales. Preanal pores arranged in L-shaped row with a small angle on scales larger than abdominal ones; around five scale rows between them and vent. Tail slightly depressed, covered by rings of smooth, oval-rec-

---

* {A new find of this rare species has been reported from three localities in the central Kuhrud Mountains, Iran, by Moravec and Cherny, 1994 – MG.}

tangular flat scales that increase in size toward the flanks; underside of tail with various combinations of enlarged plates. Finger tips of adpressed forelimb reach tip of snout, toes almost reach shoulder. Digits slightly compressed, covered above by one, from sides {p. 121} by two scale rows; base of claw covered by upper and lower smooth lamellae, which, in turn, contact one scale from above and on each side.

Key to the Species of the Subgenus Asiocolotes

1 (2). Two to five preanal pores in males; enlarged plates only on the last third
of tail . . . . *T. depressus* Minton, S. Anderson, et Jer. Anderson
2 (1). Six to seven preanal pores; subcaudal plates present along entire tail
. . . . *T. levitoni* Golubev et Szczerbak

LEVITON'S DWARF GECKO — *TROPIOCOLOTES LEVITONI* GOLUBEV ET SZCZERBAK, 1979

Type locality: Afghanistan; Kabul, Pagman River; 10 km SW Kabul; Pogar Prov., Baraki Barak.
Karyotype: not studied.

1960 – *Cyrtodactylus* (= *Gymnodactylus*) *persicus* (= *Alsophylax persicus*), Wettstein, Zool. Anz., Bd. 165, Heft 5/6, p. 190.
1963 – *Alsophylax pipiens*, Leviton and S. Anderson, Proc. Calif. Acad. Sci., ser. 4, vol. 31, no. 12, p. 334.
1979 – *Tropiocolotes levitoni* Golubev and Szczerbak, Dokl. A[kad.] N[auk] UkrSSR, no. 4, ser. "B," p. 309.

DIAGNOSIS. Six to seven preanal pores in males; subcaudal plates along entire tail.

HOLOTYPE. CAS 120283, adult male. Afghanistan: Kabul, 6 May 1968, R. and E. Clark. L = 39.5 mm; LCD = 46.2 mm; L/LCD = 0.85. Nostril between quadrangular rostral, first supralabial scales and three flat nasal scales that do not differ in size from other scales of snout. Supranasal scales flat, do not contact nostrils, do not significantly differ in size from other scales of snout, no postsupranasal scales. 18 scales across head. 10 supralabial scales, 7/8 infralabials; two pairs of postmentals, the scales of inner pair contact each other and triangular mental scale. 133 GVA, 80 [ventral] scales across midbody; 6 preanal pores; 23 subdigital lamellae.

DEFINITION (from 24 specimens in CAS, USNM, NMW, and ZML) (Fig. 56). L adult male = 27.4-44.9 (37.30 ± 1.27) mm; L adult female = 22.0-44.8 (37.91 ± 0.98) mm; L/LCD gen = 0.80-0.85; HHW 43-55 (48.00 ± 0.70); scales across head 16-21 (18.23 ± 0.29); supralabials 9-11 (9.80 ± 0.09); infralabials 6-9 (7.63 ± 0.10); GVA 123-146 (129.84 ± 1.02); [ventral] scales across midbody 71-82 (77.93 ± 0.84); subdigital lamellae 17-23 (20.02 ± 0.23); preanal pores in males 6 (11 specimens) or 7 (two specimens). Four to five supralabial scales reach frontal level of orbit; ear opening oval, 0-2 enlarged scales above ear. Mental scale pentagonal or triangular with blunt (60.9%) or pointed (39.1%) tip; it is followed by two pairs of postmentals, scales in first pair are sutured (21.8%) or separated by one (60.9%) or two (17.3%) scales. Sometimes in sacral area and on the upper parts of side of tail base, enlarged scales present, which resemble dorsal ones (some of them have keels). Along 2/3 of tail length from the base, the lower surface of each segment has a pattern of alternating one and two enlarged plates; from there to the tip of tail, a pattern of one and one plate. Enlarged tubercles, intermixed in disorder with much smaller and smooth scales present on the outer surface of crus (and a few on thigh).

COLOR AND PATTERN. In life, the main background color of the back [209] is pale yellowish brown. The rear edges of the dark transverse bands are edged with white. Viscera visible through the translucent white or grayish belly. The labial scales are covered with gray dots.

A brown stripe runs from the nostril, through the eye and to the ear opening. Small stripes of the same color may appear on top of the head and on the neck, {p. 123} and the latter are M-shaped. Seven to eight similar transverse bands are located between the neck and the sacrum, and up to 10-11 bands between the base and the tip of the tail. The bands are two to four times narrower than the interspaces, and may be broken in the middle and

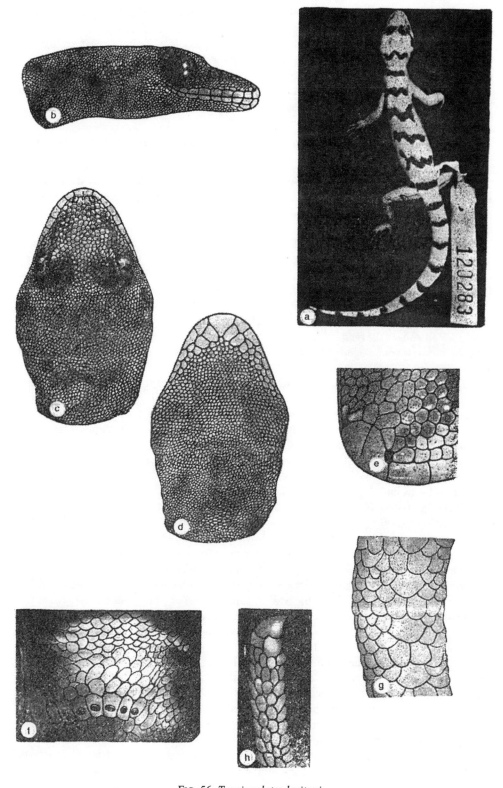

FIG. 56. *Tropiocolotes levitoni*:
a, general view from above; b, head from the side; c, head from above; d, head from below; e, nasal scales; f, preanal pores; g, tail from below; h, toe from the side (CAS 96217, Kabul, Afghanistan).

shifted along the longitudinal axis at the break. There is a poorly defined dark brown pattern on the limbs.

A small amount of data on the ecology of this species can be found almost exclusively in the work of H. Seufer [289], who observed these lizards in 1971-1973 under natural conditions and in 1977 in a terrarium.

DISTRIBUTION. The species has a limited range; it is known from several localities in the Kabul area and somewhat farther south (Fig. 52).

LEGEND to Fig. 52.
Afghanistan: 1 - Oukak [257]; 2 - Charikar [244]; 3 - Paghman River (CAS); 4 - Kabul (CAS); 5 - Baraki Barak (USNM).

SEXUAL DIMORPHISM AND AGE VARIATION. Males differ from females by the presence of preanal pores. As distinct from the smaller geckos of the genus, *T. levitoni* does not have differences between the sexes. The index EED is subject to age variation, as it is in species of *Alsophylax*.

SYSTEMATICS AND GEOGRAPHIC VARIATION. Erroneous identification of three species of northern pygmy geckos (*A. pipiens, A. laevis,* and *A. microtis* [112]) led to the listing of only the squeaky pygmy gecko for the majority of arid regions of Middle and Central Asia; this species within such limits was characterized by a wide range of variation in many features. This circumstance resulted in foreign authors [209, 243, 245, 253, 257, 263, 289, *et al.*] attributing individuals from Afghanistan to *A. pipiens*, because they did not comprehend correctly either the range or the spectrum of variation in this species. Having examined the collections, we concluded [43] that these specimens belong to a new species of the genus *Tropiocolotes*.

This species with its limited range does not show any geographical variation within the characters studied. However, the boundaries of variation for some of these characters (presence of subcaudal plates and tubercles in the sacral area, number of preanal pores, etc.) allow the assumption that Leviton's dwarf gecko and the species of the genus *Alsophylax* (*A. laevis*, in particular) had common ancestral forms.

HABITATS AND QUANTITATIVE DATA. In Afghanistan, this species is found only in areas of human activity, where it is frequently common. In Kabul, on the clay walls of houses, it sometimes forms large aggregations. It is found at altitudes to 2300 m [209].

RESPONSE TO TEMPERATURE. H. Seufer [289] has noted that in Kabul, which is located at an altitude of 1800 m, winter is cold and snow cover reaches 1 m in depth. The average temperature in January is $-1.3°$. Overnight lows can drop to $-25°$. Nevertheless, Leviton's dwarf gecko has adapted well to these conditions. The geckos that this author had in a terrarium would move about at temperatures quite low for reptiles ($+8-10°$).

Several specimens of these dwarf geckos were caught by that author on a May evening with the air temperature of $20°$ on the wall of a mud-house one meter above the ground.

DAILY AND SEASONAL ACTIVITY CYCLE. They leave shelter at early dusk and are active at night. In nature, winter hibernation lasts from November to the end of February. In some cold years it can stretch to the end of April.

FEEDING. The favorite hunting places are the outside surfaces of window mosquito netting, where many insects accumulate. They willingly eat mosquitos (*Anopheles*), which provide for this species an excellent food source into the deep autumn. In addition, beetles, spiders, night moths, caterpillars, heteropterans, ants and aphids are also in their food supply.

{p. 124} REPRODUCTION. This author [Seufer] has not had an opportunity to observe mating in these dwarf geckos. From the photographs in his work, it is evident that the females may have one or two eggs simultaneously. In captivity, eggs are laid from February to December. Their dimensions are 13×9 mm. As a rule, in the terrarium, females bury the eggs in the soil or leave them in the open on the bark of trees. The eggs were kept in an incubatror at the temperature of $20-28°$. The young emerged 44-93 days (the majority, 53-58 days) after the beginning of incubation. They fed on moth larvae, small mole crickets, and crickets, and reached adult size after 1.5 years.

BEHAVIORAL ATTRIBUTES. R. Clark et al. [209] report that these geckos have regular hunting areas, where they were observed every night, frequently sitting on sections of walls that are lighted by street lights. They are very wary. Once disturbed, they quickly disappear in the cracks or rotting planks of doors and transoms.

FLAT DWARF GECKO — *TROPIOCOLOTES DEPRESSUS* MINTON ET JER. ANDERSON, 1965

Type locality: Quetta Dist., Pakistan.
Karyotype: not studied.
1965 – *Tropiocolotes depressus* Minton et Jer. Anderson, Herpetologica, vol. 21, no. 1, p. 59.

DIAGNOSIS. Two to five preanal pores in males; subcaudal plates only on last third of tail.

HOLOTYPE. AMNH 93003 "Kach (Quetta Division), West Pakistan, on the abandoned rail line between Ziarat and Quetta at an altitude of about 2200 m on 18 May 1962, by S. A. Minton." L juvenile F = 26.8 mm; LCD about 24.5 mm; L/LCD = 1.09; nostril between quadrangular rostral, one supralabial, and two flat small nasal scales (the first insignificantly larger than second) that do not differ in size from other scales of snout. Supranasal scales flat, do not contact nostrils, do not differ in size from other scales of snout; no postsupranasal scales. 17 scales across head; 9 supralabial scales; 6/7 infralabial scales. A couple of small postmental scales, [which are] separated by two scales [that are] almost equal in size to the postmentals, following trapezoid mental scale. GVA 133, 76 scales across midbelly. No preanal pores. 16/17 subdigital lamellae.

DEFINITION from the five currently known specimens (from AMNH, RSM, and SMF), Fig. 57. L gen = 26.8-32.2 (29.82 ± 0.89) mm; L/LCD = 0.94-1.09; HHW 47-53 (49.33 ± 1.86); scales across head 16-20 (17.80 ± 0.73); supralabial scales 8-10 (8.80 ± 0.25); infralabials 6-8 (7.00 ± 0.15); GVA 129-139 (134.00 ± 1.79); [ventral] scales across midbody 74-76 (75.00 ± 0.32); subdigital lamellae 16-18 (17.30 ± 0.26); preanal pores in males 2-5 (3.67 ± 0.80). 3.7-4 supralabial scales reach front edge of orbit; ear opening roundish, no enlarged scales under it. Mental scale trapezoid with uneven posterior edge; followed by one to two pairs of small postmentals, scales of first pair do not contact each other, separated by one (20.0%) or two (80%) gular scales. No tubercles, similar to dorsal ones, in the sacral area. Tail covered underneath by rings of flat, smooth scales replaced on the last third of tail by enlarged scales, arranged on single segment at first as 2 = 1, then as 1 = 1. No enlarged tubercles on upper surfaces of limbs.

LIFE COLOR [264; 263]: the main background color on the back is saffron yellow. There are three to five dark brown transverse bands on the back and six on the tail. All the bands have clear boundaries, are densely colored, have no white edging and are narrower than the interspaces. There is a short dark brown stripe on top of the head. Stripes of the same color run from the nostril, through the eye and the ear opening on both sides of the head, coming together on the back of the neck in a horseshoe {p. 126} pattern. There is no pattern on the limbs. The ventral surfaces are pale pinkish white.

SEXUAL DIMORPHISM AND AGE VARIATION. Males differ from females by the presence of distinct preanal pores. It is impossible to judge age variation only from the five known specimens.

DISTRIBUTION. Known from three localities near Quetta (Pakistan) (Fig. 52).

LEGEND to Fig. 52.
Pakistan: 1 - Quetta District: Kach, on the abandoned rail line between Ziarat and Quetta (AMNH); 2 - Chiltan-Berges, 27 km S Quetta and 22 km N Mastung [259]; 3 - Kolpur (RSM).

SYSTEMATICS AND GEOGRAPHIC VARIATION. *T. depressus* has a local range (known from several localities of Quetta District, Pakistan), which suggests no forms within this species. Probably only individual variation can be stated: in comparison with the types, the newly found specimens [265] have more preanal pores, subdigital lamellae, and scales across head.

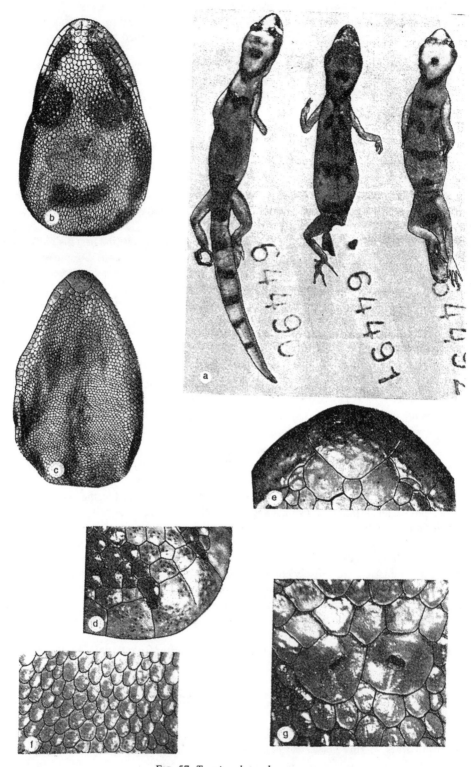

FIG. 57. *Tropiocolotes depressus*:
a, general view from above; b, head from above; c, head from below; d, nasal scales; e, chin;
f, dorsum; g, preanal pores (c, d, e – RSM 1964.58.1, Kolpur, Pakistan;
others – AMNH 93003, Ziarat to Quetta, Pakistan).

ECOLOGY. Data on this lizard are found in a few publications [263, 264, 265]. Dwarf geckos of this species were caught at early dusk between two small granite outcrops at an elevation of 2700 m, on a rocky hillside with a slope angle of 45° to the horizon and covered with thin bush and grass vegetation. They live in the old burrows of *Ochotona r. rufescens* and under stones. A male and a female were found under the same rock. They move quite rapidly on the ground and on rock fragments, onto which they frequently climb. When calm, their movements are slow: the geckos stand, bending the body in snake-like fashion, then run some distance, bending a great deal, and return to standing.

POTENTIAL ENEMIES. Scorpions (*Androchtonus astralis*, *Ruthus occitanus*), solpugids (*Galeodes caspius*), (*Skolopendra sp.*), young snakes (*Coluber rhodorachis*, *C. karelini*, *Vipera lebetina*), a lizard (*Eumeces schneideri blythianus*), Baluchistan shrews, and a hedgehog (*Hemiechinus megalotus*).

The temperatures in this region in October reach 15° in the daytime and fall to –4° at night. The researchers did not find dwarf geckos here in the summer.

Genus Keel-scaled Gecko — *Carinatogecko* Golubev et Szczerbak, 1981

Type species: *Bunopus aspratilis* S. C. Anderson, 1973.
1981 - *Carinatogecko* Golubev and Szczerbak, Vestn. Zool., no. 5, p. 34.

DIAGNOSIS. All body scales, except large head scales, strongly keeled; three nasal scales contact nostril, first slightly convex, 1.5 to 2 times larger than second, in turn slightly larger than third; no supranasal and postsupranasal scales; height of first supralabial scale from nostril to edge of mouth distinctly less than its width along edge of mouth; upper edge of supralabials curves onto the upper surface of snout. Eight to 10 supralabials, 3.5-4 of them reach front edge of orbit. Pupil vertical with serrated edges. Dorsum covered by flat scales and tubercles. No preanal pores in females (males are still unknown). Tail segmented (segments slightly distinct); two semicircles on each side of segment consist of three to four tubercles; base of each tubercle located in middle of one segment and separated from [segment's] posterior edge by row of small scales; tubercles do not contact each other; segmental midventral scales may be slightly enlarged but do not reach size of subcaudal plates, which are typical for many geckos. Digits slightly angularly bent, clawed, not widened, covered underneath by one longitudinal row of transversely enlarged lamellae; no lateral fringes.

{p. 127} ADDITIONAL CHARACTERS. Small geckos (L = 20-30 mm). Rostral scale pentagonal; first nasals separated by a scale; clear postnasal dents, no frontal dent; scales of snout heterogeneous, polygonal, alternate with tubercles on neck; ciliary growths indistinct; ear opening roundish. Mental scale pentagonal; four pairs of postmentals follow, they gradually decrease so that the latter pair is difficult to distinguish from gular scales; these scales in the first pair contact each other by wide suture, last two pairs do not contact infralabials; upper gular scales slightly larger than others. Dorsal tubercles arranged in regular longitudinal (12-14) and transverse (22-24) rows among flat small dorsal scales; abdominal scales flat, rhomboid, imbricate, their border with lateral scales indistinct. Caudal tubercles reach almost to end of tail. Base of claws covered by upper and lower lamellae with several short longitudinal keels, edged by one upper scale and two from each side.

Distributed along the Zagros Mountains in eastern and western Iran. The genus includes two species.

Key to the Species of the Genus Carinatogecko

1 (2) Middorsal scales distinctly larger than abdominal ones; caudal tubercles pointed, high, with enlarged posterior edge; tubercles, analogous to dorsal ones, present on forearm; 17-18 subdigital lamellae on fourth toes . . . .
. . . . *Carinatogecko aspratilis* (S. Anderson)

2 (1) Middorsal scales almost equal to abdominal ones; caudal tubercles not pointed, flattened, their posterior edges not high; no tubercles on forearm; 15 subdigital lamellae on fourth toes . . . .
. . . . *Carinatogecko heteropholis* (Minton, S. Anderson, and Jer. Anderson)

IRANIAN KEEL-SCALED GECKO — *CARINATOGECKO ASPRATILIS*
(S. C. ANDERSON, 1973)

Type locality: 35 km east of Gach Saran (30°20′N, 50°48′E), Fars Prov., Iran.
Karyotype: not studied.
1973 – *Bunopus aspratilis* S. C. Anderson, Herpetologica, vol. 29, no. 4, p. 355 {error in original – MG}.
1981 – *Carinatogecko aspratilis*, Golubev and Szczerbak, Vestn. Zool., Kiev, no. 5, p. 35.

DIAGNOSIS. Midbody scales distinctly larger than ventral ones; tail tubercles very mucronate, sharp-edged, high; lateral digital scales of lower row (closest to the end) flat with short longitudinal keels without spine at end; 17-18 subdigital lamellae; upper forearm scales heterogeneous.

HOLOTYPE. USNM 193961, adult female. Iran, Khuzestan, 35 km E Gach Saran, J. W. Neal coll., 10 Feb. 1964. L = 27.4 mm; LCD = 29.6 mm; L/LCD = 0.93; 14 scales across head; nine upper and seven infralabials; four pairs of postmentals; 110 GVA; 28 ventral scales across midbody; 18 subdigital lamellae; no preanal pores.

DEFINITION (from the only two known females from USNM) (Fig. 58). L females = 26.2-27.4 mm; LCD = 29.6 mm; L/LCD = 0.93; HHW 55-56; EED 29-30; supralabial scales 9-10; infralabials 6-8; scales across head 14-16; DTL 23-25; scales around dorsal tubercle 10-11; GVA 110-111; ventral scales across midbody around 28; subdigital lamellae 17-18. Two to four enlarged eyebrow scales above each eye; the last supralabial scale followed by three to four scales and then a row of three enlarged tubercles; dorsal tubercles triangular-oval, contact each other in transverse rows on the flanks. Four to five scales along the middle of tail segment; three to four tail tubercles in semicircle on each side of segment, at least first and second tubercles separated from posterior edge of the segment by a scale row; each such tubercle enlarged, mucronate, with very large posterior edge, which makes an a- {p. 129} -cute angle with the upper surface of segment; two central scales on the lower surface of the segment slightly larger than others. Adpressed forelimb almost reaches nostril with fingers, [adpressed] hindlimb [to] the axilla; tubercles, similar to dorsal ones, scattered among homogeneous scales of forearm, thigh, and crus (less defined on forearm). Lower row of lateral digital scales (closest to the end) consists of thickened scales with high keels which then become flattened with a short longitudinal keel without spine on the end; every subdigital lamella has {p. 130} two to three longitudinal keels, not mucronate and less well-developed toward end of digit.

COLOR AND PATTERN. The main background color on the upper side of the body is grayish blue. There is a poorly defined pattern on the head that is formed by thin, short, dark, curved lines. Five strongly bent, M-shaped dark stripes run from the neck to the lumbar area. In places, these stripes blend together in the protruding angles, forming a longitudinal pattern. On the tail, there are 13 transverse bands of the same color. Poorly defined transverse dark bands form an indistinct pattern on the limbs. The lower surfaces are light with numerous pigment spots.

DISTRIBUTION. Known only from the type locality.

ECOLOGY has not been studied. Judging from the capture sites, this species lives in the foothills zone at elevations of 200-500 m. A specimen was caught under pebbles, next to a dry channel, in the midst of thin grassy and woody vegetation [169].

IRAQI KEEL-SCALED GECKO — *CARINATOGECKO HETEROPHOLIS*
(MINTON, S. ANDERSON, ET JER. ANDERSON, 1970)

Type locality: Iraq: Erbil Prov.: Salahedin.
Karyotype: not studied.
1959 – *Alsophylax persicus* Nik.; Reed and Marx, Transact. Kansas Acad. Sci., vol. 62, no. 1, p. 97 (err. det.).

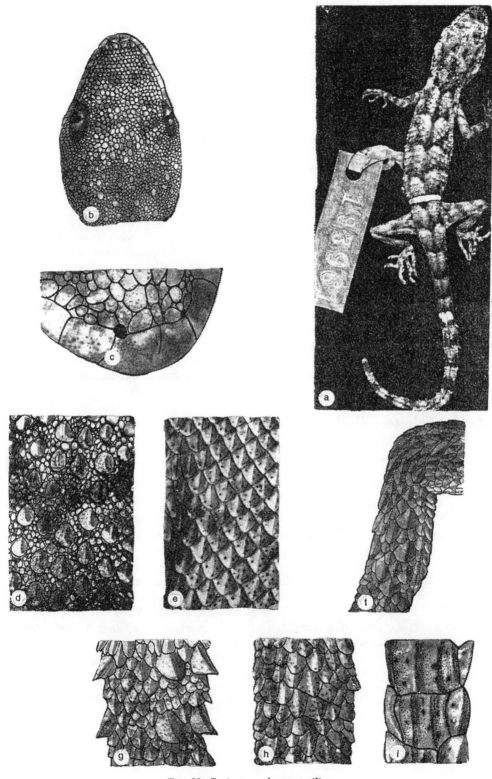

FIG. 58. *Carinatogecko aspratilis*:
a, general view from above; b, head from above; c, nasal scales; d, dorsum, e, belly; f, forearm from above; g, tail from above; h, tail from below; i, subdigital lamellae (USNM 193961, Gach Saran, Iran).

1970 – *Tropiocolotes heteropholis* Minton, S. C. Anderson, and Jer. A. Anderson, Proc. Calif. Acad. Sci., ser. 4, vol. 37, no. 9, p. 357.
1981 – *Carinatogecko heteropholis*, Golubev and Szczerbak, Vestn. Zool., no. 5, p. 37.

DIAGNOSIS. Middorsal scales slightly larger than or equal to ventrals; tail tubercles distinct but flattened, not sharp-edged, posterior plane of each such tubercle not very high; lateral digital scales in lower row with high longitudinal keels, pointed on their ends; 15 subdigital lamellae; upper forearm scales homogeneous.

DEFINITION (from holotype FMNH 74549) (Fig. 59). Adult (?) female. L = 20.4 mm; LCD = 19 mm; HHW 56; EED 19; 8 supralabial scales; 6 infralabials; 15 scales across head; DTL 19; 8-9 scales around dorsal tubercle; about 100 GVA; about 28-30 ventral scales across midbody; 15 subdigital lamellae.

No enlarged eyebrow scales; a row of three large tubercles follows one scale after the last supralabial; dorsal scales triangular, do not contact each other. Low tubercles are characteristic of caudal segments. Caudal tubercles located in the middle of segment and separated from its posterior edge by a row of scales. Three to five scales along segment, caudal tubercles are not arranged in precise semicircles and do not contact each other. Adpressed forelimb reaches front edge of orbit with fingers, hindlimb reaches mid-shoulder; upper surfaces of thigh and crus with tubercles similar to dorsal [ones]; shoulder and forearm covered by homogeneous scales. Lateral digital scales flat, [each] with low but distinct keel with no spine on end.

COLOR AND PATTERN. The pattern of the upper side of the body was not preserved due to poor fixation. The scales of the lower surfaces have numerous pigment spots.

DISTRIBUTION. Known only from the type locality.

ECOLOGY has not been studied. A specimen of this species [280] was caught under a piece of tree bark on a hotel floor. The author suspects that it was brought in with the firewood.

Genus Thin-toed Geckos — *Tenuidactylus* Szczerbak et Golubev, 1984

Type species: *Gymnodactylus caspius* Eichwald, 1831.
1825 – *Gymnodactylus* Spix (part.), Spec. nov. Lacert. Brasil. p. 17.
1827 – *Cyrtodactylus* Gray (part.), Philos. Mag. (2), [vol.] 3, p. 55.
1843 – *Cyrtopodion* Fitzinger, Syst. Rept., p. 18, 93*.

{p. 131} DIAGNOSIS. Digits slender, clawed, basal phalanges almost do not differ [from each other] in thickness, gradual transition distally (some thickening can be present on joints), covered underneath by one row of transversely enlarged, flat lamellae; second distal phalanx [on each digit] makes an angle with proximal part of digit; ends of digits feebly or absolutely not compressed; no fringes or other projections on lateral digital scales. Pupil vertical with serrated edges. As a rule, longitudinal concavity on frontal area [of head] absent or poorly developed; usually no more than 30 scales across head; preanal and (more rarely) femoral pores present only in males. Tail segmentation well developed.

ADDITIONAL CHARACTERS. Small and midsize geckos (L = 60-80 mm). As a rule, dorsal tubercles large, triangular, or trihedral to round, usually keeled, arranged in 8 to 14 longitudinal and 20 to 28 transverse (neck to sacrum) rows; three (very rarely two) distinctly enlarged nasal scales, the first one sometimes slightly larger than others; usually knobby tubercles on upper posterior part of orbit well developed; 35 ventral scales across midbody; 150 GVA (usually no more than 130); preanal pores not separated from femoral ones and arranged in almost a straight line; large tubercles that reach lower sides of tail (three to five on each side) reach its end; usually a row of enlarged scales (two or two pairs per segment) present on tail lower surface. Color pattern not bright, consists of 4-7 transverse body and 7-12 tail bands.

---

* The last name cannot be used as a generic name because it is in contradiction with the diagnosis of *Tenuidactylus*. {For discussions of this nomenclatural problem, see Böhme, 1985; Kluge, 1985, 1991, 1993; Szczerbak, 1986, 1988, 1989, 1991; Bauer, 1987; Borkin, et al., 1990; Leviton, et al., 1992; Zhao and Adler, 1993 – MG}

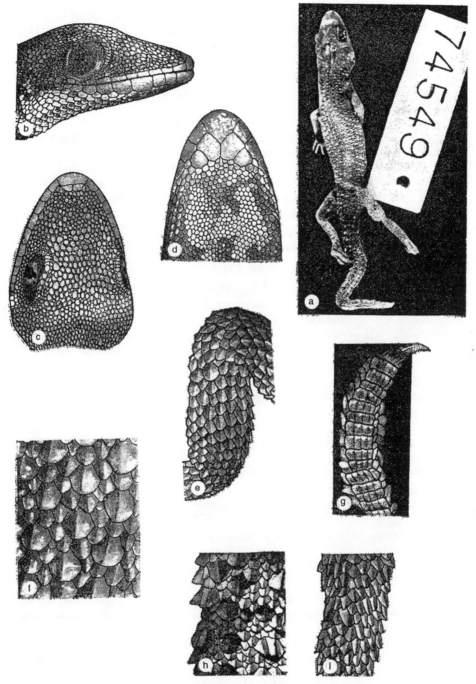

FIG. 59. *Carinatogecko heteropholis*:
a, general view from above; b, head from the side; c, head from above; d, head from below; e, forearm from above; f, dorsum; g, toe from below; h, tail from above (FNHM 74549, Salahedin, Iraq).

[The genus] includes 16 species that occur west of Indus River, in the Himalayas and north of them; consists of four subgenera.

Key to the Subgenera of Tenuidactylus

1 (2). Spine-shaped caudal tubercles do not contact each other in the semicircles of a segment, surrounded by uniform scales; height of first labial scale from nostril to edge

of mouth distinctly less than its width along edge of mouth; background color and pattern light and dark gray . . . . . *Mediodactylus* Szczerbak et Golubev

2 (1). Low or moderately high tail tubercles widely contact each other in semicircles and surrounded by one to two smaller tubercles and much smaller scales; height of first labial scale from nostril to edge of mouth distinctly longer or slightly shorter than its width along edge of mouth; background color and pattern light and dark ochre brown.

3 (4). No fewer than 20 femoral [and preanal] pores in males; row of plates covers almost entire subcaudal width; if there is suture between postmental scales in first pair, its length is no more than half the length of mental scale
. . . . *Tenuidactylus* Szczerbak et Golubev

4 (3). No more than 10 preanal pores; row of subcaudal plates distinctly narrower than subcaudal surface; suture between postmental scales of first pair longer than half the length of mental scale . . . . *Cyrtopodion* Fitz.

Subgenus *Tenuidactylus* Szczerbak et Golubev, 1984

1984 – *Tenuidactylus* Szczerbak and Golubev, Vestn. Zool., Kiev, no. 2, p. 53 (type species *G. caspius* Eichw.).

DIAGNOSIS. Males with preanal and femoral pores (no fewer than 20) arranged in almost straight line; if suture between postmental scales in first pair, its length is no more than half the length of mental scale; shape of dorsal tubercles varies from trihedral to round-triangular; dorsal tubercles widely contact each other on their lateral edges; in front of base of such tubercle, there is one to two smaller tubercles, larger than surrounding scales; row of subcaudal plates covers almost entire lower caudal surface (length of plate less than half its width).

ADDITIONAL CHARACTERS. Height of first labial scale from nostril to edge of mouth distinctly greater or slightly smaller than its {p. 132} length along edge of mouth; flattened trihedral ciliary projections in posterior part of orbit poorly developed; 2-10 small trihedral or conical tubercles usually present among small scales on posterior surface of thigh. Tail thin. Background color and pattern light and dark brown.

Includes four species distributed in Middle Asia; partially penetrate into Transcaucasus Region and central part of Southwest Asia.

Key to the Species of the Nominate Subgenus Tenuidactylus

1 (2). No more than 30 femoral pores; no more than 115 GVA; no more than 11 scales across head . . . . *T. caspius* (Eichw.)

2 (1). Usually no fewer than 30 femoral pores; usually no fewer than 120 GVA; usually more than 11 scales across head.

3 (4). Scales of first pair of postmentals separated by gular scales or contact each other by a short suture (90% of specimens have length of this suture less than 10% of length of postmental scale) . . . . *T. longipes* (Nik.)

4 (3). Scales of first pair of postmentals contact each other by wide suture (85% of specimens have length of this suture within the limits of 10-40% of postmental scale length).

5 (6). Dorsal tubercles strongly keeled, cross area between forelimbs; usually three postmental scales; usually two rows of scales between vent and row of femoral pores (or corresponding scales in females) . . . . *T. turcmenicus* (Szczerbak)

6 (5). Usually dorsal keeled tubercles do not penetrate area between forelimbs; usually two {pairs of – MG} postmental scales; usually three rows of scales between vent and row of femoral pores . . . . *T. fedtschenkoi* (Strauch)

CASPIAN THIN-TOED GECKO — *TENUIDACTYLUS CASPIUS* (EICHW., 1831)

Type locality: Baku [Azerbaijan].
Karyotype: $2n = 42$, two submetacentrics, 38 acrocentrics, two subtelocentrics, NF = 46.

1831 – *Gymnodactylus caspius* Eichwald, Zool. Spec. Ross. Polon., vol. 3, p. 181.
1954 – *Cyrtodactylus caspius*, Underwood, Proc. zool. Soc. London, p. 475.
1984 – *Tenuidactylus caspius*, Szczerbak and Golubev, Vestn. Zool., Kiev, no. 2, p. 53.

DIAGNOSIS. No more than 32 femoral pores in males; no more than 115 GVA and 31 ventral scales across midbody; no more than 11 scales across head; dorsal tubercles large, trihedral, almost always one to four smaller additional tubercles contact them; two to three pairs of postmentals, scales in first pair contact each other or separated by one or two scales; usually two rows of scales between row of femoral pores and vent; adpressed forelimb reaches tip of snout with tips of fingers, total length of forearm and shoulder comprise 29-37% (31.8 ± 0.1) of body length.

LECTOTYPE. ZIL 3182, adult male. "Baku, 1830, Ménétriés." L = 61.0 mm; LCD = 81.9 mm; L/LCD = 0.74; 9 scales across head; 11 upper and 8 infralabials; 96 GVA, 28 ventral scales across midbody, 13+1 scales surround dorsal tubercle {'+1' indicates additional tubercle – MG}; 24 femoral pores; 26/24 subdigital lamellae; 2/3 postmental scales.

DEFINITION (Fig. 60) (from 565 specimens in ZIK, ZIL, MLSU, ZDKU, NMW, MNHP [= MNHN]). L adult male (n = 249) 40.6-72.00 (56.06 ± 0.48) mm; L adult female (n = 293) 40.2-68.5 (55.33 ± 0.49) mm; L/LCD 0.62-0.95 (0.76 ± 0.004); HHW (n = 100) 55-67 (61.09 ± 0.26); EED (n = 60) 41-68 (55.88 ± 0.95); supralabial scales (n = 1121) 9-14 (10.43 ± 0.02); infralabials (n = 1117) 6-10 (8.04 ± 0.02); scales across head (n = 559) 7-11 (8.68 ± 0.04); DTL (n = 439) 23-40 (32.11 ± 0.13); scales around dorsal tubercle (n = 549) 13-22 (17.39 ± 0.06); GVA (n = 501) 89-113 (103.44 ± 0.20); ventral scales across midbody (n = 464) 22-31 (26.60 ± 0.10); femoral pores (n = 298) 22-32 (26.65 ± 0.11); femoral scales {*i.e.*, enlarged scales without pores – MG} in females (n = 185) 25-33 (30.86 ± 0.14); subdigital lamellae (n = 1095) 20-29 (23.85 ± 0.05). {p. 133} Mental scale always triangular or pentagonal; two (56.7%) or three (43.2%) pairs of postmental scales, scales of the first pair can be separated by one to two gular scales; first nasal scales separated by a scale (37.7%) or contact each other (62.3%); one (41.3%), two (32.5%), three 13.3%), or four (2.5%) additional tubercles might contact dorsal tubercle in the middle of the dorsum, 10.4% of specimens lack this contact.

{p. 134} COLOR AND PATTERN. This species is lighter than the other species of this subgenus. The main background is pale sandy gray. There are dark transverse M-shaped bands on the body (5-6) and tail (8-12). On the limbs, there are poorly developed small transverse bands. There is a pattern on the head, formed by thin dark stripes, the stripe on the neck connecting the tops of the ear openings. Parallel to this stripe, small stripes run from the rear edge of each eye. They are interrupted at the top of the head. A dark spot is located between their free edges. Above this spot, almost between the eyes, there are one to three small spots. Three parallel small stripes that extend to the infralabial scales run from the front upper edge of the orbit to the nostril, from the front edge of the orbit to the first supralabial scale and from the front lower edge of the orbit to the fifth-sixth supralabial scale. These stripes are significantly better developed in the Caspian thin-toed gecko than in other species of the genus, and can serve as one of its identifying characters in the field. The lower surfaces are white. [Body] mass to 6.5 g.

DISTRIBUTION. A line that runs from the northeastern part of Caspian Sea (Komsomolets Bay) to the northern coast of Aral Sea and Syr-Darya, is the northern border of the range. It is found along the northern shore of the Caspian Sea to Kalmykiya (Naryn-Huduk), in Transcaucasus; extreme southwestern localities are Ackera River (natural habitat) and Julfa in Azerbaijan, Tbilisi (synanthropic). We found an isolated population in the eastern part of the range, in Pherghan Valley. An increase in the range has been observed in recent years through the dispersion with transported goods on water (Caspian Sea) and by railway transportation (Caucacus, Middle Asia). It can be stated with a fair degree of confidence that the appearance of this species in the European part of the range is the result of human activity {*fide* Szczerbak – MG}. In an article [147] on new finds in Kyzylkums, an error crept in:

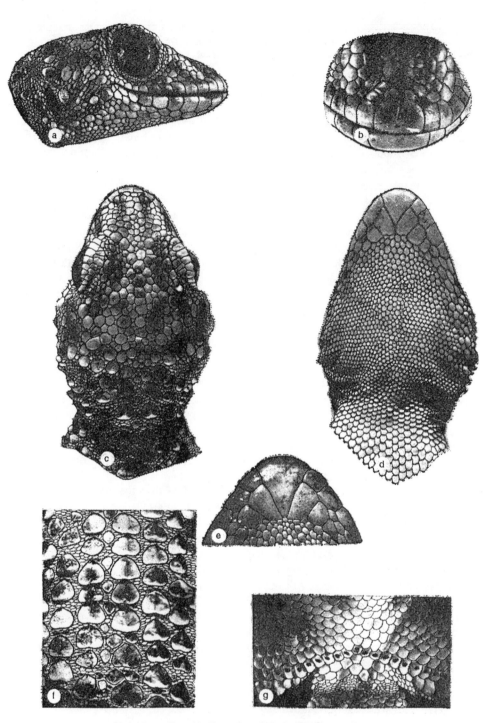

FIG. 60. *Tenuidactylus caspius*:
a, head from the side; b, snout in front; c, head from above; d, head from below; e, chin; f, dorsum, g, preanal pores (e – ZIK Re 7.0, Vulf Island, Azerbaijan; others – ZIK SR 378, Meshkhed-i-Misrian, Turkmenia).

the phrase "we found it to the north of Bukhara near Mynbulak and Uchkuduk" relates to the Caspian rather than the armored pygmy gecko (p. 69).

{p. 135} The southern border in Iran and Afghanistan is not clearly defined. The localities in central and western Iran and Syria need confirmation, because confusion with close species is quite possible (in particular, until recently the Caspian thin-toed gecko had been confused with the Turkmenian gecko) (Fig. 61).

LEGEND to Fig. 61.
U.S.S.R., R.S.F.S.R.: 1 - Kalmyk SSR: Lyman Dist.: Naryn Huduk village (ZDKU); 2 - Daghestan ASSR: Mahachkala; Buynaksk (ZIK); 3 - Derbent (ZIK).
Georgia: 1 - Tbilisi [82].
Azerbaijan: 1 - Nahichevan, Julfa (ZIK); 2 - Shemaha (ZIL); 3 - Sumghait-chaj River, 35 km N Baku (ZIK) 14 km N Sumghait (ZDKU); 4 - Apsheron Peninsula: Baku, Mardakyan, Zagulba, Vishnevka, Bilgya (ZIK); 5 - Primorsk; Kobystan (ZIK) Sangachaly (ZIL); 6 - Island Vulf (ZIK); 7 - Island Oblivnoj (ZIK); 8 - Salyany (ZIL); 9 - Fizulinsky Dist.: Akera River, Kerimbeyli (ZIK).
TSSR {Turkmenistan – MG}: 1 -Kunya-Urgench (ZIK); 2 - Deu-Kesken-Kala (ZIK); 3 - Fortresses Doyarbekir and Shahsenem [130]; 4 - Sarykamish Lake, Daryalyk River (ZIK); N shore of Sarykamish [130]; 5 - Kunyadarya Plain [130]; Kunyadarya: 3 km S Chaprak-Kala [23]; 6 - 27 km from Tashauz to Kunya-Urgench: Fortress Guldumsaz (ZIK); 7 - W edge Sarykamish Hollow (ZIK); 8 - SE chink of Sarykamish Hollow (ZIK); Kapyllarkyr Spit [130]; 9 - Tahta Dist.: Fortress Izmykshir [130]; 10 - 25 km S Tahta: Fortress Bederkend (ZIK); 11 - 90 km E of Sarykamish Hollow border (ZIK); 12 - SE shore Kara-Bogaz-gol Bay: Chagy-Bey-Bulak Well [23]; 13 - Zaunguz Karakums: Doudir-Eneje Well (ZIL); 14 - Kumsebshen [130]; 15 - Dahly Well [130]; 16 - N chink {loess cliffs – MG} of Kaplankyr {Kaplank is a region – MG} [130]; 17 - Sulmen Village (ZDKU); 18 - 25 km N Serny Zavod (ZIK); 19 - Krasnovodsk (ZIK); 20 - Nargen, Island Svyatoj [130]; 21 - Djebel Station [130]; 22 - Bolshoj Balkhan Ridge [130]; 23 - W Uzboj: Yilgynly and Togolak Wells [130]; 24 - Serny Zavod; 10 to 15 km W Serny Zavod (Zik); 25 - Dordja (ZIL); 26 - left bank of Uzboj River, 4 km W Molla-Kara (ZIL); 27 - Akhcha-Kuyma [130]; 28 - Yaskhan (ZIL); 29 - Kirkily [130]; 30 - 180 km N Bakharden: Kyzul-Gaty (ZIK); 31 - Maly Balkhan [130]; 32 - Kazandjik: 20 km S Iskander Station (ZIK); Iskander Station [130]; 33 - 44 km NE Kyzyl-Arvat, near Aul Demerdjan (ZIL); 34 - Kultakyr [130]; 35 - Danata (ZIK); 36 - Kyzyl-Arvat (ZIK); 15 km S Kyzyl-Arvat (ZIL); 14 km N Kyzyl-Arvat near Takyrabad [23]; 37 - Cheleken [130]; 38 - Bakharden Dist.: Shehr-Islam (ZIK); 39 - Meshkhed (ZIK); Messerian Plain (ZIK); 40 - Kazandjik (our data); Kyurendagh [130]; 41 - Karagez [130]; 42 - 15-20 km N Bamy (ZIK); Bamy [130]; 43 - 53 km N Bakharden [130]; 44 - 40 km S Madau [130]; 45 - Kara-Kala;

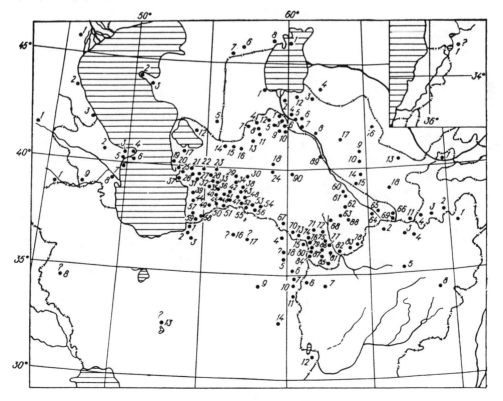

FIG. 61. Distribution of *Tenuidactylus caspius*.

between Kara-Kala and Chandyr; 20 km S Kara-Kala (ZIK); Kara-Kala Dist.: collective farm Lenina (ZDKU); Chandyr Valley (ZIL); Yoldere Gorge [130]; 46 - Aydere (ZIK); 47 - Archman [130]; 48 - Sinekly [130]; 49 - 50 km S Kara-Imam (ZIK); 50 - Sharlouk (ZIK); 51 - Chandyr River Valley: 20 km S Kyzylimam ZIK); 52 - Bakharden; 15 km S Bakharden: Arvaz River (ZIK); between Bakharden and Durun [23]; villages of Durun, Karagan, Yaradji [130]; 53 - Ashkhabad: Berzengy; Kury-Goudan; 20 km S Ashkhabad: road to Goudan (ZIK); Nah-Duin (Goudan) Mountain [23]; 54 - 57 km from Ashkhabad to Tedjen (ZIK); 55 - Chuly; 15 km W Ashkhabad: Bagyr (ZMMSU); Village Verhne-Skobelevsky [23]; Germab; Kelendjar; Phiryuza; N Geok-Tepe; N Bezmein; Cheltek Well; Porsykuyu Well [130]; 56 - Kalininsky (ZIK); 57 - near Bolshoye and Maloye Delily [130]; 58 - Bayat-Khodja locality (ZIL); 59 - Gasan-Kuly (ZIK); Ak-Patlauk Hill [23]; 60 - Denau village [80]; 61 - Chardjou [130]; 62 - 20 km N Repetek (ZIK); Repetek Station [130]; 63 - Mary Dist.: Ravnina station (ZIK); 64 - Kerky [130]; 65 - Kerkichi; Dostluk [130]; Talimardjan (ZMMSU); 66 - Hodjepil (ZIK); 67 - site of ancient town Kaahka(ZIK); 68 - Bayram-Aly (ZIK); 69 - former Bazardepe village, belt of lower hills of Kughitangh [130]; 70 - Kaahka Dist.: Meana village (ZIK); N Meana [130]; 71 - Karabata [130]; 72 - Sultanbent; Yolotan [130]; 73 - Kopetdagh: S Chaacaha village (ZIK); 74 - Takyr Station [130]; near Tedjen [130]; 20 km Tedjen Station upstream the River [23]; 75 - Imambaba (ZIK); 76 - Serakhs (ZIK); 77 - Tashkepry [130]; 78 - Karabil: Khumly Settlement (ZIK); 79 - Pul-i-Khatum; Nauruz-Abad locality; Shirdepe (ZIK); Tedjen River near Iran-Afghan border [23]; 80 - Serakhs Dist.: Adamulen locality (ZIK); 81 - N Kaka-i-Mor Station (ZIK); 82 - 40 km Takhta-Bazar; Orna-Daulet; Ravat-Kashan River, Tarashek locality (ZIK); Kashan N Tarashek; Pende Oasis [23]; 83 - Pelengoveli locality (ZIK); 84 - Badghyz: Gez-Gyadyk Ridge (ZIK); 85 - Kushka; Agashly (ZIK); 5 km N Morgunovsky Settlement (ZIL); Chemen-i-Bit [130]; 86 - Badghyz: Er-Oylon-Duz (ZIK); 87 - Badghyz: Akar-Cheshme; Penhan-Cheshma; Shakmakly (ZIK); Akrabat; Kepely; Kyzyl-Djar [130]; 88 - Nychka; 65 km SW Nychka [130]; 89 - Chardjou Dist.: Darghanata (ZIK); 90 - Ak-Molla, 150 km E Serny Zavod (our data).

Uzbekistan: 1 - SW coast of Aral (Urgha) [222]; 2 - Karakalpakia: 30-40 km NE Tahta-Kupir, foothills of Beltau (ZIL); 3 - Karakalpakia: 25-30 km N Chabankazghan (ZIK); 4 - Karakalpakia: 45 km from Nukus to Turtkul (ZIK); 5 - Karakalpakia: Boday-Tugay Reserve: Djampyk-Kala fortress (ZIK); 6 - Karakalpakia: 25 km NE Biruni: Kyzyl-Kala fortress (ZIK); 7 - Khorezm Dist.: Hiva; 30 km NE Biruni, Toprak-kala fortress; 8 - Turtkul [22]; 9 - Bukhara Dist.: 100 km N Shafrikan; Mynbulak; Uchkuduk (ZIK); 10 - Ayakagytme Kishlak (ZIK); 11 - Termez (ZIK); 12 - Beltau [22]; 13 Nuratau Ridge: Temirkauk Well (ZIL); 14 - Bukhara [22]; 40 km SE Bukhara (ZMMSU); 15 - Karakul Reserve: 20 km S Alat Station (ZMMSU); 16 - NE {p. 136} part of Bukhara Dist.: old, destroyed hills (ZMMSU); 17 - Bukhara Dist.: Djingildy (ZMMSU); 18 - 60 km N Karshi: Uch-tepe locality (ZMMSU).

Tadjikistan: 1 - Leninabad Dist. {our data – MG}; 2 - Kulyab: near Kalimtaj Kishlak [107]; 3 - Vakhsh River Valley: near Djilikul and Kumsanghir villages [107]; 4 - Beshkent Valley: near Tylkhar, Kabadian [107].

Kazakhstan: 1 - Kockaral Island [2]; 2 - Shevchenko (former Fort-Alexandrovsk) (ZIL); 3 - Mangyshlak (ZIL); Tauchik {98]; 4 - Kzyl-Orda Dist.: 40 km NW Kekreky Settlement, remnants of Chirik-Robat fortress (ZIK); 5 - Shah-Pahty Hollow and ridge of Assake-Audan Hollow [22]; 6 - northern chink [= loess hills] of Ustyurt [98]; 7 - Asmatay-Matay Sor (Shor) [98]; 8 - NW coast of Aral Sea [98].

Iran: 1 - Gorgan, NE Iran (MNHP, NMW); 2 - Astrabad (Damghan); Ak-Kala, 15 km from Astrabad; Nasirabad (ZIL); 3 - Shahrud (MLSU); 4 - 40 km M Meshed (ZIL); 5 - Boz-Houz-Pain (ZIL); 6 - near old Bahars, Keriz-i-Nou (ZIL); 7 - Terra Hoshtadan, Kerat (ZIL); 8 - Kermanshah (ZIL), the locality needs confirmation; 9 - Mirinduz (ZIL); 10 - Houz in Terra Zirkuh (ZIN); 11 - Khorasan: Chah-Isi Well (ZIL); 12 - Sistan: Neizar (ZIL); 13 - Shiraz (NMW), the locality needs confirmation; 14 - Birdjan (ZIN); 15 - Serakhs, NE Iran (MNHP); 16 - Zyurabad [80], *Gymnodactylus sp.,* the locality needs confirmation; 17 - Soleabad (Soltanabad) [80], the locality needs confirmation; 18 - Meshed [80], the locality needs confirmation.

Afghanistan: 1 - near 100 km E Fayzabad (CAS); 2 - 25 km SW Akcha (CAS); 3 - Mazari-Sharif, 36°43′N, 67°05′E [172]; 4 - 10 km W Tashkurghan (Holm) (CAS); 5 - Sari-pul, Vagieh (ZML); 6 - Rd from Herat to Islam-Qala (CAS); 7 - Herat (MLSU); 8 - Paghman, 34°36′N, 68°56′E [172].

Syria: (see inset) 1 - Heme, S Haleb (ZMMGU - locality needs confirmation).

SEXUAL DIMORPHISM AND AGE VARIATION. Unlike the females, the males have femoral pores; their heads are somewhat more massive. Males are a little larger than females in maximum size (see above); they have longer tails (LCD M = 95 mm, LCD F = 89.1 mm).

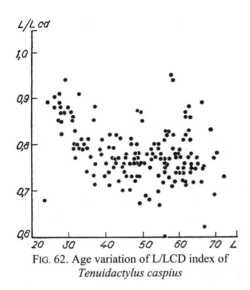

FIG. 62. Age variation of L/LCD index of *Tenuidactylus caspius*

During the mating period, males are brighter and have better definition than females; their background color is yellowish; it is grayish in females. We determined this by injecting a gonadotropic hormone. In one day, the males got sharply lighter and began to differ distinctly in color from females.

The head and limbs of juvenile specimens are relatively larger and the tail shorter than in adults (Fig. 62).

There are no differences in color and pattern associated with the area of habitation, but we observed many times [that] the color of a lizard becomes darker (to a complete disappearance of pattern) or lighter in connection with its physiological condition.

SYSTEMATICS AND GEOGRAPHIC VARIATION. Fifteen mainland samples from all parts of the range did not show any significant differences when compared in the main scalation characters (even from such a remote population as the Fergan Valley). The most variable were the number of subdigital lamellae, femoral pores, supra- and infralabials, ventral scales across midbody, GVA, scales across the head, as well as DTL. In all these characters, there is clinal variation, defined to varying degrees, in the direction of west to east. However, the largest specimens were found in almost the very center of the range, near Kushka (southern Turkmenia).

The geckos from Vulf Island in the Caspian Sea were significantly different from all the aforementioned samples and were described as a separate subspecies [14]. Almost or complete absence of contact between scales in the first pair of postmentals is characteristic for this form (the character is common with a very close species, *T. lon-* {p. 137} *-gipes*). It is characteristic that this feature is absent in 93-100% of specimens from mainland areas, such as Azerbaijan, Karakalpakia, Iran, Turkmenia, while it is absent only in 70% of the Apsheron Peninsula population, which is the closest locality to the island on the coast. The causes of this population's peculiarity are probably not in the conditions of the island's formation (tectonic origin) but in the workings of genetic mechanisms.

The data presented allow for the conclusion that this species penetrated the Caucasus from southern Turkmenia through northern Iran, which is supported by minimal differences in samples corresponding [to the just named localities].

Key to the Subspecies of T[enuidactylus] caspius

1 (2). Scales of the first pair of postmentals contact each other across a wide
    suture . . . . *T. c. caspius* (Eichwald)
2 (1). Scales in first pair of postmentals do not contact one another or have a
    narrow suture . . . . *T. c. insularis* (Akhmedow et Szczerbak)

CASPIAN THIN-TOED GECKO — *TENUIDACTYLUS CASPIUS CASPIUS* (EICHWALD, 1831)

Type locality: as in species.

DIAGNOSIS. Scales of first pair of postmentals contact each other across a wide suture. Other characters are contained in the species description.

TYPE SPECIES. [see species, p. 129 {eds.}]

DISTRIBUTION. [Same as] the species range except for Vulf Island in the Caspian Sea, about 12 km from Apsheron Peninsula.

ISLAND ROCK GECKO — *TENUIDACTYLUS CASPIUS INSULARIS* (AKHMEDOW ET SZCZERBAK, 1978)

Type locality: Vulf Island (Baku Bay of Caspian Sea).
1978 – *Gymnodactylus caspius insularis* Akhmedow and Szczerbak, Vestn. Zool., Kiev, no. 2, p. 80.

DIAGNOSIS. Scales of first pair of postmentals do not contact [one another] or have a narrow suture (65.5%) or a point suture (34.5%).

HOLOTYPE. ZIK Re no. 7, adult female. Vulf Island, 28 March 1976, M. I. Akhmedov coll. L = 55.7 mm; LCD = 69.0 mm; 10/9 supralabials; 9/9 infralabials. 105 GVA, 28 ventral scales across midbody, 8 scales across head, Scales of first pair of postmentals do not contact or have point suture.

DEFINITION (n = 30 specimens). Only characters which have distinct differences from other samples (4-13) are presented. L/LCD = 0.81-0.88 (0.83 ± 0.01). Supralabial scales 9-13 (11.0 ± 0.1); infralabials 7-10 (8.45 ± 0.11). DTL 26-32 (28.66 ± 0.29) (this and the following are most divergent); subdigital lamellae 24-29 (26.3 ± 0.2).

Other characters are in accordance with the species description.

DISTRIBUTION. Known only from the type locality.

HABITAT AND QUANTITATIVE DATA (Figs. 31, 40). The Caspian thin-toed gecko, in the Caucasus part of its range, lives, as a rule, on the vertical surfaces of rocks, cliffs, and buildings. The walls, rocky cliffs of Uzboi, Unguz and Ustiurt are also the usual biotopes of this species in western Turkmenia. However, in the plains portion of the Karakums they are found in the burrows of *Rhombomys opimus*, *Meriones libycus* and *M. meridianus*, and other burrow dwellers. This trait is most apparent toward the south and east (Er-Oilan-Duz, Fergan Valley), except in the loess walls of ancient buildings, such as in the Merv Oasis (Bairam-Ali). Because in the eastern part of the range the Caspian thin-toed gecko is found with related species (Turkmenian {*T. turcmenicus* – MG}, Turkestan {*T. fedtschenkoi* – MG}, long-legged {*T. longipes* – MG}), a suspicion arises that the cause of this is the competitive relationship with them [24].

{p. 138} In Badkhyz, rocky outcrops and cliffs of the Ghez-Gyadyk Crest are occupied by the long-legged thin-toed gecko, the exposed rocks in the Kushka-Kushan Divide – by the Turkmenian thin-toed gecko and the mid-section of the Amudaria course – by Fedchenko's gecko. However, in places where there are no aforementioned "competing" species (C and E Kopetdagh), the Caspian thin-toed gecko, as a rule, also avoids exposed rock and can be found in the foothill zone in karizes, burrows, crevices, and clay rain channels. It is even difficult to imagine that the eurytopic Caspian thin-toed gecko can be displaced by the stenotopic long-legged thin-toed gecko. Apparently, there is a different cause here. In the range of the Caspian thin-toed gecko, the aridization of the climate grows from west to east.

As geographers report [30, *et al.*], the relative humidity in May in the Transcaucasian part of the range is 50-70%, in western Turkmenia, approximately up to Kazandjik, it is 40-50%, and further, beyond the line Ashkhabad-Nukus, it is 35%. This is also confirmed by data on the distribution of precipitation during the warm part of the year: in the Transcaucasus it is no less than 200 mm, in western Turkmenia – from 100 to 50 mm, in the central Karakums and in Kyzylkum – less than 50 mm. The annual precipitation in the Transcaucasus (relating specifically to biotopes of the Caspian thin-toed gecko) is no lower than 300 mm, in western Turkmenia – 100-200 mm, farther east – less than 100 mm. Consequently, Caspian thin-toed geckos live on rocks in places where the May relative humidity is no lower than 40%, the precipitation during the warm time of year is around 100 mm and the annual precipitation exceeds 100 mm. It has also been established that relative humidity is higher in clay and sand substrates than in rocky ones (the richer vegetation is in sand deserts, the most meager – in pebble deserts). Thus, we suppose that the biotopes of the Caspian thin-toed gecko are delimited by humidity. The closely related Iran-Afghan species, which have more southern origins, — the long-toed and Turkmenian thin-toed geckos — are better adapted to the low humidity, which is confirmed by laboratory observations.

In the mountains, Caspian thin-toed geckos have been found at elevations of up to 900 m (Kopetdagh). S. Anderson and A. Leviton [172] reported records of this species in Afghanistan to 1653 m above sea level.

On the Apsheron Peninsula, up to 12-15 specimens are found on 10 m² of limestone outcrops. In Turkmenia, in the brick ruins of the medival town Messerian, one individual is encountered on 1-2 m². On the loess walls of the ancient Merv (in the vicinity of the town Bairam-Ali), in a 2000 m² area of walls with southern exposure, 35 geckos were counted, or 185 specimens per hectare (on 15 April 1977; also here, on 22 May 1982, at dusk, on a cloudy and rainy day, nearly 125 individuals per hectare were noted). In Badkhyz, on the unprotected part of the Er-Oilan-Duz shor, on a 5 m-wide strip of a 1 km transect only single individuals were encountered; but in an area of sand with *Salsola* sp. {MG} bushes, near *Meriones sp.* {MG} burrows, up to 18 lizards per hectare were recorded (16 May 1976, 10 April 1977, 8 October 1981, 24 April 1984). On stabilized sands interspersed with takyrs, between Serakhs and Khauskhan, on 28 April 1984, one gecko per 1 km of the transect was encountered on the average. The same numbers were noted in Kyzylkum on a salinated takyr in the vicinity of Ayak-Agytma village (14 June 1976).

Single individuals were encountered here in winter sheep enclosures, on turanga trees {*Populus* sp. — MG}; in the Karakums they are found on the foundations and in the ruins of loess buildings, on rock outcrops along the Unguz, on the walls of wells among the sand barkhans, on hillocks on a takyr with the remnants of *Salsola* {MG} and *Haloxylon* {MG} bushes, in clay flood channels, on loess cliffs on rock slides, under a chink cliff to the southwest of Sarykamysh and Ustyurt; in Badkhyz on a loess plateau with sagebrush, in telephone-pole cracks; on Karabil (beside the *Meriones* {MG} burrows) on loess cliffs; in the Beshkent Valley (Tadjikistan) in rodent burrows in the ephemerous desert. In the vicinity of Repetek, in the black saxaul forest {*Haloxylon aphyllum* — MG} there are three individuals per hectare, and in the village, 6.5 individuals [120]. Single individuals also occur in the extreme east of the range, in the Fergan Valley. Thus, Caspian thin-toed geckos reach their highest numbers in the western part of the range and in buildings (expecially in the ruins).

RESPONSE TO TEMPERATURE. We have noted active geckos when the air temperature was 15-30°, 21° on the average (n = 14). O. P. Bogdanov {**p. 139**} [23] had noted lizards of this species active at air temperatures of 17-36°. Simultaneously, we would determine the substrate temperatures of the places where they were encountered. It turned out that the geckos were observed on the surface with temperatures of 16-22°, 19.6° on the average (n = 14). The greatest portion of the measurements was done in the spring. In the fall (8 October 1981), at the air temperature of 20°, the lizards were not observed. A direct relationship between the air and soil temperatures is often absent in the continental climate of Middle Asia (the soil may be warmer than the air and vice versa). Thus, geckos would not emerge from shelter at the air temperature of 17° and the soil temperature of 21.5° (30 April 1979) or with both temperatures at 18° (16 April 1977, 24 April 1977), however, they were on the surface on 3 May 1984 in the overcast weather at the air and soil temperature of 16°. In May, lizards were observed on the surface even during drizzle. When studied in a thermal gradient apparatus, Caspian thin-toed geckos (n = 42) would select the spots with temperatures in the range of 13-41°, 32.7° ± 1.1° on the average.

In winter shelters, at the air temperature of 6.5°, on 6 February 1971, the geckos' body temperatures reached 8-13° [12].

DAILY ACTIVITY CYCLE. In the spring, the geckos frequently appear in the daytime (in February — at 1300-1700 hours [23]). First, they warm themselves in the sun and then forage in the shade. Frequently, in southern Turkmenia, we observed the geckos catching insects, as they tried to hide from the hot sun in burrows where they sat near the entrance. In the summer, they are active at dusk and at night. In the fall, on cool days, they once again are frequently seen in the morning and during the day. They are also active in the daytime in the scattered light on buildings and in caves. At dusk and at night, they hunt from 1900 hrs until dawn, around 0500-0600 hrs; however, the peak of activity is around 1900-2000 hrs [130]. The first appearance in the springtime (April) we noted between 2000 and 2100 hrs, and in the fall (October) — at 1920 hrs.

SEASONAL ACTIVITY CYCLE. In southern Turkmenia, in some years, Caspian thin-toed geckos would appear on warm days even in the winter (25 January 1964, 7 February 1971 [11]; 25 February 1977, 27 February 1964 [130]). However, the appearance *en masse* here is noted for the end of March, the beginning of April (from our observations on 31 March 1977, near Serkhas; on 6 April 1978, near the villlage Berzenghi; on 7 April 1979, at the Kury-Khoudan locality in the vicinity of Ashkhabad; on 13 April 1983, near the village Berzenghi and on 14 April 1963, by the town of Kazandjik). Close dates of springtime gecko emergence are known from Azerbaijan (26 March 1975, from Sumghait and 28 March 1976, from Vulf Island {Azerbaijan coast of Caspian Sea – MG}) and Kalmykia (3 April 1951). The last records of Caspian thin-toed geckos from our Turkmenian collections are: 26 October 1966, near Yaskha Lake; 28 October 1966, vicinity of Kazandjik Station; 4 October 1981, south of the town of Kara-Kala; 8 October 1981, from Er-Oilan-Duz; 18 October 1981, in the vicinity of Uch-Adji station. From summarizing the above data and data from the literature [1, 23, 98, 107, 130, *et al.*], it can be concluded that, in a wide portion of the range, single Caspian thin-toed geckos re-appear after hibernation in the second half of March, emergence *en masse* happens in the first half of April, disappearance for winter is at the end of October, single individuals may be encountered into November and, in the southern portion of the range, during warm winters, they may leave their shelters for a short time.

SHEDDING. According to data in the literature data [1, 8, 107, 130], shedding individuals are encountered from April until June and, occasionally, at the beginning of August. We saw shedding in Er-Oilan-Duz on 8 October 1981. From observations in captivity, they shed three times: in the spring, summer, and fall; this picture may lose definition due to more frequent shedding among malnourished and sick individuals. Usually, shedding begins with the dulling of the cornea and whitening of the main body color. Then, the keratinous layer splits in the area under the jaw; in the abdominal area it separates almost completely; large shreds also come off the back. The lizard pulls the old skin off by biting it. As a rule, the shed skin is eaten.

FEEDING. When dissecting the stomachs of 27 Caspian thin-toed geckos from Azerbaijan [1], we discovered the remnants of locusts, snout beetles, tiger beetles, crickets, hemipterans [true bugs], ants, bees, mantises, and caterpillars, as well as spiders.

The analysis of 154 stomachs from Turkmenia, obtained at different times [130], revealed that in first place in the diet of these lizards are beetles (31.2% frequency rate), followed by termites (28%), orthopterans (24.6%), crickets {p.140} (17.5%), spiders (13%), butterflies and their larvae, as well as isopods (7.8% each). In April and May (n = 20), the frequency rates are 50% for beetles, 25% for butterflies, 20% each for orthopterans and spiders. In June (n = 120), in the stomachs of geckos caught in old earthen buildings, termites made up 32.5% of the contents, beetles were 30% and orthopterans, primarily crickets, were 25.8% of the contents. In September and October (n = 14), spiders are encountered most frequently (35.7%), orthopterans and termites (28.6% each), and isopods (7.1%). In Turkmenia [23], there was reported a case of a gecko eating a small lizard (*Eremias velox*).

According to the data from Uzbekistan (n = 15 [22]), they most frequently eat beetles (80% encounter rate), snout beetles predominate at 33.3%, darkling beetles and lamellibranchs at 12 and 6%), dipterans (26.3%), isopods (20%). In captivity, Caspian thin-toed geckos eat grasshoppers, small cockroaches, mealworm larvae, and young individuals eat fruit flies and the larvae of *Galleria mellonella* {MG}.

REPRODUCTION. Caspian thin-toed geckos become sexually mature at body lengths over 40 mm. Usually females predominate in the population (from 1.4 to 2.1 for every male [130]). Mating begins shortly after emergence from winter shelters. Evidence for this are the early finds of gravid females: at the end of May in Azerbaijan [1]; in Turkmenia, as early as late April (26 April 1979, Karakums, 50 km north of Bakharden, our data) and in May (15 May 1979, Khumly, Karabil, our data [130]); in Uzbekistan, also at the end of April [22]; in Kazakhstan, at the end of May, beginning of June [98]. Eggs are laid twice and, in rare cases, possibly thrice. This lasts until August [130] and even September [1]. From observations in captivity, egg laying was seen in the fall and in February. There are reliable, known dates for clutches: 30 April 1979, central Karakums, Djarli Well and 5 May 1975 south of Kara-Kala, our data; 5 May 1952, Murghab Valley, Turkmenia [23]; 15 May 1979, Khumly, Karabil, our data; 6 June 1931, vicinity of Iskander Station [8]; 13 June 1967, north of Geok-Tepe [130]; 14 June and 19 June 1979, vicinity of Berzenghy, our data; 16 June 1947, Mangyshlak Peninsula [98]; 26-28 June 1979, Kyzylkum, Pustinnaya Station; 26 June 1983, Karakalpakia, vicinity of Chabankazghan Village; 6 July 1979, Kyzylkum, Pustinnaya Station, our data; 27-30 August and 7-16 September 1947, Apsheron Peninsula [1]. Somewhat smaller Caspian thin-toed gecko egg dimensions are cited in the literature [1, 22, 130], probably because they were extracted from the oviducts before full maturity. We obtained, from captured lizards and from finds in the field, 16 normally laid eggs. Their dimensions were 10.9-12.5 × 9.0-10.2 mm, 11.1 × 9.1 mm on average. Usually, two eggs are laid, less frequently it is one (40% [130]). The eggs are white, oval, with a light pink hue and weigh 380-670 mg. Several females may lay eggs in the same place (we have found groups of six or eight eggs). In captivity, the incubation period lasts 68-81 days. Known dates of finds of newly hatched geckos are: in Turkmenia, 8 July 1963, 24 July 1931, 8 August 1962, 20-27 September 1968, 13 October 1967, 26 October 1969 [130]; in Azerbaijan, 5-26 September 1952 and 17 October 1951; on 14-15 September 1948 eggs with fully formed embryos were found [1].

GROWTH RATE. Newly hatched geckos have the dimensions L = 17-23 mm; LCD = 22-28 mm. Before disappearing for the winter, the current year's hatchlings grow a minimum 7-8 mm. By the middle of the following summer, they reach the size of 40-45 mm [130] and become sexually mature. According to observations in captivity, sexual maturity is reached in less than two years. In two years, all the lizards are larger than 46-48 mm [130]. The maximum body size (68-72 mm) is reached no earlier than at five years of age.

ENEMIES. In the Transcaucasus, the Caspian thin-toed gecko is attacked by the lizard snake (*Malpolon monspessulanus*) and gyurza (*Vipera lebetina*) [1]. In Turkmenia, they are eaten by the Karelin racer (*Coluber karelini*), cliff racer (*C. rhodorhachis*), steppe ribbon snake (*Psammophis lineolatum*), saw-scaled viper (*Echis carinatus*) {*Echis multisquamatus*, *fide* Cherlin, 1981 – MG}, glass lizard (*Ophisaurus* {= *Pseudopus* Merrem – MG} *apodus*), barn owl (*Athene noctua*), black kite (*Milvus korschun*), and fox (*Vulpes vulpes*) [8, 23, 52, 53, 130; our observations]. Data on Caspian thin-toed gecko enemies from other areas are not available.

As we discovered, many geckos of this species die in winter {p. 141} shelters and on the surface during frosts. Thus, on 27 June 1976, we discovered 14 Caspian thin-toed gecko skeletons in a hollow in a loess wall, near Kunya-Urgench. In the same region, in a similar situation, 18 skeletons and the remains of two more lizards nearby were found in a crack on 6 June 1979. In a burrow near Berezenghi village, on 11 April 1984, two mummified geckos were found. This was, undoubtedly, the result of the deep and sudden frosts (down to −13°) at the end of February. A. M. Alekperov [1] reported finding 5-17 skeletons of these lizards in a single place.

BEHAVIORAL ATTRIBUTES. Caspian thin-toed geckos move about easily on vertical surfaces and ceilings. They run fast on rocks, loess cliffs, and in vertical burrows dug in loose soil. They are capable of moving about in grass near rocks during foraging, migrating over fairly great distances, and are found in steppes and on takyrs. Some adults and sexually immature individuals constantly migrate in search of suitable places. On the plain near Ashkhabad, all lizards were captured in a 2 m^3 concrete pit located in a wormwood steppe, and in ten days, with cool weather and frequent rains (April 1978), two geckos appeared. In the following ten days, ten more appeared. The geckos live individually or in groups of 10-15 individuals in suitable places [1] (we have observed congregations of up to 50). Once a gecko notices its prey, it slowly moves toward it, grabs it with a lunge, and eats it. Individual lizards are attached to their shelter: having been released 20-30 m away, they returned [98]. They hide in cracks in rocks and cliffs, hollows between stones in taluses and buildings, and in rodent burrows. They also winter in these. In hilly areas near Ashkhabad, Ch. Atayev [12] dug out Caspian thin-toed geckos from burrows at depths of 8-10 cm (16 females and 12 males were discovered together), and 80 cm long – five more individuals. During the mating season, these geckos fight frequently, emit metallic sounds (the squeak is quieter and of lower tone than that of the gray thin-toed gecko). In daylight, they sit no closer than 2 m apart. Caspian thin-toed gecko egg clutches have been found in hollows in old loess buildings, where the eggs were slightly buried in loess or sand.

PRACTICAL IMPORTANCE AND PROTECTION. The Caspian thin-toed geckos are especially beneficial living alongside humans. They willingly eat Transcaspian termites {*Acanthotermes ahngerianus* – MG}, which damage buildings, as well as snout beetles, flies and other harmful insects. The wide range and stable numbers assure normal existence for this species. Thus, there is no need for any special protection measures.

TURKMENIAN THIN-TOED GECKO — *TENUIDACTYLUS TURCMENICUS* (SZCZERBAK, 1978)

Type locality: T[urkmen]SSR: Badghyz: near Kushka, Agashly locality.
Karyotype: 2n = 42; 4 subtelocentrics, 38 acrocentrics, NF = 46.
1978 – *Gymnodactylus turcmenicus* Szczerbak, Vestn. Zool., Kiev, no. 3, p. 39.
1984 – *Tenuidactylus turcmenicus*, Szczerbak and Golubev, Vestn. Zool., Kiev, no. 2, p. 53.

DIAGNOSIS. No fewer than 30 femoral pores in males; GVA usually no more than 115 (to 130); 25-32 ventral scales across midbody; 8-12 scales across head; dorsal tubercles triangular roundish or trihedral, usually large, usually one to three additional tubercles present (in about 94%); two to three (usually three) pairs of postmentals, the scales of first pair contact each other by wide suture (92%); usually two rows of scales between femoral pore row and vent; adpressed forelimb reaches tip of snout with finger tips (total length of upper- and forearm comprises 33-38% (35.4 ± 0.2) of body length).

HOLOTYPE. ZIK Re No. 10, adult male. "Turkmenia: near Kushka: Agashly locality, 21 Sept., 1977," L = 78.3 mm. LCD (half regenerated) 97.7 mm; 10 scales across head; 11 supralabial scales; 8/9 infralabials; 128 GVA; 30 ventral scales across midbody; 12 scales around {p. 142} tubercle + 2 additional tubercles; 35 femoral pores; 25/27 subdigital lamellae; two pairs of postmentals, scales in first pair contact each other across wide suture.

DEFINITION (Fig. 63) (from 132 specimens in ZIK, ZIL, NMW, and FMNH). L adult male (n = 51) 41.4-80.0 (70.76 ± 1.09) mm; L adult female (n = 51) 40.6-71.0 (60.76 ± 0.79) mm; mass 16.3 g; L/LCD (n = 31) 0.67-0.89 (0.73 ± 0.008); HHW (n = 21) 50-66 (58.10 ± 1.00); EED (n = 24) 40-62 (50.54 ± 1.20); supralabial scales (n = 235) 10-14 {p. 143} (11.43 ± 0.05); infralabials (n = 238) 6-10 (8.32 ± 0.05); scales across head (n = 120) 8-12 (9.81 ± 0.07); DTL (n = 118) 21-32 (27.79 ± 0.22); scales surrounding dorsal tubercles (n = 132) 14-21 (16.76 ± 0.13); GVA (n = 108) 106-133 (119.31 ± 0.65); ventral scales across midbody (n = 102) 25-32 (28.20 ± 0.18); femoral pores (n = 55) 31-39 (35.04 ± 0.23); femoral scales in females (n = 49) 35-41 (37.63 ± 0.23); subdigital lamellae (n = 222) 23-30 (25.75 ± 0.10). Mental scale triangular or pentagonal; two (28.4%) or three (71.2%) pairs of postmental scales (four pairs recorded once, equals 0.4%); scales in first pair contact each other across wide suture (92.2%), this suture is point like (5.2%), 2.6% specimens lack it; first nasal scales contact each other (50%) or separated by a scale (50%); one (30.3%), two (45.5%), three (18.1%) or four (1.5%) additional small tubercles contact central dorsal tubercle, 4.6% specimens lack such contact; row of femoral pores in males is not interrupted in the middle.

COLOR AND PATTERN. The color on the back is ochrish. There are five washed out, poorly defined, brownish transverse bands on the body, and around ten such dark bands on the tail. The abdominal side is light and without spots.

DISTRIBUTION (Fig. 64). Extreme southern U.S.S.R. (Turkmenia) and Afghanistan from the divide between Kushka and Murghab Rivers and southern extreme of the Karabil Upland eastward through Paropamisus to the western hills of the Hindukush. Finds in Pakistan are

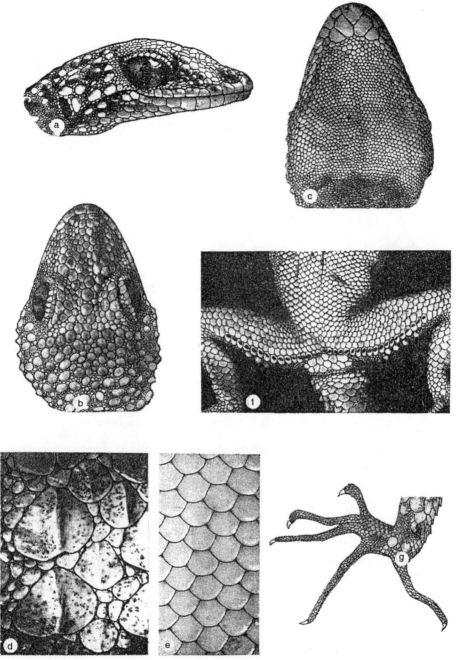

FIG. 63. *Tenuidactylus turcmenicus*:
a, head from the side; b, head from above; c, head from below; d, dorsum;
e, belly; f, femoral pores; g, hindfoot from above (ZIK SR 561).

possible. Specimens from Afghanistan were determined as *Cyrtodactylus caspius* and *C. fedtschenkoi* (foreign collections).

LEGEND to Fig. 64.
U.S.S.R.: Turkmenia: 1 - Karabil Upland: Pelenghovely Gorge (ZIK); 2 - Chemen-i-Bit (ZIK); 3 - Agashly Gorge (ZIK).
Afghanistan: 1 - 60 km N Talukhan (CAS); 2 - Mazar-i-Sharif (FMNH); 3 - 24 km E Khanabad (CAS); 4 - E of Faizabad, 36°32′N, 71°21′E (FMNH); 5 - **{p. 144}** Vaqieh, 10 km Sar-Pol [317]; 6 -

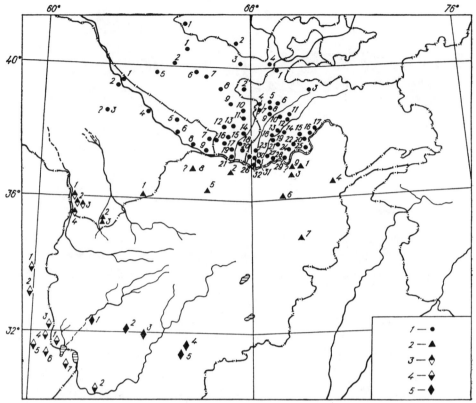

FIG. 64. Distribution of *Tenuidactylus fedtschenkoi* (1), *T. turcmenicus* (2), *T. longipes microlepis* (3), *T. l. longipes* (4), *T. l. voraginosus* (5).

Cheshmeh-Shehr, 17 km Pul-i-Humri (NMW); 7 - Paghman (FMNH); 8 - Sheberghan (ZFMK); 9 - Kunduz (ZFMK).

SEXUAL DIMORPHISM AND AGE VARIATION. The males of this species are larger than females. Other features are similar to those in close species (see *T. caspius*).

SYSTEMATICS AND GEOGRAPHIC VARIATION. Four samples from various parts of the range were compared to find geographical variation. The results show that compared with nearby Agashly populations, the specimens from Chemen-i-Bit have a lower number of ventral scales across midbody, [lower] GVA and [fewer] scales across the head; the mean values of such characters as the number of femoral pores and supralabial scales also gradually increase toward the east (Mazar-i-Sharif, Afghanistan). The number of scales that surround the dorsal tubercle is greatest in Karabil. There were no characters detected that would clearly distinguish any of the examined samples.

ECOLOGY. This is the first such study and all data are published for the first time.

HABITAT AND QUANTITATIVE DATA. The first discovery of this species was at the Agashly locality near Morgunovka (Badkhyz) in hill country with pistachio trees and well developed grass cover on sandstone cliffs. Here, 27 geckos were noted on a 400 m-long, 3 m-high cliff, 216 specimens per hectare.

At the Pelengoveli (Karabil) locality, the Turkmenian thin-toed geckos were encountered only on vertical rocky (limestone) cliffs; they were absent from taluses and individual rocky boulders. Here, the cliffs are without lichens, have cracks, and the slope has a southern exposure. On the slope with a northern exposure, ten times fewer animals were encountered. It is curious that Caspian thin-toed geckos have been found here only at the bases of cliffs and in rain channels in loess, but not even a single individual was noted on rocks. Here, 20 geckos were counted over the distance of 500 m at the height of 2.5 m, or 160 specimens per hectare of rock area.

On the left bank of the Kushka River, near the spot where it flows into the Islim River (Fig. 65), we found these geckos on conglomerate rock cliffs. Here, a narrow strip of rocky cliffs (about 2 m) on sandy hill slope runs for 33 km from the town of Kushka (with breaks). Individual geckos are encountered 1-10 m from one another. Over a distance of 1 km and {p. 145} with a 2-m-high strip for the count, we encountered 60 individuals, or 300 specimens per hectare. The elevation of Turkmenian gecko habitats is 570-720 m.

RESPONSE TO TEMPERATURE. In the spring, the geckos appeared when the rock temperature was 17° (9 May 1979, Agashly locality; 12 May 1979, locality Pelengoveli), and their greatest numbers were seen with a substrate temperature of 25°. In the fall, these lizards would appear with the substrate and body temperature of 22° (9 October 1981, Chemen-i-Bit). With a rock temperature of 21°, they would disappear and then, 45 minutes later, we would observe a temperature drop down to 15° for air and 18° for rock. The geckos would not emerge from the cracks in strong wind. In studies of Turkmenian thin-toed geckos in a thermal gradient apparatus (n = 46), lizards were seen at temperatures of 9-44° (29.5°±1.5°).

DAILY ACTIVITY CYCLE. They are nocturnal. On 13 May 1978, at the Agashly locality, they appeared one hour after dusk, at 2130 hrs. Also here, on 9 May 1979, they came out at 2200 hrs. At the Pelengoveli locality, on 11 May 1979, the first lizard was noted at 2100 hrs and they emerged *en masse* – at 2300 hrs. The last geckos remained on the surface until 0200 hrs. In the fall (9 October 1981, vicinity of Chemen-i-Bit St.), lizards of this species appeared at 2045 hrs, and at 2115 hrs they disappeared into their shelters. Thus, Turkmenian thin-toed geckos, unlike the Caspian ones, emerge from cover later.

SEASONAL ACTIVITY CYCLE. The earliest capture date for this species is 11 April 1977 and the latest is 9 October 1981. Probably these accurately reflect its period of activity (there are possible shifts of 10-15 days in either direction, depending on the weather conditions).

SHEDDING. From terrarium observations, the lizards shed during the periods of 17-27 April, 13-20 May, and 1-10 September. They shed soon after hatching. Shedding *en masse* was noted on 9 October 1981 in the vicinity of Chemen-i-Bit.

FEEDING. These data on the Turkmenian thin-toed gecko feeding are published here for the first time. Fifty Turkmenian thin-toed gecko stomachs were dissected for analysis (food remnants discovered in only 17 of them) (Table 5). These geckos eat snout beetles (28% frequency rate), butterflies (22%); in the cracks in the rocks there were many noctuid moths at this time (on 11 May 1979, we observed a lizard grabbing one of them), hemipterans

FIG. 65. Biotope of *Tenuidactylus turcmenicus* (sandstone slabs along Kushka River bed, Chemen-i-Bid Station, Turkmenia).

TABLE 5. Analysis of stomach contents of the Turkmenian thin-toed gecko
(*T[enuidactylus] turcmenicus*, n = 50) (n = 30, locality Pelenghovely, 11 May 1979;
n = 20, locality Agashly, 14 May 1978)

Food Item	Constitutents		Number food items found	
	Absolute (number)	%	Absolute (number)	%
Crustacea				
Isopoda	2	4	2	1.6
Arachnoidea				
Phalangidae	4	8	6	4.8
Araneidae	2	4	2	1.6
Hylopoda	1	2	1	0.8
Insecta				
Orthoptera	2	4	3	2.4
Tettigonidae	1	2	1	0.8
Acrididae	2	4	2	1.6
Blattoptera	2	4	2	1.6
Homoptera	4	8	5	4.0
Hemiptera	6	12	9	7.2
Reduviidae	2	4	3	2.4
Coleoptera	1	2	1	0.8
Carabidae	1	2	1	0.8
Chrysomelidae	2	4	4	3.2
Curculionidae	14	28	42	30.2
Hymenoptera	1	2	1	0.8
Formicidae	5	10	21	21.6
Diptera	2	4	2	1.6
Tipulidae	2	4	2	1.6
Lepidoptera	11	22	13	10.6

(12%), ants (10%), cicadas and solpugids (8% ea). During {p. 146} foraging, the geckos descend to the base of the rocks. We even observed them in the grass.

REPRODUCTION. The sex ratio in the population is 1:1. The mating period was not observed. On 13 May 1978, well developed eggs were noted in the females from Agashly locality. A gecko, caught near Chemen-i-Bit on 18 May 1982, laid a pair of eggs 12.4-13.0 × 8.5-9.0 mm. On 9 October 1981, we caught the current year's hatchlings. In a rock crack occupied by a sparrow's nest, we found a Turkmenian thin-toed gecko egg. It was accidentally crushed and found to contain a live developing gecko.

GROWTH RATE has not been studied.

ENEMIES. Near Chemen-i-Bit, in the biotope of the Turkmenian thin-toed gecko, we caught a tree snake (*Boiga trigonatum*), which can, given an opportunity, attack the geckos.

BEHAVIORAL ATTRIBUTES. In the manner of movement, this lizard does not differ from the Caspian and Turkestan thin-toed geckos, although it is a larger lizard and runs somewhat faster. Summer and winter shelters are cracks in sandstone, conglomerate, and limestone rocks. In windy weather, they hunt insects that hide in the cracks. On quiet days, they come down to the bases of the rocks and forage in the grass. Judging by the discovery of old egg shell fragments, they lay the eggs in the rock cracks. Captured geckos emit quiet squeaks.

PRACTICAL IMPORTANCE AND PROTECTION. Within the territory of our country, the Turkmenian thin-toed gecko is found only in three places. Its overall numbers are small, not more than about 500 individuals. In order to protect the gene pool of this rare species, it is included in the *Red Books* of the U.S.S.R. and T[urkmen] S.S.R. on our recommendation. Fortunately, the biotopes of the Turkmenian thin-toed gecko are not in areas of human activity and so, for now, the species is not threatened. However, conservation agencies of Turkmenia should, in the future, create a preserve in at least one area where geckos of this species are found (for example, the Agashly locality).

TURKESTAN THIN-TOED GECKO — *TENUIDACTYLUS FEDTSCHENKOI* (STRAUCH, 1887)

Type locality: Samarkand, Uzbekistan.
Karyotype: 2n = 42; two submetacentrics, 38 acrocentrics, two subtelocentrics, NF = 46.
{p. 147} 1870 – *Gymnodactylus scaber*, Fedtschenko, Izv. Obshch. lubit. anthropol., etnogr., vol. 8, no. 1, p. 165.
1887 – *Gymnodactylus fedtschenkoi* Strauch, Mem. Acad. Sci. St. Petersbourg, ser. 7, vol. 35, no. 2, p. 46.
1954 – *Cyrtodactylus fedtschenkoi*, Underwood, Proc. zool. Soc. London, p. 475.
1984 – *Tenuidactylus fedtschenkoi*, Szczerbak and Golubev, Vestn. Zool., Kiev, no. 2, p. 53.

DIAGNOSIS. No fewer than 30 femoral pores in males (26-29 femoral pores were found in less than 5% of males examined). 105-130 GVA (only in sample from Nurek GVA reached 144, other samples (around 3%) have more than 130, within limits 131-136). 24-37 ventral scales across midbody, 9-18 scales across head; dorsal tubercles triangular-roundish, usually of medium size; {p. 148} as a rule additional tubercles do not contact them. Two to four (usually two) pairs of postmentals, scales of first pair contact each other across wide suture. Usually three rows of scales between preanal pore row and vent. Adpressed forelimb reaches tip of snout with finger tips (total length of forearm and upper arm is 31-38 (35.1 ± 0.1) % of body length).

LECTOTYPE. ZIL 6354, adult male. "Samarkand, 1874, V. Russow." L = 60.7 mm; tail regenerated; 13 scales across head; 12/10 supralabial scales; 8/9 infralabials; 127 GVA; 28 ventral scales across midbody; 13 scales around dorsal tubercle and one slightly enlarged additional tubercle; 33 femoral pores; 26 subdigital lamellae; two pairs of postmentals, the scales in first pair contact each other by a wide suture.

DEFINITION (Fig. 67) (from 461 specimens in ZIK, ZIL, ZMMGU, and DIZP). L adult male (n = 162) 48.2-76.7 (59.66 ± 0.53) mm; L adult female (n = 166) 48.0-70.0 (58.79 ± 0.42) mm; L/LCD (n = 100) 0.68-0.91 (0.76 ± 0.004); HHW (n = 100) 48-60 (54.35 ± 0.29); EED (n = 100) 33-67 (45.53 ± 0.57); supralabial scales (n = 919) 9-14 (11.51 ± 0.03); infralabials (n = 921) 7-11 (8.28 ± 0.02); scales across head (n = 461) 9-18 (12.36 ± 0.07); DTL (n = 442) 15-32 (23.39 ± 0.14); scales around dorsal tubercle (n = 434) 10-18 (13.92 ± 0.09); GVA (n = 428) 105-144 (120.73 ± 0.31); ventral scales across midbody (n = 403) 24-37 (29.47 ± 0.10); femoral pores (n = 217) 26-44 (33.86 ± 0.21); femoral scales in females (n = 204) 28-48 (37.11 ± 0.27); subdigital lamellae (n = 893) 19-31 (25.02 ± 0.08).

FIG. 66. A solpugid attacking a Turkestan thin-toed gecko [*Tenuidactylus fedtschenkoi*], Beshkent Valley, Tadjikistan. (Photo by A. D. Bautin.)

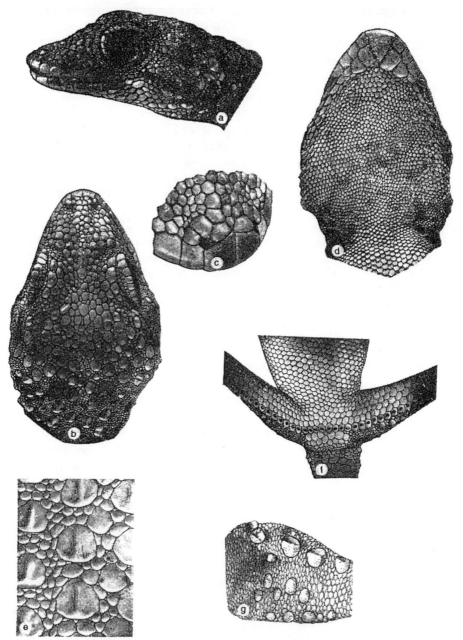

FIG. 67. *Tenuidactylus fedtschenkoi*:
a, head from the side; b, head from above; c, nasal scales; d, head from below; e, dorsum;
f, femoral pores; g, thigh from the rear (ZIK SR 1081, Karatau, Tadjikistan).

Mental scale triangular or pentagonal, two (27.1%), three (69.1%) or four (3.8%) pairs, the scales in first pair contact each other by wide (88.8%) or point (8.5%) suture or separated by a scale (2.7%); first nasal scales separated from each other by one (71.0%) or two (0.7%) scales or in contact (28.3%); one (32.4%), two (10.6%) or three (0.7%) additional tubercles contact central dorsal tubercle, 56.3% of specimens lack such contact; row of femoral pores is not interrupted in the middle in 95% of males. Mass to 9.0 g.

COLOR AND PATTERN. Unlike that of the Caspian thin-toed gecko, the main background color on the back is light or coffee brown, and the color of the body and transverse bands on the tail is dark brown. Another distinction from the preceding species is the nearly complete absence of a pattern on the head. Individuals, caught in the vicinity of Nurek on dark rocks, had a rare dark brown color with a violet or reddish hue.

DISTRIBUTION (Fig. 64). Widely distributed in the mountain systems of western Pamir and adjoining plains; penetrates southern Kyzylkums and eastern Karakums, probably brought there by man; Amu-Darya and Pyandz Rivers probably form southern border of species' range.

LEGEND for Fig. 64.

U.S.S.R.: Turkmenia: 1 - Farab (ZMMSU); 2 - Chardjou fortress (ZIK); 3 - Repetek (ZIL); 4 - 85 km S Chardjou, Karabek-aul (ZIK); 5 - 80 km N Kerkichy, Talimardzhan (ZIK); 6 - Kerkichy (ZIK); Dostluk [130]; 7 - Kugitangh Ridge: Sherabad; Khodjepil; Kugitangh Village (ZIK); Bazar-Tepe [23]; Kyzyl-Alma: Shalkan, Zarabak, Maydan (kyshlaks in Kugitangh foothills) [22]; Svintsovy Rudnik, Yurek-Tepe Hills, Kaynarata Village [130]; Ak-Bulak (ZIL); 8 - 80 km SE Kerkichy (ZIK); 9 - Kelif Village (ZIL).

Uzbekistan: 1 - Western Aktau: Kockcha Village [161]; 2 - Yllandy Village; Kermine Station [22]; 3 - Pystelitau Mountain (NE foothills of Nuratau Ridge) [98]; near Djizak (ZIL); 4 - Khavast; Djulanghar (ZIL); 5 - Bukhara (ZIK); Kaghan (ZDKU); 6 - Aktash Settlement (ZIK); 7 - Katta-Kurghan Reservoir [22]; 8 - Samarkand (ZIK); Kara-Tybe (ZIL); 9 - Sanghardakdarya River (ZIL); 10 - near Padakhan Kishlak, Tulalangah River; Chinary Settlement; Babatagh Ridhe (ZDKU); 11 - Denau on Surkhandarya River (ZIL); 12 - Surkhandarya Dist.: Derbent Settlement (ZIL); 13 - Kashkadarya Dist.: Ak-Rabat (ZIK); 14 - Shurchy Settlement (ZIK); 15 - Leylyakan (ZDKU); 16 - Kugitanghtau: Vandob (ZDKU); 17 - Karasu Settlement (between Termez and Shirabad) [22]; 18 - Kockaity Settlement (ZIL); 19 - W Djarkurghan Settlement; 11 km on road from Djarkurghan {p. 149} to Anghor (ZDKU); Anghor (ZMMSU); 20 - Surkhan Settlement; 8 km W Surkhan Station (ZDKU); 21 - near Termez, Sultan-Saadat fortress (ZIK).

Kazakhstan: 1 - C. Kyzylkum: Ayakkuduk Well [161]; 2 - Tuzkan Lake, 100 km N Nuratau Reservoir [161].

Tadjikistan: 1 - Ura-Tyube Settlement [161]; 2 - Shingh Settlement (ZIK); 3 - Garm Dist.: Shinglich locality on Sarbach River (ZIK); 4 - S slope of Ghissar Ridge: Sary-Pul and Shurhok Kishlaks (DIZP); 5 - Varzob Settlement [125]; 6 - Ramit Gorge: Chuyangharon (ZIK); 7 - Reghar (ZIK); 8 - Dushanbe (ZIK); 9 - 90 km SW Dushanbe: Isambay Settlement (ZIK); near Ghissar Town (ZIL); 10 - Yavan Town (ZIL); 11 - Chermazak Cross (1670 m elevation); Nurek (ZIK); 12 - S Nurek: road to Turtkoul [125]; 13 - Karatau: N of Pyandzh, 860 m elevation (ZIK); 14 - Surkhob (ZIK); 15 - Kulyab on road to Shurchy (ZIK); 16 - Dashtidjum (DIZP); 17 - Darvaz Ridge [96]; 18 - 12 to 20 km S Kyzyl-Kala on road to Shaartuz (ZIK); 19 - Kurghan-Tjube Town; Uialy Kishlak on road Dushanbe - Kurghan-Tyube (ZIL); 20 - near Shurabad (ZIK); 21 - Vakhshstroy [125]; 22 - Khodja-Mumin (ZIK); 23 - Shaartuz Dist.: Beshkent fortress (ZIK); Chil-Chor-Chashma [125]; 24 - Parkhar between Saray and Chubek [119]; 25 - near Yol between Chubek and Toghmay [119]; 26 - Tuyun-Tau Ridge (ZIK); 27 - 12 km S Djilikul Settlement (ZIL); 28 - 26 km from Pyandz to Parkhar (ZIK); 29 - Kirovabad (DIZP); 30 - Shaartuz (ZIK); 31 - 18 km from Chirik Settlement (ZIK); 32 - Ayvadz Kishlak [125].

Kirghiziya: 1 - Frunze {now Beshkek – MG} (ZIK).

SEXUAL DIMORPHISM AND AGE VARIATION corresponds to those of the Caspian thin-toed gecko (see above).

SYSTEMATICS AND GEOGRAPHIC VARIATION. To study geographic variation, the main scale characters of sixteen samples from all parts of the range were compared. The comparison showed some differences in the western lowland populations. Geckos from Charjou, Karabekaul, Termez, Shurchy, Kerkichy, and 80 km east of the latter locality, have lower mean values for such characters as the number of femoral pores and femoral scales in females, number of additional tubercles that contact the central dorsal tubercles, and the relative length of these tubercles and the scales around them. Increases in the averages for the number of subdigital lamellae and scales across the head are [also] observed in these populations. The opposite trend for these features is characteristic of the lowland populations of southern Tadjikistan and upland geckos. Samples such as the Samarkand or Karlyuk ones (i.e., those that inhabit foothills) have intermediate values for many characters. The presence of a vertical zonation of the variations in some characters is apparent for this species. A sharp increase in the number of ventrals across midbody and GVA in geckos near the Nurek Reservoir attracts attention as a distinctiveness among the populations, but we believe that it is insufficient to separate them into a different subspecies.

HABITAT AND QUANTITATIVE DATA. From published sources [22, 23, 98, 107, 125, 130 et al.], as well as from our data, it is known that Turkestan thin-toed geckos exclusively inhabit vertical surfaces, such as rock cliffs, loess ravines, walls of ruins, on duvals, on houses and other buildings, and are sometimes found on tree trunks. We found the greatest numbers of these lizards on loess cliffs in the foothills. Thus, in the vicinity of Karlyuk village (Kugitangh), in an area of schist outcrops 20 m long and 2 m tall, thirty lizards were caught (frequently, two or more geckos would be looking out of a crack). Near here, in an abandoned $2 \times 1 \times 1$ m concrete pit (total [vertical surface] area of 6 m^2), 11 geckos were noted. In the vicinity of Talimardzhan village (80 km northeast of Kerkichy), on loess cliffs along a brackish stream, the highest density was recorded: six to eight individuals per m^2.

Near the Karatau crest, on a porous gypsum rock cliff (100×2 m), six lizards were caught; 12 geckos were caught in the vicinity of Surkhob Settlement (Tadjikistan), on a saline-clay cliff with a 10 m^2 area; in the vicinity of Nurek, on a ferriferous sandstone cliff, one gecko was encountered per 10 m of the route; on loess cliffs near Dushanbe, in a 10 m^2 area, 7-13 individuals were counted. Single individuals were noted on the rocks in the vicinity of Aktash station, in karst craters and on rock-salt and gypsum outcoppings in the vicinity of Moskovsky Village (Khodja-Mumin), between the settlements Shingh and Kostarash. These lizards are numerous (1 specimen every 1-5 m of route) on ancient loess buildings and fortress walls (Chardjou, Karabekaul, Turkmenia), {p. 150} and their ruins (Beshkent Fortress near Nasir Khirsov Settlement, Tadjikistan; Kyrk-Kyz and Sultan-Saadat, Uzbekistan). Very rarely, the Turkestan thin-toed geckos are found on rocky slopes in the pistachio savannah (Karatau Crest, Tadjikistan): here, a total count was conducted on a 900 m long, 6 m wide transect and only one gecko was found under a rock. Literature data on the numbers of Turkestan thin-toed geckos [107, 130] frequently cannot be compared with ours, because the lizards were being counted for 1 hour (at unknown speeds, unidentified biotope) or over imprecisely determined distances (6-8 km, 7-9 km). However, the maximum numbers of geckos in Tadjikistan [107] is given for the vicinity of Gandjin Settlement: 50-60 individuals were counted on an 8-10 km route, or six individuals per km, which is a significantly lower result. A different source [327] cites the average density of 11.3 individuals per 100 km^2 (in the area of the Nurek Reservoir). This species is found at elevations up to 2300 m [125].

RESPONSE TO TEMPERATURE. K. P. Paraskiv [98] reported the appearance of these geckos on the surface after rock temperatures reached 23-25°. On 23 April 1980, in the vicinity of Shurab (Tadjikistan), they were seen in the daytime with the air temperature of 13° (at an elevation of 1520 m, there was snow still in the ravines). With the air temperature of 19°, at dusk, active geckos were seen on 8 April 1980 at an elevation of 580 m in the vicinity of Sumbula settlement, and with an air temperature of 16° at dusk – in the vicinity of Sukhrob settlement. A. D. Bautin (written report) incubated the Turkestan thin-toed gecko eggs at the temperature of 36° and with 60-80% humidity.

DAILY ACTIVITY CYCLE. In early spring and late fall, geckos of this species are active in the daytime. In summer, they are also frequently seen during the day, during cool hours or in darkened places, but they are most active at dusk [22, 23, 98, 125, our observations]. They emerge from cover no earlier than 0800-0900 hrs. In April-May, active lizards were encountered at night at 2200-0100 hrs (8 April 1979, vicinity of Sumbula settlement; 10 April 1980, southern Tuyun-Tau; 20 May 1979, vicinity of the town Chardjou).

SEASONAL ACTIVITY CYCLE. In the foothills of Kugitangh, in 1954, the first individuals appeared after hibernation in the middle of February, and on the Zeravshan Crest they were noted at the beginning of April [22]. Near Dushanbe, the geckos appeared after hibernation in the second half of March, and in the plain of the Vakhsh River mid-section – even in the first half of the month [125]. S. A. Said-Aliyev [107] observed single individuals on 7 February 1963 in Dushanbe and in Chor-Teppa village, on 23 February 1963 in the vicinity of Zighar settlement, but *en masse* emergence was observed in March. In southern Kazakhstan, Turkestan thin-toed geckos appear in the first half of March [98]. The specific timing of these lizards' appearance depends on the weather conditions and elevation. Geckos, caught on 8 April 1980 in the vicinity of Sumbula settlement, had fed already (cool spring, 580 m elevation), but in the vicinity of Shurabad (1520 m elevation), on 23 April 1980, they had just left the winter shelters. Undoubtedly, lizards living in the high mountain regions (western Pamir) may emerge from hibernation at the end of April or the beginning of May. The disappearance for the winter in Tadjikistan (near Dushanbe) was noted in December [125]. At the end of November and the beginning of December of 1963, geckos were still encountered in the vicinity of Gulbista, and during the periods of 15-18 and 29-30 November 1964-1968, in the Kafirnighan River plain [107]. In southern Kazakhstan, these lizards disappear in October-November [98].

SHEDDING. In Turkmenia, shedding individuals were encountered in April-May. *En masse* shedding in Tadjikistan has also been noted in April-May [125, 130]. During observations in the terrarium, shedding was noted for various individuals in the periods of 8-14 April 1982 and 13-25 May 1982. Undoubtedly, fall shedding also occurs.

FEEDING. In the stomachs of 35 geckos, caught in Kugitangh in August [23], termites predominated (74.4% encounter rate), followed by beetles (34.3%, among them were tiger beetles, darkling beetles, scarabids, [and] snout beetles). In addition, there were homopterans (14.3%), stink bugs, hymenopterans, and grasshoppers (11.4% each), as well as butterflies (5.7%). In insignificant amounts, they feed on earwigs, neuropterans, butterfly larvae, spiders, centipedes {*Scolopendra* sp. – MG}, isopods and solpugids. In the diet of 10 geckos, caught in August in the Surkhandarya River valley [22], coleopterans (70%), and hymenopterans (50%) predominated. Secondary food components include hemipterans, dipterans, {p. 151} stridulating orthopterans, lepidopterans, termites, homopterans and spiders.

The diet of 12 geckos, caught the same month in Babatagh, was made up of homopterans and lepidopterans (33.3% frequency of occurrence) and hemipterans (25%). The secondary components were beetles, hymenopterans, orthopterans, dipterans, cockroaches, isopods, scorpions, spiders, [and] solpugids. According to M. V. Kaluzhina's data [63], the stomachs of 50 geckos, caught in the rice field region in Uzbekistan, were full of mosquitoes. As the author determined, each gecko eats up to 73 mosquitoes in one night.

The analysis of 35 stomachs from geckos collected in the vicinity of Dushanbe in April-May [125] demonstrated that the arachnids are the main food for them at this time of year (100% frequency of occurrence). Other invertebrates, such as millipeds and insects and their larvae, are of secondary importance (around 40%). In the summer, spiders disappear from the diet and insects and their larvae become the main prey source. The most frequent prey for these geckos are beetles, small grasshoppers, dipterans, and hymenopterans, in particular ants and bees.

REPRODUCTION. The sex ratio in our samples is close to 1:1. Sexual maturity is reached when the body length is 48 mm and greater.

According to the data from Uzbekistan [22], the largest testes were discovered in the first ten days of April. Probably that is the time of mating. The author, cited above, found eggs that were ready to be laid in mid-May, in the vicinity of Termez (he supposes that approximately 1.5 months elapses between the beginning of egg development and laying). In Turkmenia, eggs, ready to be laid, had been found around the same time [130]. The time of reproduction, in the vicinity of Termez, comes to a close by mid-July (laying of the second clutches is done in two to three weeks). During this period, in the foothills and mountains as well as in areas further to the north, reproduction continues. The last females with eggs were caught in Babatagh, near Kashka-Bulak, on 6 August, and in Illias on 12 August. Most often, females lay two eggs (108 cases), less frequently – only one (32 cases). Very rarely, three eggs are laid (one case). The last circumstance is given only for the Turkestan thin-toed gecko. According to observations in Tadjikistan [125], females most often lay two eggs.

The beginning of egg laying, in the vicinity of Dushanbe, has been noted in the second half of May (in particularly cold years this is delayed until mid-June). The subsequent clutches are laid, probably, at the end of June or even in July. Females, ready to lay eggs, were still encountered in the second half of July. There is a possibility of a third laying. The egg dimensions vary within 11-12.6 × 8.8-10.1 mm limits (n = 20). Frequently, several females lay their eggs together. Ten to 14 eggs had been found at the end of July and beginning of August, and the embryos were at different stages of development: from the barely formed to the ready to emerge. The supposed incubation period is 60-65 days.

According to our observations, the first clutch was found on 17 May 1979, in the vicinity of Kerkichi (also here, nearly 20 old eggs were found; colonial nest). A pair of fresh eggs, laid by a gecko of this species, was also found on 18 May 1979, in a crack in a loess cliff 80 km northeast of Kerkichi village. We noted a massive colonial nest in a loess cliff, in the vicinity of Shurchi station, on 24 May 1964. On 30 May 1976, a gecko caught in the vicinity of the Karlyuk settlement laid a single egg. We found a colonial nest (around 10 previous year's eggs) in a loess building in the vicinity of Sambula settlement on 8 April 1980. The egg dimensions were 10.5-12 × 8.6-9.6 mm (n = 5). According to the data from the Leninsky region of Tadjikistan, in 1976, the eggs of Turkestan thin-toed geckos were found from the end of May to the end of September. In captivity, these lizards started mating at the end of November, the first egg was laid at the end of December and the second one two days later. Its dimensions were 13.3 × 10.2 mm with a mass of 0.69 g. The average egg dimensions were 12.8 × 9.9 mm and the average mass was 0.6 g (n = 5). Incubation lasted 62 days and the hatchling's mass was 0.45 g (A. D. Bautin, *in litt.*). The first young lizard was caught in Bukhara on 2 July 1955 [22]. Newly hatched lizards begin appearing in the vicinity of Dushanbe at the end of July and beginning of August [125].

GROWTH RATE. The bodies of the hatchlings are 20-24 mm in length [107, 125]. According to S. A. Chernov's observations in Tadjikistan [125], {**p. 152**} 1.5 months after hatching the geckos grow approximately 5 mm, and by the end of May of the following year, that is after four to five months of active life (not counting the quiet winter period), they grow another 12-13 mm and reach an average length of 42-45 mm. By the end of July, another 9.5-10 mm are added, and toward the winter they become near adult [in size]. Thus, the full size is reached in the middle of the second year of life. Once the body reaches 65-68 mm, growth stops, although, very rarely, some individuals grow to 70-71 mm. The females that hatch from the year's first clutches reach sexual maturity by the following spring. Individuals from later clutches reach puberty later.

The authors note the geckos' tail growth rate in the first month of life: in newly hatched geckos, the tail is barely longer than the body and head (L/LCD = 0.93 on average), but just one month later, it is significantly longer and the ratio L/LCD is equal to 0.75-0.78, on average. This ratio does not change with further growth.

From the observations in the terrarium, Turkestan thin-toed geckos reached sexual maturity and engaged in reproduction 15-16 months after hatching (A. D. Bautin, *in litt.*).

ENEMIES. In Uzbekistan, these lizards are eaten by spotted desert racer (*Coluber karelini*) and saw-scaled viper [22]. We found a Turkestan thin-toed gecko in the stomach of a spotted desert racer that was caught on 24 May 1964, in the vicinity of Shurchi Station. On 21 May 1979, in the ruins of Karabekaul Fortress (100 km south of Chardjou), we observed a *Coracias garrulus* {MG} picking off geckos as they emerged from cover in the morning. In Beshkent Valley, A. D. Bautin photographed geckos being attacked by a large solpugid (Fig. 66) and by spotted desert racers.

BEHAVIORAL ATTRIBUTES. In the modes of motion and hunting, Turkestan thin-toed geckos do not differ from the Caspian thin-toed geckos. Ten or more geckos may live in a single crack. Fights among them occur constantly; sexually mature males are the only ones [that] fight. Even if only one gecko moves, the others start moving also. Here and there, confrontations arise among them. When an insect lands on a wall of the crack, two to three individuals head for it. The squeak of a disturbed gecko is easily heard [98].

Based on observations in Uzbekistan [22] and from our observations in Turkmenia, they hide in various cracks and hollows in rocks, loess cliffs and between bricks in buildings.

The gecko hibernates in the same cracks by going to depths of 20-30 cm. According to

M. V. Kaluzhina's observations [63], in the vicinity of Kattakurghan, 5-11 individuals spend the winter in the same place.

As S. A. Chernov reported [125], in Tadjikistan the Turkestan thin-toed gecko is particularly numerous on loess cliffs. Here, in the cracks, hollows, and insect and bird burrows, the geckos find cover, places to lay their eggs, and hibernate. The geckos do not go far from these places; this is confirmed by specially conducted tagging.

In the vicinity of Dushanbe, the eggs are found in the same burrows in the loess ciffs that these animals use for shelter. As a rule, the eggs are found predominantly in those burrows in which the entrance is as large as that of a *Coracias garrulus* {MG} burrow. The eggs are 12-20 cm from the entrance and are covered by a thin layer of loess. In the daytime, the temperature in that part of the burrow is about 33-34°.

Mating can occur on a vertical wall. In some terrarium observations, a male approached a motionless female, walking on highly extended legs, swinging slightly from side to side and with its waving tail lifted high, so that his motion resembled a peculiar dance. He grabbed the female at the base of a hindlimb (in the pelvic region), lifted her slightly, turned her over and began copulating (A. D. Bautin, *in litt.*, Dushanbe). He also observed egglaying: the female, having selected cover in the terrarium (crevice in a snag), rested, [then] 20 minutes later slightly lifted herself, and turned her head, as if watching the emergence of the egg. The egg emerged {p. 153}, soft, and rolled to the bottom of the crevice. It was pink. A short time later, right before his eyes, the shell began to harden and turn white. At the end of incubation, the newborn made an almost round burrow 1.5 mm in diameter, rested 15 min and started to enlarge it with sharp, snapping motions. This continued until the gecko was completely free of the shell. Once out, it immediately disappeared in the shelter. Four hours after hatching it started taking food.

PRACTICAL IMPORTANCE AND PROTECTION. The Turkestan thin-toed gecko is a beneficial animal because of its eating habits. This species has a wide range and high density, transforming to a synanthropic lifestyle fairly easily, and is found within the confines of many preserves (Kugitangh in T[urkmen] S.S.R.; Kyzylsui, Mirakin and Nuratin in Uzbek S.S.R.; Ramit in Tadjik S.S.R. *et al.*). Thus, no special preservation measures are needed.

LONG-LEGGED THIN-TOED GECKO — *TENUIDACTYLUS LONGIPES* (NIK., 1896)

Type locality: Neh, Iran.
Karyotype: 2n = 42; 30 acrocentrics, 12 not elucidated {error in original – MG}; NF = 42
1896 – *Gymnodactylus longipes* Nikolsky, Ann. Zool. Mus. Imp. Acad. Sci., v. 1, no. 4, p. 369.
1918 – *Gymnodactylus microlepis* Lantz, Proc. Zool. Soc. London, no. 11-17, p. 11.
1963 – *Cyrtodactylus longipes*, S. Anderson, Proc. Calif. Acad. Sci., ser. 4, vol. 31, no. 16, p. 474, San Francisco [*sic*].
1984 – *Cyrtodactylus voraginosus* Leviton and Anderson, J. Herpetol., vol. 18, no. 3, p. 270.
1984 – *Tenuidactylus longipes*, Szczerbak and Golubev, Vestn. Zool., Kiev, no. 2, p. 53.

DIAGNOSIS. No fewer than 30 femoral pores; usually more than 130 GVA; 30-40 ventral scales across midbody; 12-19 scales across head; dorsal tubercles triangular-roundish, usually small additional tubercles do not contact central dorsal tubercles more often than not {than in other species of this group – MG}; two to three pairs of postmental scales; usually two to three scale rows between row of femoral pores and vent; appressed forelimb reaches beyond tip of snout with fingertips.

LECTOTYPE. ZIL 8810, adult female. "E Persia, Neh, 1896, N. Zarudny." L = 62.7 mm; no tail; 16 scales across head; 13 supralabial scales; 9/8 infralabials; 153 GVA; 33 ventral scales across midbody; 12 scales around dorsal tubercle, additional tubercles do not contact central dorsal tubercles; 40 femoral pores; 22/23 subdigital lamellae; two pairs of postmentals, scales of first pair separated from each other by two scales.

DEFINITION. (Fig. 68) (from 117 specimens in ZIK, ZIL, ZMMSU, NMW, FMNH, ZML). L adult males (n = 40) 45.3-68.8 (60.41 ± 1.48) mm; L adult females (n = 35) 45.0-66.4 (59.05 ± 1.91); L/LCD (n = 27) 0.71-0.82 (0.78 ± 0.02); mass 3.75 g; HHW (n = 95) 39-58 (49.89 ± 0.57); EED (n = 55) 31-56 (44.68 ± 0.78); supralabial scales (n = 234)

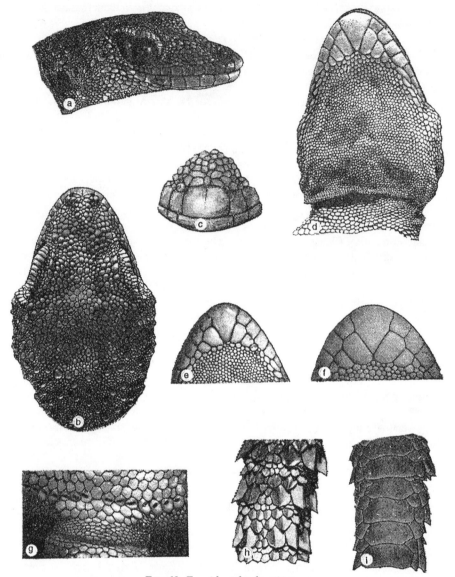

FIG. 68. *Tenuidactylus longipes*:
a, head from the side; b, head from above; c, snout in front; d, head from below; e-f, chin; g, femoral pores; h, tail from above; i, tail from below (f – ZIL 8810, Neh, Iran; others – ZIK SR 307, Akarcheshme, Turkmenia).

11-16 (13.27 ± 0.08); infralabials (n = 234) 7-11 (8.84 ± 0.05); scales across head (n = 115) 12-19 (15.95 ± 0.27); DTL (n = 80) 15-22 (18.82 ± 0.20); scales around dorsal tubercle (n = 117) 10-15 (11.92 ± 0.13); GVA (n = 98) 119-169 (141.67 ± 1.27); ventral scales across midbody (n = 106) 30-40 (34.05 ± 0.39); femoral pores (n = 58) 31-44 (35.86 ± 0.29); [enlarged] femoral scales in females (n = 21) 35-44 (39.45 ± 0.77); subdigital lamellae (n = 220) 20-30 (25.17 ± 0.19). Mental scale triangular or pentagonal; two (58.9%) or three (40.2%) (four scales were found in one case, or 0.9%), scales in first pair contact each other by wide (19.7%) or pointed (19.7%) suture or separated by one (57.3%) or two (3.3%) scales; first nasal scales separated from each other by a scale (96.6%), two or none of these scales may be present (1.7% each); one (30.2%) or two (8.6 %) additional tubercles contact central dorsal ones but more often (61.2%) they do not contact each other. A row of femoral pores in males is interrupted in the middle.

COLOR AND PATTERN. The main background is brownish ochre, occasionally with olive or lilac hue. The pattern is darker than the main {p. 154} color, but is slightly washed out, indistinct. The dorsal tubercles are a little lighter. All of this creates an impression of a kind of velvety texture. Overall, this gecko is lighter [in color] than the Turkmenian and Turkestan thin-toed geckos, and not as bright as the Caspian one. There are seven transverse bands on the back, 8-12 on the tail, and six to nine on the legs. The width of the bands is equal to or greater than the width of the spaces between them. Frequently, the stripes do not reach the sides of the body. The lower surface is white.

SEXUAL DIMORPHISM. Males are distinguished from females by femoral pores.

AGE VARIATION corresponds to that in Caspian thin-toed gecko (see above).

DISTRIBUTION (Fig. 64). Extreme southern U.S.S.R. (Turkmenia), eastern Iran, Afghanistan from western part of the Badghyz Plateau southward through the Kayen Mountains to Pelenghan Ridge and eastward from the latter through the southern foothills of Hindukush to Kandagar. Finds in Pakistan are possible.

{p. 155} LEGEND to Fig. 64.

T. longipes microlepis. U.S.S.R.: Turkmenia: 1 - Puli-Khatum locality (ZIK); 2 - Badkhyz: Akar-Cheshme Cordon {cordon = reservation frontier post – MG} (ZIK); 3 - Badghyz: Kerelek locality: "Kamenny" spring (ZKDU); 4 - Zulfaghar locality (ZIK) {all "localities" here are frontier posts – MG}.

T. l. longipes. Iran: 1 - Baaza Bed in Terra Zirkuh [89]; 2 - Birdjand [80]; 3 - Chah-i-Ziru in Terra Zirkuh (ZIL); 4 - Chah-i-Gyuishe in Terra Zirkuh [89]; 5 - Neh Settlement (ZIL); 6 - Aliabad Settlement [59]; 7 - Neizar in Sistan [89]; 10 km SE Rud-E Hirmand, road from Zabol to Dust-E Mohammed Khan.

Afghanistan (in foreign museums, all identified as *T. fedtschenkoi*): 1 - Koh-Siah-Pochteh, Distr. Naouzar (ZML); 2 - Husseinabad [59].

T. l. voraginosus: Afghanistan: 1 - Farah (NMW); 2 - Kandagar: 30 mi. of Dilaram (FMNH); 3 - 55 and 64 km W Girishk (CAS); 4 - 32 km W Kandahar (CAS); 5 - Pandjavan (ZML); Kandahar 31°36′N, 65°47′E [172].

SYSTEMATICS AND GEOGRAPHIC VARIATION. It was ascertained recently [48] that *T. microlepis* (Lantz) is a subspecies of *T. longipes* Nik. The first of the named forms occurs in the south of Turkmenia (Badghyz), the second one inhabits most of the range, in eastern Iran, and southern Afghanistan. The above is contradicted by a publication [97], whose author asserted that he had found in the ZIL collections four geckos of the nominate form that were supposedly obtained by V. Ya. Lasdin on 15 May 1915, in the vicinity of Termez, "on the way to Ghissar Crest." The results of the special research we conducted contradict this opinion:

1. On 15 May 1915, Mr. V. Ya. Lasdin was not "on the way to Ghissar," as Mr. N. L. Orlov states, but in Patta-Ghissar (south of Termez), which is mentioned on the original label and in the journal of the investigator [71].

2. The series ZIN 19215 consists of three adult *T. longipes* geckos (a, b, d), in varying conditions of preservation, and a juvenile gecko 19215 c, which was redetermined by us as *T. fedtschenkoi*. This indicates that these specimens were put into the container at different times. Undoubtely, three long-legged thin-toed geckos were put there by mistake not in Patta-Ghissar in 1915, but later, in the laboratory. Lasdin's preserved original label relates to Turkestan thin-toed gecko 19215 c.

3. Altogether, 21 specimens of long-legged thin-toed gecko were received by the ZIL collection (according to the catalog) but only 11 remain there now (by the way, a series ZIL 9877, consisting of four geckos, previously identified as *G. longipes*, was found to belong to *T. fedtschenkoi*; judging by its locality, "Khorasan: Terra Bakhars, ruins of Pesh-Robat," the previous identification was an obvious mistake). The fate of only one specimen out of 10 is known so far (NMW 17393). Probably these three geckos, 19215 a, b and d, came from among the nine missing specimens that were caught by Mr. N. A. Zarudny in eastern Persia.

In addition, this publication contains two other disappointing examples of negligence. The description of Neh Settlement in eastern Persia as "terra typica restricta" was groundless because the restriction of type locality was made by A. M. Nikolsky [87] 80 years prior to Orlov's publication. The restriction of type series to three females led to this situation, where such a significant diagnostic character in this group as the number of femoral pores is not represented in the series, even though there was a real opportunity to find a proper specimen from the remnants of the 8809 series (currently – "former types").

Thus, the point of view that the long-legged thin-toed gecko inhabits southern Uzbekistan is based on a museum error. In conclusion, it should be mentioned that N. N. Szczerbak investigated the territory of southern Tadjikistan and southern Uzbekistan along the Soviet - Afghan border in detail in 1980 but, as expected, no specimens of *T. longipes* were found.

A confused situation has arisen in connection with the description of a new species from Afghanistan [325]. The types of this form from the Chicago museum that were available to us convinced us that geckos from the southeastern foothills of Hindukush can be distinguished under this name. However, they can only be given the subspecies status. But the authors of the description included a specimen from eastern Iran (Sistan Province) in the supplementary materials. We disagree with that, because it is the nominate form that is found in the foothill system of the eastern Iranian crests of Kayen and Pelenghan. In addition, as it follows from the table in the original description, it is this very specimen that possesses the characters, which prevent {p. 156} its unequivocal assignment to the new subspecies. Possibly, the absence of significant physical barriers could have led to the formation of a wide intergradation zone between these forms in the plains portion of southwestern Afghanistan and the contiguous regions of eastern Iran in the recent times.

Key to the Subspecies of T. longipes

1 (2). Length of forelimb (forearm and upper arm) is 36-40% of body length including head; scales of first pair of postmentals usually (65%) separated by gular scales
. . . . *T. longipes microlepis* (Lantz)
2 (1). Length of forelimb (forearm and upper arm) is 38-45% of body length including head; scales of first pair of postmentals usually (75%) contact each other 3
3 (4). 135 to 158 GVA; 13 to 17 scales across head . . . . . . *T. longipes longipes* (Nik.)
4 (3). 156 to 169 GVA; 17 to 19 scales across head . . . .
. . . . *T. longipes voraginosus* (Leviton et Anderson)

LONG-LEGGED THIN-TOED GECKO — *TENUIDACTYLUS LONGIPES LONGIPES* (NIK., 1896)

Type locality: Neh, E Iran, 31°30′N, 60°E.
1974 – *Gymnodactylus longipes longipes*, Gorelov, Darevsky, and Szczerbak, Vestn. Zool., Kiev, no. 4, p. 35.

DIAGNOSIS (from 16 specimens in ZIL, NMW, and ZML). Length of forelimb (forearm and upper arm) is 38-45% of body length including head; scales of first pair of postmentals usually contact each other by relatively wide (50%) or point (37.6%) suture or separated by one to two scales (12.4%); GVA (n = 16) 135-158 (146.64 ± 1.64); scales across head (n = 16) 13-17 (15.93 ± 0.32); ventral scales across midbody (n = 16) 30-36 (32.79 ± 0.47); femoral pores (n = 9) 32-38 (34.56 ± 0.67). Other characters are contained in species description.

TYPE OF SPECIES.
DISTRIBUTION. Eastern Iran, southwestern Afghanistan: Kayen Mountains and Pelenghan Ridge.

SMALL-SCALED THIN-TOED GECKO — *TENUIDACTYLUS LONGIPES MICROLEPIS* (LANTZ, 1918)

Type locality: Tedjen Valley near Iran frontier.
1918 – *Gymnodactylus microlepis* Lantz, Proc. Zool. Soc. London, p. 11.
1975 – *Gymnodactylus (Cyrtodactylus) longipes microlepis*, Gorelov, Darevsky, and Szczerbak, Vestn. Zool., Kiev, no. 4, p. 35.

DIAGNOSIS (from 96 specimens in ZIK, ZIL, and ZMMSU). Length of forelimb (forearm and upper arm) is 34-40% of body length including head; scales of first pair of postmentals separated by one (65.1%) or two (1.8%) gular scales or contact each other by point (16.1%) or relatively wide (17%) suture; GVA (n = 78) 119-150 (137.61 ± 0.83); scales across head (n = 95) 12-16 (13.86 ± 0.09); ventral scales across midbody (n = 87) 30-40 (34.15 ± 0.22); femoral pores (n = 28) 32-39 (35.79 ± 0.35). Other characters are contained in species description.

LECTOTYPE. ZMMGU 151a, adult male. "Tedjen River valley near Persian frontier, Apr. - Sept., 1914, N. B. Mariacri." L = 60.2 mm; LCD = 78.6 mm; L/LCD = 0.77; 12 scales across head; 12/13 supralabial scales; 9 infralabials; 136 GVA; 32 ventral scales across midbody; 10 scales around dorsal tubercle and two additional ones; 34 femoral pores; 23 subdigital lamellae; three pairs of postmentals, scales in first pair are in contact.

DISTRIBUTION. Western part of the Badghyz Plateau and, probably, adjoining parts of Iran and Afghanistan.

{p. 157} SOUTHWEST THIN-TOED GECKO — *TENUIDACTYLUS LONGIPES VORAGINOSUS* (LEVITON ET ANDERSON, 1984) COMB. ET STAT. NOV.

Type locality: Afghanistan: 55 km W Girishk.

1969 – *Cyrtodactylus fedtschenkoi*, Clark, Clark, Anderson, and Leviton, Proc. Calif. Acad. Sci., ser. 4, vol. 36, p. 300.
1970 – *Cyrtodactylus sp.*, Leviton and Anderson, Proc. Calif. Acad. Sci., ser 4, vol. 38, p. 185.
1984 – *Cyrtodactylus voraginosus* Leviton and Anderson, J. Herpetol., vol. 18, no. 3, p. 270.

DIAGNOSIS (from 5 specimens in FMNH, NMW, and ZML). Length of forelimb (forearm and upper arm) is 38-45% of body length including head; scales of first pair of postmentals contact each other by point (40%) or relatively wide (20%) suture, often (40%) separated by gular scales; GVA (n = 5) 156-169 (161.40 ± 2.32); scales across head (n = 5) 17-19 (17.60 ± 0.40) (according to [325], 16-19); ventral scales across midbody (n = 5) 34-39 (37.0 ± 1.05) (according to [325], 36-40); femoral pores (n = 3) 35-44 (according to [325], 35-46). Other characters are contained in species description.

HOLOTYPE [325]. CAS 120332, adult female. "55 km W Girishk, Afghanistan, 10 May 1968, R. and E. Clark coll." L = 60 mm; LCD = 85 mm; L/LCD = 0.71; 17 supralabial scales; 11/9 infralabials; 36 ventral scales across midbody; no femoral pores; 25 subdigital lamellae; one to two postmentals, scales of first pair contact each other by point suture.

DISTRIBUTION. Southwestern foothills of Hindukush in Afghanistan.

There are no data on ecology in the literature. In a section of the "Guide to the Amphibians and Reptiles of the U.S.S.R." [17], written by N. N. Szczerbak, it was reported that, in Badghyz, this species lives on rock outcroppings and is encountered on stone and earthen fences and building walls.

HABITAT AND QUANTITATIVE DATA. As we have determined, the long-legged thin-toed gecko penetrates the territory of the U.S.S.R. from Afghanistan and Iran along the Gez-Gyadyk and Zulfagar Ridges, which are oriented longitudinally. In the former, areas of exposed rock are found in the Tedjen River valley. They are composed of limestone, and the geckos are encountered here as single individuals. Farther north, in the hill country of Badghyz with pistachio savannah, there is a group of outcrops and rocky cliffs in the vicinity of the Akar-Cheshme Preserve boundary (Fig. 69). The rocks here are sandstone and volcanic conglomerates cemented by sandstone. Even though there is less cover here than in the Tedjen River valley, geckos are more numerous. Thus, on the rock "Tooth," at the height of 1.5 m from the base and on a strip 20 m long, on 13 May 1976, 13 individuals were seen. Also here, on 22 April 1984, on a 2 × 50 m transect, seven geckos were counted. All were found on the slopes with southwestern exposure.

In the area of the Zulfagar Ridge, they are also encountered (single individuals) on the cliffs of Tedjen {River – MG}, in the rain channels on a plateau with gypsum outcops further to the north (in a 2 × 300 m area, three geckos were caught on 18 May 1982), and, finally,

FIG. 69. Biotope of *T[enuidactylus] longipes* (sandstone relicts near Akar-Cheshme, Turkmenia {pistachio trees in foreground – MG}).

also on a new ridge of hills with a southwestern exposure, on the cliffs of the plateau and on the outcrops of dark volcanic rocks and conglomerates without vegetation. Here, they are fairly rare, one for every 7-10 m of transect (18 May 1982; 25 April 1983). The elevation of the biotopes is 760-900 m above sea level. On the outcrops of Akar-Cheshme, for every 50 long-legged geckos, two Caspian thin-toed geckos were encountered.

RESPONSE TO TEMPERATURE. Sluggish geckos were first observed on 4-7 April 1977 in Akar-Cheshme at the minimum air temperature of 14° and substrate temperature of 17°. The highest rock temperature that we had noted at dusk, the geckos' active period, was 25°. In the fall, on 7 October 1981, they were on the surface with the air temperature of 18.5° and rock temperature of 25.0°. The lizards' temperatures were 25-26°. The long-legged thin-toed geckos are, probably, more heat-loving than the Caspian species. When being photographed in the sun, the former was calm, but the latter attempted to escape from the heat (they were kept in identical conditions prior to this).

DAILY ACTIVITY CYCLE. In the spring and fall, long-legged thin-toed geckos can be observed sunning themselves. On 6 April 1977, the temperature at night was {p. 158} 8-11°, but they would appear in the morning, in sunlit places in their cracks. In the fall, when the sun had warmed the rocks, the geckos began to emerge on the surface after 1400 hrs. Overall, this is a crepuscular animal. In the spring, active lizards were noted from 2200 until 0100 hrs. In the fall, the times were from 1900 until 2300 hrs. On moonlit nights, they take cover in the shadows.

SEASONAL ACTIVITY CYCLE. We encountered the first individuals near winter shelters on 4 April 1977, and the last ones on 7 Octobrer 1981 (Akar-Cheshme). It would appear that they are active from the end of March or beginning of April through the middle of October, and in some years possibly later, until November.

SHEDDING. In the field, shedding individuals were encountered on 13 May 1976 and 24 April 1984. In the terrarium, these lizards shed on 8-14 April and 13-23 May.

FEEDING. Analysis of the gastro-intestinal tract contents of 25 lizards, caught in April in southern Turkmenia, showed (Table 6) a predominance of insects in their diet: butterflies and their larvae (100%), flies (24%), tiger beetles and snout beetles (20% each), as well as spiders (16%). From among other invertebrates, they eat, in small numbers (4-8%), solpugids, scorpions and ticks. Moths predominate in the long-legged thin-toed geckos' diet, which distinguishes them from other species.

TABLE 6. Analysis of stomach contents of *T[enuidactylus] longipes microlepis* from Turkmenia April-May (n = 25)

Food Items	Frequency		Number Eaten	
	Total	%	Total	%
Crustacea				
Isopoda	1	4	1	0.9
Arachnoidea				
Solpugidae	1	4	1	0.9
Scorpionidae	2	8	2	1.8
Araneinae	4	16	5	4.5
Acarina				
Ixodidae	1	4	1	0.9
Hylopoda	1	4	1	0.9
Insecta				
Orthoptera	2	8	2	1.8
Coleoptera	4	16	10	9
Carabidae	5	20	6	5.4
Staphylinidae	1	4	2	1.8
Dermestidae	1	4	1	0.9
Curculionidae	5	20	6	5.4
Neuroptera				
Mirmeleonidae	1	4	1	0.9
Plasmodea	1	4	1	0.9
Diptera	6	24	18	17.5
Lepidoptera				
Caterpillar	15	60	24	26.1
Imago	10	40	21	20.4

REPRODUCTION. The sex ratio in our sample is close to 1:1. These geckos reach sexual maturity when the body length is 45 mm and greater. Mating occurs, probably, shortly after the lizards emerge from hibernation in April because on 12 May 1978 well-developed eggs were clearly visible in the majority of the females. They lay one or two (more frequently) eggs at the end of May. Judging by the shells of old eggs found in rock cracks in Zulfagar locality, their diameter is 10 mm. The times of hatching have not been determined.

GROWTH RATE has not been studied. The smallest dimensions of hatchlings of the current year available to us were L = 27.1 mm, LCD = 28.4 mm. Because of similarity to closely related species, it can be supposed that they reach sexual maturity at the age of 15-16 months, that is by the middle of their second year of life. The maximum size is reached by the age of four or five.

ENEMIES. There are no concrete data. It is possible that they are hunted by hawks {*Falco tinnunculus* – MG}, which nest on the same outcrops inhabited by the geckos.

BEHAVIORAL ATTRIBUTES. Judging from observations in nature and in the terrarium, long-legged thin-toed geckos move and hunt like the Caspian species, although the long legs give them a chance to protect against overheating; {p. 159} these lizards raise their bodies higher when moving or stopping on a rock heated by the sun. They are found only on vertical surfaces. Their summer and winter shelters are located in cracks in the rocks, where they also lay their eggs. When caught, they emit quiet squeaks. Mating behavior has not been observed.

PRACTICAL IMPORTANCE AND PROTECTION. As a rare species in the U.S.S.R. fauna having a limited distribution, this gecko is protected by law and has been included in the *Red Books* of the U.S.S.R. and T[urkmen] S.S.R. It is protected in the Badghyz Preserve. Introduction of this species in compatible areas of this preserve is possible.

Subgenus *Mediodactylus* Szczerbak et Golubev, 1977

Type species: *Gymnodactylus kotschyi* Steindachner, 1870.
1977 – *Mediodactylus* Szczerbak and Golubev, Herpetol. Sborn. Tr. ZIN AN SSSR, vol. 74, Leningrad, p. 130.

DIAGNOSIS. No more than 10 preanal pores in males; if suture between scales of first pair of postmentals present, then its length usually is equal to no more than half length of mental scale; dorsal tubercles oval or roundish; spiny caudal tubercles intermixed with homogeneous scales and do not contact each other in semicircles; row of subcaudal plates does not occupy entire ventral surface of tail (height of one such scale is less than half its width).

ADDITIONAL CHARACTERS. Height of first supralabial scale from nostril to edge of mouth distinctly smaller than its width along edge of mouth; conical growths well developed in upper posterior part of orbit; supracaudal scales homogeneous, no additional tubercles at bases of regular ones. Tail slightly swollen, row of subcaudal plates may be replaced by two rows or just regular scales. Main color and pattern usually gray.

This subgenus includes six to seven species distributed in the Mediterranean region; it penetrates Middle Asia.

{p. 160} *Key to the Species of the Subgenus Mediodactylus*

1 (2). No dorsal tubercles . . . . *T. amictopholis* (Hoofien)
2 (1). Dorsal tubercles present 3
3 (4). Dorsal tubercles roundish, smooth . . . . *T. spinicauda* (Strauch)
4 (3). Dorsal tubercles oval or oval-triangular, keeled 5
5 (6). Diameter of ear opening in adults more than half the longitudinal
    diameter of eye . . . . *T. russowi* (Strauch)
6 (5). Diameter of ear opening less than half the eye diameter 7
7 (8). All scales and plates of tail (except tubercles) smooth; upper front
    surfaces of arms smooth or with weakly defined tracks of keels . . . .
    . . . . *T. kotschyi danilewski* (Steindachner)
8 (7). All scales and plates of tail, as well as upper front surfaces of arms,
    distinctly keeled 9
9 (10). No more than 90 GVA; enlarged scales (two pairs on each segment)
    along [ventral] midline of unregenerated tail . . . . *T. sagittifer* (Nik.)
10 (9). No fewer than 90 GVA; no noticeably enlarged scales on ventral
    tail surface . . . . *T. heterocercus* (Blanford)

MEDITERRANEAN THIN-TOED GECKO — *TENUIDACTYLUS KOTSCHYI* (STEINDACHNER, 1870)

Type locality (Mertens and L. Müller, 1928): Siros Island, Cycladen.
Karyotype: 2n = 42, 42 acrocentrics, NF = 42.
1870 – *Gymnodactylus kotschyi* Steindachner, Sber. Akad. Wiss. Wien. math-naturwiss. Kl., pt. 1, vol. 62, p. 329.
1981 – *Cyrtodactylus kotschyi*, Beutler, Handb. Amph. Rept. Europ., vol. 1, 1, p. 53.
1984 – *Tenuidactylus kotschyi*, Szczerbak and Golubev, Vestn. Zool., Kiev, no. 2, p. 54.

DIAGNOSIS. Shape and size of dorsal tubercles are highly variable; they can be keeled or smooth, oval, heart-shaped, or elongated. Ear opening diameter less than half (0.3) of eye diameter. No less than 90 GVA. Upper front surfaces of arms smooth or keeled.

LECTOTYPE. Male, no.10868, Insel Syra (= Siros) and 7 paratypes labelled as "Sira, 1866, CD 8a, don. Steind." are in NMW. The locality in description [294] was mentioned as "Goree, Senegal," but Siros Island was also mentioned there. Later, the type locality was restricted to Siros Island [260; 314].

DESCRIPTION (from 231 specimens [183]). Male max. L = 51 mm ($\overline{X}$ = 42.6 mm); female max. L = 56 mm ($\overline{X}$ = 44.4 mm). L/LCD = 0.71-1.0. Three to four, rarely two or more than four supranasals; 8-10 supralabials, 6-8 infralabials. Head covered above with granular scales intermixed with separate or groups of enlarged, slightly keeled scales. Dorsum covered by convex scales and 8-14 longitudinal rows of enlarged tubercles. Dorsal tubercles separated from each other by one to three granular scales in longitudinal row and by two to four in transverse row. Abdominal scales large, flat, with serrated posterior edge, in 20-40 longitudinal rows. Males with two to nine preanal pores or without them. Usually one to three postanal tubercles from each side behind vent. Unregenerated tail covered above by small scales, segmented, with four to six enlarged spiny tubercles form semicircles on each segment. Transverse rings on tail, 17-24. Scalation of lower tail surface very variable. Regenerated tail covered by homogeneous or cycloid scales, or irregular scales in some forms. Often a lateral skin fold can be distinguished.

COLOR AND PATTERN. The main background varies from yellowish, orangish or reddish (*T. k. stepaneki, T. k. adelphiensis*), gray to dark brown and almost black. There are usually six or seven dark M-shaped bands with the angle pointing toward the tail. From the end of the snout to the ear drum runs a small, regular, dark stripe. The intensity of color and pattern in the same individual can change, depending on the animal's physiological condition. The abdomin is {p. 161} white or grayish, slightly yellowish or orange. The cloacal area is often spotted.

DISTRIBUTION. Southern Italy, Balkan Peninsula, Aegean Islands, Crete, Cyprus, southern Crimea, Asia Minor, southwestern Transcaucasus region, Syria and Palestine (map and legend are given only for the form from the U.S.S.R. fauna, *T. k. danilewskii*, see below).

SYSTEMATICS AND GEOGRAPHIC VARIATION. Exceptionally variable species. According to A. Beutler's data [183], maximum body size (56 mm) is found in specimens from the southern Cyclades, whereas South Aegean forms barely reach 40 mm. Regenerated tails with irregular subcaudal plates were found in geckos from Greece, Italy, and Macedonia, and with small scales, from Crete, Bulgaria, and Crimea. Lower surface of unregenerated tail also varies significantly. The forms with divided subcaudal plates or with only cycloid scales (Crete, Bulgaria, Crimea) occur along with the forms with one row of enlarged scales. Specimens from Crimea, Bulgaria, and southeastern Aegean Islands have one postanal tubercle behind the vent, whereas [among specimens from elsewhere] there can be two or three [tubercles]. Adult males from the northern Sporades, Crete, and Siros do not have preanal pores at all. Males from the southeastern Aegeas have only two preanal pores, males from Crimea and Bulgaria – six to nine, males from other parts of range – three to five. Most forms have 26-32 longitudinal abdominal scale rows, the lower number of these scales are in specimens from Crete, southeastern Aegeas, Yugoslavia, and mainland Greece. The populations from the northern Sporades have 34-38 longitudinal rows.

Usually, dorsal tubercles form 12 longitudinal rows; however, lizards from the southeastern Aegean, eastern Crete, Yugoslavia, and Siros Island have, as a rule, only 8-10. The size of the dorsal tubercles in the forms from the mainland and northern Cyclades equals four to five surrounding scales; specimens from northern Sporades, southern Cyclades, and Crete have significantly smaller tubercles; tubercles reach the size of 1.5-2.5 surrounding scales in forms from southeastern Aegeas and western Crete, [and] they are smooth and hard to distinguish. The same situation is noted with the scalation on the upper and lower surfaces of the thighs; specimens of *T. k. solerii, T. k. bartoni*, and lizards from Gavdos and southeastern Aegeas lack these tubercles. Especially well developed tubercles on upper thighs are found in *T. k. fuchsi, T. k. schultzewestrumi*, and *T. k. tinesis*. Usually, there are six longitudinal rows of tubercles on the upper surface of the base of the tail; but *T. k. schultzewestrumi* and the specimens from the southeastern Aegeas more often have seven to nine [165, 184, 296].

Body color also varies geographically. Combinations of the above mentioned characters allows for the grouping of various populations into forms and groups of forms. Many forms of Mediterranean gecko were described at different times but only 25 [183] are currently recognized as valid. Their diagnoses, which can be used to define separate specimens and samples, are given below. Owing to the great variation, it is not yet possible to create reliable keys to the subspecies. Differentiation of some forms is often based on insignificant characters and on a limited number of specimens.

I group "*kotschyi*"

T. k. kotschyi (Steindachner, 1870): No preanal pores in males, up to 33 longitudinal rows of abdominal scales; 12 [rows of] dorsal tubercles; 6 [rows of] tail tubercles; three separately situated tubercles on upper thigh surface (Siros Archipelago).

T. k. fuchsi (Beutler et Gruber, 1977) (= *C. k. gruberi*): Closely related to *T. k. kotschyi* but 34 to 40 longitudinal rows of [ventral] abdominal scales, six separately situated tubercles on upper thigh surface (N chain of northern Sporades Islands).

T. k. schultzewestrumi (Beutler et Gruber, 1977): Closely related to *T. k. kotschyi*, but tail tubercles arranged in 7 to 8 longitudinal rows, six separately situated tubercles on upper thigh surface (Valaxa Island, turns into *T. k. tinensis* on the other islands of Siros Archipelago).

{p. 162} *T. k. tinensis* (Beutler et Frör, 1980): Three to five preanal pores in males; upper thigh tubercles four times wider, and dorsal ones four to five times wider than granular dorsal scales; six separately situated tubercles on upper thigh surface, 12 longitudinal rows of dorsal tubercles, 30 ventrals (northern Cyclades, turns into *T. k. saronicus* on Seriphos, Mykonos, Naxos and Ikaria Islands).

T. k. saronicus (Werner, 1939): Closely related to *T. k. tinensis* but with separately situated tubercles on upper thigh surface (western and central Cyclades, Saron Islands).

T. k. buchholzi (Beutler et Gruber, 1977): Closely related to *T. k. saronicus* but with 10 longitudinal dorsal tubercles (Siphnos Archipelago).

T. k. solerii (Wettstein, 1937): Closely related to *T. k. saronicus* but dorsal and upper thigh tubercles three times longer than granular dorsal tubercles (southern Cyclades, Astipalaia; turns into *T. k. saronicus* on southwestern Cyclades). *T. k. christinae* is connected via a zone of intergradation (Buchholz, 1955, nomen nudum).

T. k. rumelicus (L. Müller, 1939): Closely related to *T. k. saronicus* but subcaudal plates on regenerated tail divided (southwestern Bulgaria).

T. k. bibroni (Beutler et Gruber, 1977): Closely related to *T. k. saronicus* but 24 longitudinal abdominal scales (Apulia, south of Balkan Peninsula, Kithira and Kithnos Islands).

T. k. skopjensis (Karaman, 1965): As *T. k. saronicus* but dorsal tubercles in 10 longitudinal rows (Northern Macedonia).

II group "*danilewskii*"

T. k. danilewskii (Strauch, 1887) (= *C. k. bureschi* (Stepanek, 1937) (Crimea, Bulgaria, Frakia): Detailed description is given below. The single form representing this species in the U.S.S.R. fauna.

Subspecies *T. k. steindachneri* (Stepanek, 1937) from southwestern Anatolia and *T. k. lycaonicus* (Mertens, 1952) from Anatolia synonymized recently [181] with *T. k. danilewskii*.

T. k. colchicus (Nik., 1902): 12 longitudinal rows of dorsal tubercles; 26-29 longitudinal rows of ventral scales [across abdomen]; 3-4 longitudinal scale rows below tail; 2-4 preanal pores. Belly yellowish, lower tail surface orange (probably, in juveniles) (northeastern Anatolia. Finds in Adjaria, U.S.S.R. are possible).

T. k. syriacus (Stepanek, 1937): Large keeled dorsal tubercles in 10-14 (more often 12) longitudinal rows; 4-5 preanal pores (4.5 average), 24-26 ventral scales across midbody, lower surface of unregenerated tail covered below by 2 scales close to vent and one row of enlarged scales in other 3/4 of tail (northern Syria).

T. k. orientalis (Stepanek, 1937): Dorsal tubercles in 10 somewhat irregular longitudinal rows, keeled, heart-shaped; 26-30 ventral scales across midbody (Syria, Lebanon, Palestine).

T. k. fitzingeri (Stepanek, 1937): 10-12 longitudinal rows of strongly keeled and elongated dorsal tubercles; 20-25 ventral scales across midbody. Two to three preanal pores or none. 1-1 postanal tubercles. Unregenerated tail covered by two rows of cycloid scales, regenerated tail – by small scales (Cyprus).

T. k. karabagi (Baran et Gruber, 1982): 12 or more longitudinal rows of tubercles, 30-34 (32 average) ventral scales across midbody, 3-4 preanal pores, 2-2 postanal tubercles (Fener Island in Marmara Sea).

T. k. beutleri (Baran et Gruber, 1981): 10-12 longitudinal rows of dorsal tubercles, 25-33 (30 average) ventral scales across midbody, 1-6 (average 3.8) preanal pores, 2-2 postanal tubercles (southwestern Turkey and adjacent islands in Aegean Sea).

T. k. ciliciensis (Baran et Gruber, 1982): Keeled scales on head (flat in *T. k. danilewskii*), 10-12 longitudinal rows of dorsal tubercles, 23-26 ventral scales across midbody (23.8 average), 2 preanal pores, 1-1 postanal tubercles, {p. 163} 14-18 (average 15.5) subdigital lamellae, (18-24 in *T. k. danilewskii*) (southern coast of Mediterranean Sea in Turkey).

T. k. ponticus (Baran et Gruber, 1982): 10-14 (average 12) longitudinal rows of dorsal tubercles, 26-31 (average 29.2) ventral scales across midbody; 2-5 (average 3.5) preanal pores; 1-2 postanal tubercles; 17-21 (average 19.9) subdigital lamellae (known from Tokat, Amasia, and Sivas in northern and central Turkey).

III group *"bartoni"*

T. k. bartoni (Stepanek, 1934): Dorsal tubercles twice as long as granular dorsal scales (western Crete).

T. k. wettsteini (Stepanek, 1937) (= *G. k. rarus* Wettstein, 1952, = *G. k. stubbei* Wettstein, 1952): Dorsal tubercles three times longer than granular dorsal scales (coast of eastern Crete).

IV group *"oertzeni"*

T. k. kalypsae (Stepanek, 1939): Six longitudinal rows [of tubercles] on tail base, 26-29 longitudinal rows of ventral tubercles (Gavdos Island near Crete).

T. k. oertzeni (Boettger, 1888) (= *G. k. unicolor* Wettstein, 1937): 7-8 longitudinal rows of tail tubercles, 23-27 ventral scales across midbody, gray background color (Karpathos Archipelago).

T. k. stepaneki (Wettstein, 1937): Closely related to *T. k. oertzeni*, but red background color (Megali Sofranow Island near Sirna).

T. k. adelphiensis (Beutler et Gruber, 1978): Closely related to *T. k. stepaneki* but more than 28 longitudinal rows of tubercles across abdomen (coast of Sirna Island).

DANILEWSKI'S MEDITERRANEAN THIN-TOED GECKO — *TENUIDACTYLUS KOTSCHYI DANILEWSKII* (STRAUCH, 1887)

Type locality: Yalta, Crimea.

1887 – *Gymnodactylus danilewskii* Strauch, Mem. Acad. Sci. St. Petersbourg, ser. 7, vol. 35, no. 2, p. 48.
1940 – *Gymnodactylus kotschyi danilewskii*, Mertens and Müller, Abh. Senckenberg. naturf. Ges., Frankfurt am Main, vol. 451, p. 23.
1966 – *Gymnodactylus kotschyi bureschi*, Szczerbak, Herpetologica Taurica, Kiev, p. 80.
1981 – *Cyrtodactylus kotschyi danilewskii*, Beutler, Handb. Amph. Rept. Europ. vol. 1, 1, p. 63.
1982 – *Cyrtodactylus kotschyi steindachneri, C. kotschyi lycaonicus*, Baran and Gruber, Spixiana, Munich, vol. 5, 2, p. 132.
1983 – *Tenuidactylus kotschyi danilewskii*, Szczerbak and Golubev, Vestn. Zool., Kiev, no. 2, p. 54.

DIAGNOSIS. 10 to 13 (more often, 12) longitudinal rows of oval keeled dorsal tubercles, 22-30 ventrals across midbody; 2-15 preanal pores, one postanal tubercle on each side; all dorsal and caudal scales (except for dorsal tubercles) flat; upper front surfaces of limbs smooth or with barely noticeable keels.

LECTOTYPE. In Zoological Institute, Academy of Sciences of USSR, Leningrad ZIL 3688. "Crimea: Yalta, 1868, Danilewsky."

DEFINITION (Fig. 70) (from 75 specimens in ZIK from Crimea). L adult male (n = 25) 32.9-43.0 (38.68 ± 0.64) mm; L adult female (n = 30) 31.5-45.0 (39.79 ± 0.78) mm; L/LCD (n = 20) 0.86-0.98 (0.91 ± 0.0018); HHW (n = 29) 52-74 (54.41 ± 0.75); EED (n = 32) 23-48 (34.2 ± 1.00); supralabial scales (n = 146) 5-8 (6.15 ± 0.06); infralabials (n = 148) 7-10 (7.87 ± 0.08); scales across head (n = 75) 12-19 (14.41 ± 0.17); DTL (n = 70) 14-25 (18.67 ± 0.26); scales around dorsal tubercle (n = 73) 9-15 (10.85 ± 0.14); GVA (n = 48) 89-112 (100.75 ± 0.81); ventral scales across midbody (n = 54) 22-30 (25.67 ± 0.27); subdigital lamellae (n = 138) 14-21 (18.93 ± 0.13), up to 24 in Asia Minor [181]; preanal pores (n = 27) 2-9 (6.44 ± 0.31) (6-15, average 8, in Asia Minor and adjacent islands [181]). Mental scale triangular (73.6%) or pentagonal {p. 164} (26.4%); usually three pairs of postmentals (96.4%), very rarely two (1.2%) or four (2.4%) pairs; scales in first pair contact each other (96.2%) or separated by a scale (3.6%); three nasal scales; first nasal scales contact each other (34.3%) or separated by one (60.0%) or two (5.7%) scales; additional dorsal tubercles almost always absent, defined in 6.8% and contact regular {p. 165} ones. Subcaudal scales well developed, smooth, make combinations 1-1 or 1-2 on single segment; other tail scales, except tail tubercles, smooth; regenerated tail covered by small cycloid scales; on upper front surfaces of forelimbs feebly keeled scales may appear, these keels may be better developed on thighs.

COLOR AND PATTERN. The main background color on the upper surfaces is ash or sandy

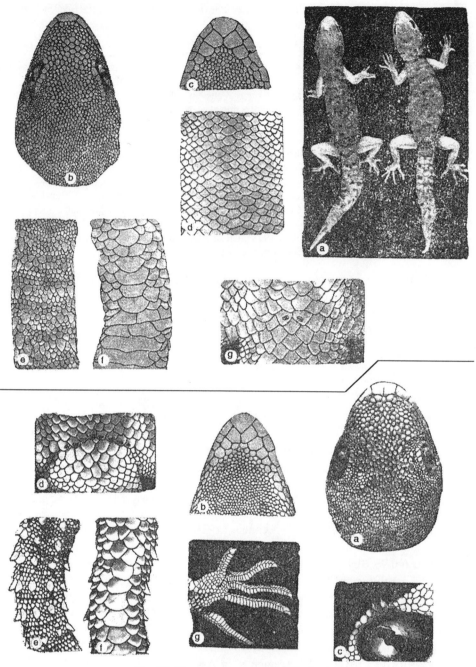

Fig. 70. *Tenuidactylus amictopholis* (top):
a, general view from above; b, head from above; c, chin; d, belly; e, tail from above;
f, tail from below; g, preanal pores (TAU R 10351, Herman Mountain, Lebanon).
Tenuidactylus kotschyi danilewskii (bottom):
a, head from above; b, chin; c, upper edge of orbit; d, preanal pores; e, tail from above;
f, tail from below; g, hindfoot from below (ZIK SR 794, Khersones, Crimea).

gray. The body's upper surface has 5-8, usually 6, dark M-shaped bands that look like an angle pointed to the rear in the middle of the dorsum. There are 10-12 such bands on the tail, and in juveniles they have an ochrish or light orange piping. The sides of the abdomen are light yellowish green or dirty white. The underside of the tail is reddish ochre or orange, brighter in juveniles. The regenerated tail is gray-bluish or gray brown. When the regeneration is only begining (to 0.5 cm), [the stub] is almost black.

DISTRIBUTION (Fig. 71). Southern coast of Crimea (not found above 200 m elevation) from Sevastopol to Alushta, Black Sea coast of Bulgaria south of Varna, and western and southwestern Turkey.

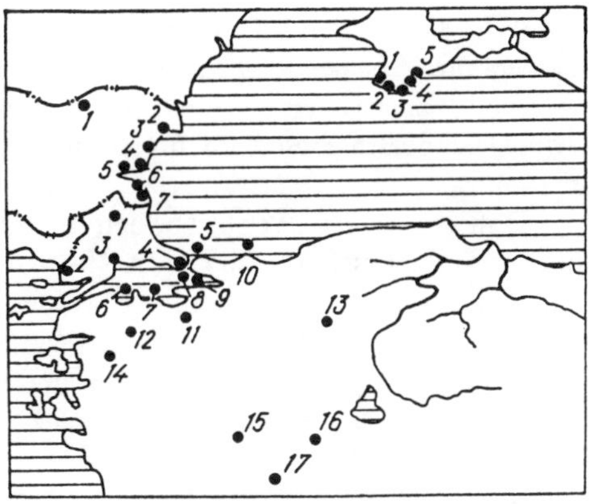

LEGEND to Fig. 71.
U.S.S.R.: 1 - Crimea: Sevastopol, ruins of Khersones (ZIK); 2 - Batiliman; 3 - Symeiz; 4 - Yalta; Gurzuf; 5 - Karabakh; Alushta (2-5 [141]).
Bulgaria: 1 - Ruse [317]; 2 - Varna [66]; 3 - Kamchiya [66]; 4 - Nesebyr [66]; 5 - Burghas [66]; 6 - Sozopol; Svyatoj Ivan Island [207]; 7 - near Primorsko, Maslen Nos Cape [181].

FIG. 71. Distribution of *Tenuidactylus kotschyi danilewskii.*

Turkey: 1 - Kirklareli [180{error in original – MG}]; 2 - Enez [181]; 3 - Tekirdag [181]; 4 - Istanbul [181]; 5 - Ezek Adazi Island [181]; 6 - Pashalimani Island [179]; 7 - Imrali Island [179]; 8 - Prinz Islands [179]; 9 - Hayirsiz Island [179]; 10 - Kefken Island [179]; 11 - Bursa {error in original – MG} [180{error in original – MG}]; 12 - Balikesir: 50 km SE Bursa [181]; 13 - Ankara [180{error in original – MG}]; 14 - Savashtepe [180{error in original – MG}]; 15 - Egridir [180{error in original – MG}]; 16 - Konya [180{error in original – MG}]; 17 - Akseki [181].

ECOLOGY was first studied by one of the authors [139, 141].

HABITAT AND QUANTITATIVE DATA. It inhabits the ruins of ancient buildings, modern residential and utility buildings, both stone and wooden ones, the stonework of road supports, stone fences, and rocks, and it is occasionally found under the bark of old juniper trees and stumps near ruins and buildings (Fig. 72). According to our observatons, in areas where building maintenance (repairs, restorations, plastering and painting of walls, etc.) is not done, the numbers of these geckos has been stable for the last 25 years. In the areas where they are found, the distribution is not uniform. The highest density is seen in Khersones, near the lights on building walls, where there are three to four individuals per m^2, but among the ancient ruins in the Khersones Museum, during the period of 1958-1981, there was one individual per 100-1600 m^2.

RESPONSE TO TEMPERATURE. In the summer in the Crimea, the geckos were not encountered at air temperatures below 18°. In April-May, they were noted at air temperatures

FIG. 72. Biotope of *Tenuidactylus kotschyi danilewskii* (ruins of Khersones town, Crimea, Ukraine).

of 14.5°, 13° and even 12°. A rise of 1° in body temperature, in comparison with the ambient temperature, has been noted in a moving lizard. The largest numbers of geckos were observed at an air temperature of 26°. Research in a thermal gradient apparatus showed the optimal temperature range for this species to be 34-40°. Eggs were incubated at temperatures of 18-35°.

DAILY ACTIVITY CYCLE. The activity is nocturnal. This species is active in the evening and during the first half of the night. The first individuals emerge {p. 166} from cover before sunset (in the shady places) at 1910 hrs. They come out in large numbers around 2100 hrs, and the highest numbers were noted by 2200 hrs. By midnight they already begin to disappear except for the individuals around street lights. The latter leave for cover at dawn, when the lights are turned off. In the early spring and late fall, basking geckos were also seen in the daytime.

SEASONAL ACTIVITY CYCLE. The earliest times of appearance after hibernation are on 24 April 1959 (Batiliman) and 25 March 1960 (Khersones). In Khersones, the last times the lizards were seen on the surface were [on] 3 November 1958, [in] mid-October 1959, and [at] the end of September 1960.

SHEDDING. Shedding *en masse* has not been observed. Shedding individuals were encountered on 27 May 1958 and 8 July 1960. In captivity, the geckos shed on 24 July and 15 August 1961.

FEEDING. The contents of 75 stomachs were analyzed [141]. Among the preferred diet items are spiders (38.8% frequency of occurrence), butterflies (45.3 %), flies (36.9%), hymenopterans (32.7%), and beetles (28.0%). In addition, insignificant numbers of orthopterans, homopterans, neuropterans, and isopods were found.

REPRODUCTION. The sex ratio in our samples is close to 1:1. The maximum body size of a mature male is 36.5 mm and 41 mm for a gravid female. Mating occurs in May. Egg laying was observed on 1 and 7 June 1961, 17 and 20 July 1960, 25, 26, and 27 July 1963 (Khersones). Fully formed eggs were found in oviducts on 27 May 1958, 7 and 8 July 1960, and 15 May 1961 (Khersones). One or two eggs are laid with the dimensions of 5.8-8.1 × 7.5-9.7 mm. The eggs are oval in shape, white with a pinkish hue. Eggs are laid only once but over an extended period. The hatchlings appear in July or August (newly hatched individuals were caught on 2, 4 and 22 August 1958). In the laboratory, incubation lasted 50 days, and hatching was noted on 11 and 12 September 1963. The laying of eggs and hatching occur at night.

GROWTH RATE. The dimensions of the hatchlings are: L = 17.8-18.5 mm; LCD = 16-16.1 mm. Marking them has demonstrated that the most intensive growth is in the first year of life. In 8.5 months from the time of hatching (five of these in hibernation), the body length in these lizards increased from 18 to 23.5 mm. In two years the gecko reaches the size of an adult lizard – 39 mm. In another year it is 41.8 mm. The maximum size (L female = 50.5 mm, L male = 43.5 mm) is reached at the age of five.

ENEMIES. According to the observations in Crimea, they are attacked by domestic cats. Owls {*Athene noctua* – MG}, common hedgehog {*Erinaceus europaeus* – MG}, and rock marten {*Martes foina* – MG} inhabit the same biotopes and could be among the possible predators on this lizard.

{p. 167} BEHAVIORAL ATTRIBUTES. A vertical position is the norm for this species even when resting, and it moves well on vertical and ceiling surfaces. It moves about in short running bursts with short rests. In the places frequented by people, the geckos become easily frightened and run away when they hear approaching footsteps. They usually hunt by sitting still and waiting for prey. Near street lights, they pursue insects that bump against the wall several times, when blinded by the light. If one lizard catches something, the others squeak excitedly, scratch the snout with a hindlimb and lick the eyes. Sometimes, fights occur in the places where large numbers of them congregate. An excited gecko holds its body still and moves its tail in a snake-like fashion, squeaks, threateningly runs at the opponent and chases it off. Frequently, the opponent is another mature male. When in danger or being chased, the geckos jump off the wall. If they fall in the grass, they remain still awhile and then climb onto the wall again. If a gecko falls on uncovered soil, it quickly runs off in search of a suitable crack.

In the summer, the geckos use cracks, stone piles, and spaces under house roofs and behind signs on buildings for cover. Usually, these are found 2 to 3 m above ground. Winter shelters are in deep cracks in the walls of ruins and residential buildings. They have even been found in occupied apartments and in sheds.

Mating has not been observed. Excited geckos emit a characteristic sound, which can be expressed as "peek...peek...peek...k...k...k...." In places where they are common, these quiet but clear sounds can often be heard at night. Sometimes the geckos' squeaks can be heard coming from their shelters in the daytime.

In the main, the Mediterranean thin-toed geckos exist today as a synanthropic species. They find shelter in human buildings, adapting even to the light and heat of street lights, where they hunt for insects and look for warmth.

PRACTICAL IMPORTANCE AND PROTECTION. The Mediterranean thin-toed gecko is a species of interest to science as a sort of monument to Crimean nature. It is rare and should be protected by whatever measures necessary. Currently, it is protected in the Khersones Archeological Preserve. It has been included in the *Red Books* of the U.S.S.R. and Ukrainian S.S.R. We have conducted an experiment on attracting the geckos to artificial shelters [141]. This measure should be used especially in the places where extensive repair and restoration work is being performed. In order to increase the numbers and improve the protection of Mediterranean thin-toed geckos, they should be released in the Karadagh Preserve in Crimea.

GRAY THIN-TOED GECKO — *TENUIDACTYLUS RUSSOWI* (STRAUCH, 1887)

Type locality: ruins of old fortress at Novo-Aleksandrovskoye[1].
Karyotype: 2n = 44, acrocentrics 44, NF = 44 {error in original – MG}.
1887 – *Gymnodactylus russowi* Strauch, Mem. Acad. Sci. Petersbourg, ser. 7, vol. 35, no. 2, p. 49.
1899 – *Gymnodactylus zarudnyi* Nikolsky, Ann. Zool. Mus. Imp. Acad. Sci., vol. 4, p. 385.
1954 – *Cyrtodactylus russowi*, Underwood, Proc. Zool. Soc. London, vol. 124, no. 3, p. 475.
1984 – *Tenuidactylus russowi*, Szczerbak and Golubev, Vestn. Zool., Kiev, no. 2, p. 54.

DIAGNOSIS. Dorsal tubercles oval or oval-triangular, keeled; diameter of ear opening in adults more than half of eye diameter; 90 GVA; scales of tail (except tubercles) and limbs smooth.

LECTOTYPE. ZIL 3658, adult male. "Novo-Alexandrowsk, 1842, Lehmann." L = 44.3 mm; LCD = 52.9 mm; L/LCD = 0.84; 19 scales across head; 9 supralabial scales; 7 infralabials; 110 GVA; 29 ventral scales across midbody; three pairs of postmentals, the scales of first pair contact each other; four preanal pores; 15 subdigital lamellae.

{p. 168} DEFINITION (Fig. 73) (from 285 specimens in ZIK, ZIL, ZMMSU). L adult male (n = 104) 25.1-49.6 (39.73 ± 0.54) mm; L adult female (n = 112) 25.5-53.2 (42.0 ± 0.61); mass up to 3 g; L/LCD (n = 126) 0.69-1.02 (0.869 ± 0.01); HHW (n = 100) 53-66 (57.65 ± 0.27); EED (n = 79) 36-81 (56.60 ± 0.50); supralabial scales (n = 568) 8-11 (9.78 ± 0.03); infralabials (n = 570) 6-10 (7.89 ± 0.03); scales across head (n = 200) 15-24 (18.36 ± 0.10); DTL (n = 100) 14-28 (19.93 ± 0.22); scales around dorsal tubercle (n = 190) 11-18 (13.92 ± 0.08); GVA (n = 204) 94-130 {p. 169} (113.33 ± 0.47); ventral scales across midbody (n = 198) 23-34 (28.81 ± 0.16); subdigital lamellae (n = 554) 15-24 (19.67 ± 0.07); preanal pores (n = 119) 2-6 (3.16 ± 0.09). Mental scale triangular or pentagonal, one (10%), two (55.7%) or three (34.3%) pairs, scales of first pair contact each other (71.7%) or separated by one or two scales (28.3%); first nasal scales separated from each other by one (22.3%), two (65.0%) or three (12.7%) scales; three nasal scales; additional dorsal tubercles present (71.0%) or absent (29.0%); more or less well developed smooth subcaudal plates in

---

[1] An erroneous viewpoint is established in the literature [17] that the type locality is Fort Shevchenko (formerly Alexandrovsk). Actually, the gray thin-toed gecko was described from the ruins of the old fortress of Novo-Aleksandrovskoye, situated 30 km to the east.

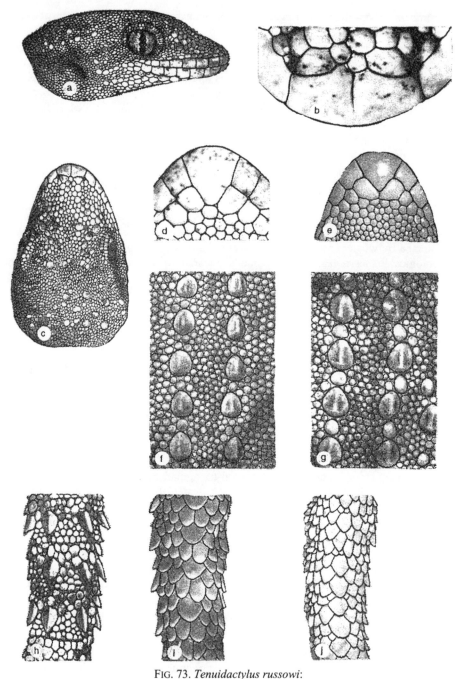

FIG. 73. *Tenuidactylus russowi*:
a, head from the side; b, nasal scales; c, head from above; d-e, chin; f-g, dorsum;
h, tail from above; i-j, tail from below (d, f, j – ZIL 9334, Sistan, Iran
{*T. russowi zarudnyi* – MG}; others – ZIK SR 971, Kirghizia).

one segment form combination 1/1 (very rarely 1/2); other tail scales (except tubercles) and limbs smooth (sometimes, separate scales of forearm may have feeble keels).

COLOR AND PATTERN. It is ashen gray or brownish gray above with narrow, dark M-shaped or transverse bands. The upper surface of the head is covered with small dark spots. On the sides of the head there is a dark longitudinal stripe, frequently with light edges, which continues onto the sides of the neck and, sometimes, onto the body. The gray thin-toed geckos change their color and pattern. Under optimal conditions, the pattern is clearly contrasted and the main background is lighter. In unfavorable conditions, their color darkens and the pattern disappears. It has also been noticed that the populations living on saxaul {*Haloxylon* sp. – MG} trunks acquire a complicated pattern (Fig. 22, color plate 3) made up of longitudinal elements, which makes them nearly unnoticable on the woody background. Individuals that inhabit buildings and loess cliffs often lose their pattern completely or possess only transverse bands on the body and tail.

DISTRIBUTION. The main range is situated east of the Caspian Sea in Middle Asia and Kazakhstan to northwestern China; to the south it occurs in northeastern and eastern Iran. [There is] one known find in the Precaucasus Region (Starogladkovskaya stanitsa {sort of large Cossack village – MG}, Checheno-Ingushetiya) (Fig. 74).

LEGEND to Fig. 74.
Kazakhstan: Paskevich Bay near Turangly Cape (ZIL); 2 - Djarbulak Kishlak [98]; 3 - Kos-Kazakh locality [98]; 4 - Chegan River (ZIL); 5 - Tyup-Karaghan Peninsula [98]; 6 - Komsomolets Bay [98]; 7 - Asmatay-Matay Shor, Ustyurt [98]; 8 - Barsakelmes Island [98]; 9 - Shevchenko Fort; Taushick; N Karatau Ridge [98]; 10 - Novo-Alexandrovsk [98]; Beke locality (ZIL); 11 - between Kungrad and Beyneu, Ustyurt (our data); 12 - Kugharal [98]; 13 - Aryskum Sands [98]; 14 - Kzyl-Orda (ZIL); 15 - Baygakum Station [98]; 16 - left bank of Chu River [98]; 17 - Karatau Ridge [98]; 18 - Chimkent [98]; 19 - Khumsan, Brichmolla [23]; 20 - ruins of old fortress Chirik-Robat (ZIK); 21 - SW shore of Lake Balkhash: near Buru-Baytal (ZIL); 22 - S Prebalkhash Region: between Ili and Karatal Rivers [137]; 23 - divide between Ili and Karatal Rivers [98]; 24 - Bakanas Settlement (ZIL); 25 - foothills of Djunghar Alatau [98]; 26 - Panfilov Settlement [98] {Djarkent – MG}; 27 - Lake Alakol (ZIK); 28 - Djunghar Gate, neal Lake Djalanash-kul [98]; 29 - divide between Charyn and Chilik Rivers [98]; 30 - the frontier, Ili River valley [98]; 31 - Tasaral village [17]; 32 - Zheltau [83; 328]; 33 - Koy-Kara, Iman-Kara, Saryniyaz, Ak-Tologhay [328].

Uzbekistan: 1 - chink of Ustyurt near Lake Sudochye (ZIK); 2 - Chabankazghan (ZIK); 3 - Urghench [23]; 4 - Khazarasp [23]; 5 - Angren River, Toy-Tyube (ZIL); 6 - Tashkent (ZIL); 7 - Chatkal Ridge [17]; 8 - Chinaz (ZIL); 9 - Golodnaya Step, near Ursatyevskaya (ZIL); 10 - Djulenghar Kishlak (ZIL); 11 - Khavast [106]; 12 - Zaamin (ZDKU); 13 - Kassansay Settlement [33]; 14 - Balikchy (ZIN); 15 - Uchkurghan near Naryn (ZIL); 16 - Buvaydy Station [33]; 17 - Andizhan (ZIN); 18 - Kokand (ZIK); 19 - Kuvech Valley (ZMMSU); 20 - Fergana (ZMMSU); 21 - 23 - S Kyzylkum.

Turkmenia[1]: 1 - ruins of Deu-Kesken-Kala (ZIK); 2 - Kunya-Urghench (ZIK); 3 - Sarykamish (ZIK); 4 - 10 km W Budzhnuyu-Daudak [130]; 5 - Charishly Well [130]; 6 - 100 km NW Darvaza Settlement (ZIL); 7 - Unghuz, 25 km N Serny Zavod (ZIK); 8 - Dardja Peninsula (ZIL); 9 - Djebel Station; Molla-Kara Station [130]; 10 - Lake Yaskha [23]; 11 - Togholak [23]; 12 - Upper Uzboy: Charishly Well [23] {no. 5 and 12 are the same locality but no. 5 referenced by [130] whereas no. 12 - by reference [23] – MG}; 13 - 180 km N Bakharden Settlement (ZIK); 14 - Ata-Kuyu Well, 40 km N Darvaza Settlement (our data); 15 - 20 km E Darvaza Settlement: Djarly Well (ZIK); 16 - 20 km E Serny Zavod: Okuz collective farm (ZIK); 17 - Kirkily [130]; 18 - Doyakhatyn (ZIK); 19 - Kultakyr [130]; 20 - Akhcha-Kuyma Station [130]; 21 - Kara Boghaz locality, 40 km N Kyzyl-Arvat (ZIL); 22 - SE Maly Balkhan Ridge: Chal-Su Well [130]; 23 - N Kyzyl-Arvat (ZIK); 24 - near Danata (ZIK); 25 - foothills of Kyurendagh Ridge [130]; 26 - Iskander Station [8]; 27 - Kyzyl-Arvat [130]; 28 - 50-100 km N Bakharden Settlement (ZIK); 29 - 100 km N Repetek, Er-Adji Well (ZIK); 30 - 30 km S Kara-Kala Settlement [130]; 31 - Aydere Gorge, Saparbakar Settlement [130]; 32 - 37 km W Repetek (ZIK); 33 - Repetek (ZIK); 34 - Bayram-Aly [130]; 35 - Karabata [130]; 36 - 20 km W Nich- {p. 171} -ka Settlement [130]; 37 - Kerky [23]; 38 - Dortkuyu Station [130]; 39 - Yolotan (ZIK); 40 - Kelif [128]; 41 - Tashkepry (ZIK).

Kirghizia: 1 - Orto-Tokoy Reservoir [22]; 2 - N Naryn River on road to Toghuz-Torou [162]; 3 - Sarygh-Bulun Settlement, 15 km before Kavak [162]; 4 - Middle Naryn River, from Ala-Buga to Naryn Town [162]; 5 - Djalal-Abad (ZIK); 6 - Osh Town (ZIL); 7 - Irka-Kashka between Aravan and Andizhan (ZIL); 8 - Gulcha Settlement (ZIK); 9 - Frunze (ZIK).

Tadjikistan: 1 - Oby-Asht Gorge (Boboiob) [106]; 2 - Asht Dist.: cliffs of old Syrdarya River bed [106]; 3 - Mohol-Tau; Murza-Robat (ZIL); 4 - S slope of Khuramin Ridge [106]; 5 - Kyrkkuduk [106]; 6 - Kurghancha Settlement (ZIK); 7 - Kock-Kurak [106]; 8 - near Kayrakkum (ZIK); 9 - Kanibadam (ZIK); 10 - near Ura-Tyube Town (ZIL); 11 - Kattacksay near Djoylyanghar Kishlak [106]; 12 - Yavan [106]; 13 - "Tigrovaya Balka" Reserve (ZIL).

---

[1] References to this species in Aydere [23, 24, 109] are in error; they relate to the spiny-tailed thin-toed gecko {*T. spinicauda* – MG}.

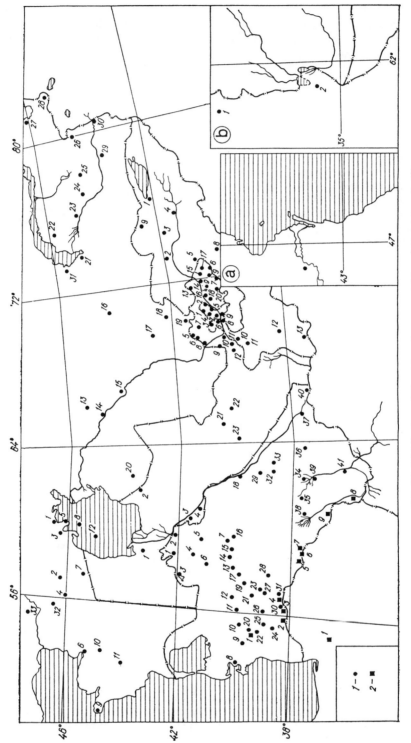

FIG. 74. Distribution of *Tenuidactylus russowi* (1) and *T. spinicauda* (2).

Caucasus: *Checheno-Ingushetiya* (inset "A"): Starogladkovskaya Stanitsa (ZIL).
Iran (inset "B"): 1 - Terra Behars (ZIL); 2 - Sistan Hollow (ZIN).

SEXUAL DIMORPHISM AND AGE VARIATION. Females are larger than males (male L = 49 mm; female L = 53.2 mm); they do not have preanal pores.

Juvenile specimens differ from adults in having a relatively larger head, thinner body, and shorter tail (L/LCD juvenile = 0.8-1.0) (Fig. 75).

SYSTEMATICS AND GEOGRAPHIC VARIATION. A study of nine samples (n = 285) with 14 characters showed that the values of some of them increase from the center of the range to the periphery (L/LCD); increase (contact between first postmental and second infralabial scales) or decrease (number of subdigital lamellae; ventral scales across midbody; GVA) to N and to S from the line Nebit-Dagh – Fergana – Djalal-Abad. The number of preanal pores, the frequency of occurrence of additional dorsal tubercles, and the maximum body size increase from south to north. Such characters as the ratio of eye diameter to length of snout and head, length of snout to length of fourth toe, as well as length of hindlimbs to L (these characters were used to describe a subspecies from Semirechye [137]) are not characteristic of any single population, and do not allow the recognition of *T*. {error in original – MG} *r*[*ussowi*] *copalensis* as a valid [subspecies]. To the contrary, the geckos from Iran (Sistan), which were described as a separate species [89] but then synonymized [125], are characterized by several original characters, which was the basis for re-establishing them at the subspecies level [152; 282].

Key to the Subspecies of T[*enuidactylus*] *russowi*

1 (2). 100-130 (114.99 ± 0.44) rows of ventral scales from chin to
vent [= GVA] . . . . *T. russowi russowi* (Str.)
2 (1). 94-105 (99.2 ± 0.99) rows of ventral scales from chin to
vent [= GVA] . . . . *T. russowi zarudnyi* (Nik.)

GRAY THIN-TOED GECKO—*TENUIDACTYLUS RUSSOWI RUSSOWI* (STRAUCH, 1887)

Type locality of species.

1887 – *Gymnodactylus russowi* Strauch, Mem. Acad. Sci. St. Petersbourg, ser. 7, vol. 35, no. 2, p. 49.
1928 – *Gymnodactylus russowi copalensis* Shnitnikov, Tr. ob-va izuch. Kazakhstana, Kzyl-Orda, vol. 8, no. 3, p. 24.

DIAGNOSIS. Usually, scales of first pair of postmentals contact each other (72%). 100-130 (114.99 ± 0.44) GVA; 25-34 (29.09 ± 0.14) ventral scales across midbody.

TYPE SPECIES.

DEFINITION. All other characters are in accordance with the description of the species.

DISTRIBUTION. From northeastern Precaucasus region through all of Middle Asia and Kazakhstan to Djungaria and northern Iran.

{p. 172} ZARUDNY'S GRAY THIN-TOED GECKO—*TENUIDACTYLUS RUSSOWI ZARUDNYI* (NIKOLSKY, 1900)

Type locality: Neizar in Sistan, eastern Iran.

1900 – *Gymnodactylus zarudnyi* Nikolsky, Ann. Zool. Mus. Imp. Acad. Sci., vol. 4, p. 385.
1981 – *Gymnodactylus russowi zarudnyi*, Szczerbak, Handbuch Rept. Amph. Europ., vol. 1, 1, p. 79.

DIAGNOSIS. Scales of first pair of postmentals always separated from each other by one to two scales; 94-105 (99.2 ± 0.99; t = 13.93) GVA; 23-26 (24.3 ± 0.35; t = 12.71) ventral scales across midbody. No additional dorsal tubercles.

LECTOTYPE. ZIL 9334 a, adult male "Neizar in Seistan, 21-24 May 1899, N. Zarudny." L = 31.4 mm; LCD = 38.9 mm; L/LCD = 0.81; 17 scales across head; 10 supralabial scales; 9/8 infralabial scales; 102 GVA; about 25 ventral scales across midbody; 11 scales around dorsal tubercle; two pairs of postmentals, scales of first pair separated by a scale; two preanal pores; 18/17 subdigital lamellae; no subcaudal plates on the basal segment, then two {on each segment – MG} in one row (1-1).

DEFINITION. All other characters are in accordance with the description of the species.
DISTRIBUTION. Inhabits Sistan Hollow (southeastern Iran).

HABITAT AND QUANTITATIVE DATA. The northern boundary of the range of *G.* [*sic*] *russowi* does not extend beyond the limits of the northern boundary of the desert zone. In the Precaucasus, it had also been found in an area of sand desert. Even though this gecko does live on vertical surfaces, it does not range far into the mountains. This is especially true of the western parts of its range. Thus, in Turkmenia, this species avoids even the foothills of Kopetdagh, even though in parts of Kirghizia, adjacent to the Pherghan Valley, it is found at elevations of 900-1200 m, and along the Naryn Depression it occurs up to 2000 m above sea level. Where it inhabits sand deserts and semi-deserts (Turkmenia, southern Tadjikistan, Prebalkhashye), *G.* [*sic*] *russowi* lives on the trunks of saxaul and other trees. In clay deserts and loess foothills (Uzbekistan, southern Kazakhstan, northern Tadjikistan), it is common on the cliffs in river and ravine plains, under rocks and (especially often) on walls of inhabited and abandoned buildings and other buildings of clay and stone. Some of the lizards have adapted to feeding near street lights, feeding on insects that are attracted to the light.

In Turkmenia, central Karakums, in the vicinity of Kirpily Well, one gecko was encountered on every fourth to sixth saxaul trunk [130]. In the west of the aforementioned desert, in the vicinity of Mollakara Settlement, on 7 May 1983, one to two individuals were noted for every old saxaul tree 5-8 m tall. In the "Tigrovaya Balka" Preserve (southern Tadjikistan), in an old saxaul grove, one to three geckos were found on each tree on 12 April 1980. In Uzbekistan and northern Tadjikistan, on loess duvals (fences) in villages, 10-15 individuals can be observed for every 10 linear meters [22, 125]. Much less frequently (single individuals), the gray thin-toed gecko is encountered here on loess and conglomerate cliffs and in tree hollows. On 12 August 1972, on the edge of the town Djalal-Abad (Kirghizia), we found 16 geckos on duvals with a total length of 1 km. On 14 April 1980, in "Tigrovaya Balka," on a wooden plank shed with a total wall area of 64 m^2, 15 geckos were observed.

Along the northern boundary of the range in the region along the lower part of Emba River (V. V. Neruchev, in litt.[1]), gray thin-toed geckos do not go beyond the zero horizontal (the lowlands boundary) and live on the outliers with steep slopes and on rock outcroppings (sand nodules). Here, up to eight individuals could be found per evening field trip.

In the extreme east of the range, in central Tjan-Shan on adyrs in the vicinity of Baygonchek Settlement, we observed two gray thin-toed geckos along a 5 km transect on 2 April 1984. On 3 June 1984, 15 km above the mouth of Alabuga River, on the upper terrace of the right bank, and on rocky cliffs and among boulders (large rock fragments), 10 geckos were recorded along a 500 m transect. In the vicinity of Konorchok Settlement, on broken rock and rocky slopes of hills, {p. 173} these geckos inhabit the base of the slopes and Tokobayev [or Tjan-Shan] pygmy geckos [*Alsophylax tokabajevi*] live near the the hill tops.

RESPONSE TO TEMPERATURE. It is known that in Uzbekistan, gray thin-toed geckos were found to be quite active in the spring at an air temperature of 17° and cliff surface temperature of 37°. A foraging lizard was observed in the summer at an air temperature of 37°. In the fall, they were seen on the cliff surface at an air temperature of 10° and a soil temperature of 24° [22]. In our observations, the voices of active geckos could be heard in windy weather with an air temperature of 20° on 13 April 1977 in saxaul thickets near Tashkepri (Turkmenia). On 30 April 1979, 25 km east of Darvaz, a gecko was caught on a rock outlier at an air temperature of 17° and soil temperature of 21.5°. When the substrate temperature was below 1°, these lizards did not leave their shelters. We observed active geckos in June of 1984 in the Alabuga River Valley (central Tjan-Shan) at dusk at an air temperature of 18°.

During experiments with gray thin-toed geckos in a thermal gradient apparatus, they would occupy sites with substrate temperatures of 19-41°. The preferred temperature is 33.3° ± 0.83° (n = 62).*

---

[1] As V. V. Neruchev had determined, it is the gray thin-toed gecko and not the Caspian one, as V. A. Kireyev had asserted [65, p. 64], that inhabits the Zheltau Mountain.

* {Data obtained by Cherlin (1988) under natural conditions in the eastern Karakums show that the preferred temperatures of gray thin-toed geckos are around 37-38° in spring and 37-40° in summer – MG.}

DAILY ACTIVITY CYCLE. In the early spring, these geckos are encountered in the daytime. They bask and forage near their shelters. On 27 February 1955, their activity was noted from 1400 to 1800 hrs, with the peak from 1600 to 1700 hrs [22]. In the summer, these lizards can bask only during morning hours, and they forage at dusk and at night, frequently until dawn. From observations in Turkmenia in April, gray thin-toed geckos, as a rule, were active after 2100 hrs. On 12 April 1980, in southern Tadjikistan, the activity of these tree-dwelling lizards was well defined in the daytime, and at dusk, after 2100 hrs, it stopped. In the cool days of September and October, gray thin-toed geckos were, once again, more often encountered during the day.

SEASONAL ACTIVITY CYCLE. The earliest discovery of a gray thin-toed gecko in Turkmenia, from our collecting, was on 27 March 1977. Their appearance after hibernation in Uzbekistan, in Angren Valley, was seen on 16 February 1955, and on 27 February 1955, in the same place, their emergence *en masse* was observed [22]. In the vicinity of Ursatyevskaya Station, they were noted somewhat later, on 15-20 March [54]. In Tadjikistan, the geckos appear in March [107] after hibernation, in southern Kazakhstan, at the beginning of March, [and] in Pribalkhashye Region, at the beginning of April [98].

In Turkmenia, these lizards disappear for the winter probably at the beginning of November, because on 15 October 1981, in our observations near Repetek, they were still quite active and common. In Uzbekistan, they disappear by mid-November [22]. In 1950 in southern Pribalkhashye, the disappearance of gray thin-toed geckos for the winter was noted at the beginning of September [96].

SHEDDING. In Uzbekistan, shedding *en masse* was observed at the end of April and in May [54]. In Kirghizia, shedding individuals have been found at the end of July [162].

FEEDING. Representatives of 34 insect families and 2 families of other arthropods have been found in the gray thin-toed gecko diet. In Kirghizia (39 gecko stomachs, obtained in July, were examined), [gray thin-toed geckos] primarily eat orthopterans (84.2% frequency of occurrence), heteropterans (36.8%), beetles (15.7%), as well as hymenopterans and butterflies (5.3% each) [162]. In Uzbekistan, in the vicinity of Ursatyevskaya Station (24 stomachs, obtained in September, were dissected), prey consisted of beetles (50%), heteropterans (29%), hymenopterans, including ants (25%), butterflies (25%), and spiders (16%) [22]. In the Tashkent area in Angren Valley, the contents of 35 gecko stomachs, obtained on 2 August 1955, and 36 stomachs, collected on 25 August 1955, were examined [28]. In the first case, beetles were encountered most frequently (46.8%), among which the best represented were the [bystryanki] beetles (28.13%) and darkling beetles (15.6%), followed by hymenopterans (25.2%, ants composed 12.5%), and spiders (15.6%). In the second case, hymenopterans were in first place (40.6%, ants made up 28.1%), followed by heteropterans (18.75%), beetles (15.6%) and flies (15.6%). In Turkmenia, 32 gecko stomachs obtained at different times were dissected [130]. Here, in first place are hymenopterans (46.8%), including ants (43.7%), butterflies and their larvae (34.4%), and beetles (15.6%). By comparing the materials on the gray thin-toed gecko feeding, and collected at different times, it can be concluded {p. 174} that the foraging intensity declines as fall nears (the number of individuals with empty stomachs increases) and the contents of the diet change somewhat. The numbers of beetles and orthopterans decreases and the numbers of hymenopterans (especially ants), and heteropterans and flies increase.

REPRODUCTION. The sex ratio in our samples is 1:1. In the Naryn Depression (Kirghizia), it is 1:1.9, male to female. Gray thin-toed geckos reach sexual maturity with a body size of 37 mm for males and 40 mm and greater for females. We observed courtship and mating on 12 April 1980 in the vicinity of Nizhnij Pyandz Village (southern Tadjikistan). According to a number of data, some time elapses between the emergence of the gray thin-toed gecko from hibernation and reproduction. Thus, in Uzbekistan [22], mating occurs from mid-May until the end of June, or two months after [their] appearance, and eggs are laid from mid-June until the end of July. Hatchlings appear here from the end of July or beginning of August until the end of October. In the western part of the range (Mangyshlak), females lay their eggs over 45 days, from the beginning of June until mid-July, and earlier in the east (Prebalkhashye Region), from the beginning of May, where hatchlings begin appearing in early August [98].

The eggs are elliptical and white. Their dimensions are: from Uzbekistan, 7.5-9 × 9-11 mm [22]; from Kazakhstan, 8 × 10-11.5 mm [98]; and from Kirghizia, 8-9 × 10-12 mm [162]. We collected clutches of the following dimensions: August 1972, in Djalal-Abad, 7.4-7.9 × 10.0-10.8 mm (n = 7); in Tadjikistan (Kurgancha Settlement) on 26 June 1979, 7.5 × 10.0-10.5 (n = 2); and in central Tjan-Shan (Kirghizia, Konorchok Settlement) on 29 June 1984, 7.7-7.8 × 11.2-11.8 mm (n = 2). It follows from the above data that the largest eggs are from geckos from central Tjan-Shan. There can be one or two eggs in a clutch, and two clutches per season are possible. Several females often lay their eggs in a single place, which led some authors [54, 107] to suppose a larger number of eggs in a clutch. This is refuted by O. P. Bogdanov's research [22] and by our materials. Incubation lasts 50-54 days [28] and 45-55 days [17]. We caught the first current year's hatchling on 21 August 1972, in Djalal-Abad. From the eggs we collected there, a hatchling gecko hatched in the laboratory on 21 September 1972.

GROWTH RATE. The dimensions of hatchlings that hatched in our laboratory from eggs we collected in Djalal-Abad were: L = 16.6 mm, LCD = 16.9 mm. After the winter, young geckos, caught in March and April in Tadjikistan, were 20.1-22.2 mm [107], 21-25 mm [98] in Kazakhstan, [and] 27-34 mm [162] in central Tjan-Shan in July. These data show that these lizards reach sexual maturity at an age of one year, but they do not begin breeding until after the second winter, at the age of 20 months. The geckos reach their maximum size by the age of four years (Fig. 75).

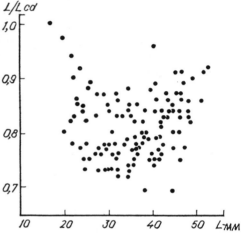

FIG. 75. Age variation of L/LCD index in *Tenuidactylus russowi*.

ENEMIES. According to our observations, geckos of this species are attacked by the spotted desert racer (*Coluber karelini*) and steppe ribbon snake (*Psammophis lineolatum*). There is a record of this gecko in the stomach of a northern wolf snake (*Lycodon striatus*) [54]. Gray thin-toed geckos have been found in the stomachs of the gull-billed tern {*Gelochelidon nilotica* – MG} (2.2% encounter rate) [155]. Among the possible predators are several other species of bird (shrikes {*Lanius* sp. – MG}, the little owl {*Athene noctua* – MG}), snakes, as well as some invertebrates (the eggs are sometimes eaten by ants, [and] solpugids attack geckos). Other factors have a much greater impact on gecko populations. Populations of this species (from many years of observations in Middle Asia) are more or less stable. There is a possibility of death in winter shelters during especially hard winters [22]. Their numbers decrease sharply after repairs on earthen structures (especially fences), and often they dis- {p. 175} -appear completely. According to reports of the same author [24], conversion of virgin and long-fallow lands to agricultural use leads to a sharp decline in the numbers of reptiles. In only one year, the gray thin-toed gecko completely disappeared from converted areas.

BEHAVIORAL ATTRIBUTES. The gray thin-toed gecko is agile and quick. It climbs and moves on tree trunks and branches with great skill. When disturbed, it runs around a trunk in a spiral and can run quite rapidly down a vertical surface, but when in danger, immediately jumps. It also runs along the underside surfaces of branches and rocks. O. P. Bogdanov [22] saw a gray thin-toed gecko jump from a cliff 1 m away and down, and catch a locust larva.

These geckos hide in cracks of saxaul trunks, loess cliffs and buildings, in the hollows of rock piles and between bricks of buildings. Less often, they will use rodent burrows, abandoned ant and termite hills, and hollows in rotting bush trunks. In the sand desert near

Kazandjik in April 1977, we dug a lizard out of a rodent burrow in a barkhan, at a depth of 40 cm. Also here, the shells of its eggs were found. These geckos winter in burrows in the sand desert, cracks in loess cliffs, and in the remnants of loess buildings and walls. Torpid lizards have been found at depths of 8-25 cm [22]. Three to four immature lizards live in a single shelter, and adults are solitary [98]. They do not go far from their cover. According to observations in Uzbekistan [22] in the summer, these geckos descend from the cliffs at dusk (in kishlaks, they come off the fences onto the roads) and run in search of food. In ravines, they will sometimes go as far as 100-200 m from their shelter. The geckos run from stone to stone, catching insects that hide under rocks. By morning, the geckos once again return to their burrows in the cliffs.

During the mating period, males engage in fierce fighting, and they frequently fall off a wall with their jaws locked on each other. When excited, the lizards thrash their tails and emit loud sounds. These sounds can be expressed as rapid fire "ek...ek...ek". In the Krakum Desert in the spring, these voices are clearly heard coming from many saxaul trees. Captured geckos also emit weak squeaks. In diffused daylight, their vision extends up to 2 m [282].

PRACTICAL IMPORTANCE AND PROTECTION. This is a common and abundant species. It is beneficial and is not in need of special protective measures in nature. Large populations are protected already in the Repetek (Turkmenia) and "Tigrovaya Balka" (Tadjikistan) Preserves.

SPINY-TAILED THIN-TOED GECKO — *TENUIDACTYLUS SPINICAUDA* (STRAUCH, 1887)

Type locality: Shahrud, Iran.
Karyotype: $2n = 42$; 2 metacentrics, 38 acrocentrics, 2 subtelocentrics, NF = 46.
1887 — *Alsophylax spinicauda* Strauch, Mem. Acad. Sci. St. Petersburg, ser. 7, vol. 35, no. 2, p. 58.
1977 — *Gymnodactylus spinicauda*, Szczerbak and Golubev, Tr. Zool. In-ta, vol. 74, Leningrad, p. 130.
1984 — *Tenuidactylus spinicauda*, Szczerbak and Golubev, Vestn. Zool., Kiev, no. 2, p. 54.

DIAGNOSIS. Dorsal tubercles roundish; diameter of ear opening smaller than half of longitudinal eye diameter; no less than 90 GVA; scales of tail (except tubercles) and limbs smooth.

HOLOTYPE. ZIL 4047, adult female. "Shahrud, 1875, Khristof." L = 37.2 mm; LCD = 30.5 mm; L/LCD = 1.21; 17 scales across head; 8/9 supralabial scales; 7 infralabials; 107 GVA; 27 ventral scales across midbody; scales in the single pair of postmentals separated from each other by a scale; no preanal pores; 14/15 subdigital lamellae. One (on the left side) and two (right) nasal scales.

DEFINITION (Fig. 76) (from 41 specimens in ZIK, ZIL, IZT, ZMKU, and ZDKU). L adult male (n = 10) 26.2-43.1 (34.22 ± 1.95) mm; L adult female (n = 19) 28.7-48.2 (39.46 ± 0.95) mm; mass to 2.5 g; L/LCD (n = 16) 1.02-1.21 (1.09 ± 0.01); HHW (n = 31) 51-68 (59.87 ± 0.86); EED (n = 31) 14-33 (25.16 ± 0.71); supralabial scales (n = 80) {p. 176} 8-11 (9.50 ± 0.09); infralabials (n = 80) 6-9 (7.81 ± 0.08); scales across head (n = 41) 16-21 (17.90 ± 0.20); DTL (n = 41) 10-16 (13.27 ± 0.25); scales around dorsal tubercle (n = 41) 8-10 (8.80 ± 0.09); GVA (n = 32) 103-127 (113.41 ± 1.34); ventral scales across midbody (n = 23) 26-32 (28.22 ± 0.33); subdigital lamellae (n = 80) 13-18 (15.45 ± 0.14); preanal pores (n = 12) 2-3 (2.17 ± 0.11).

{p. 177} Mental scale pentagonal (2.4%), hexagonal (65.9%), or trapezoid with curved posterior edge (31.7%); no additional dorsal tubercles (100%); poorly developed subcaudal plates (1/2, 1/1) sometimes (about 5%) can be present; other caudal scales smooth. Usually, first nasal scales separated from each other by one (66.4%) or two (21.6%) scales, rarely three of them can be present (4.8%), or none (7.2%). Usually three nasals on each side; none (1.7%), one (49.0%), two (34.0%), or three (15.3%) {pairs of – MG} postmentals. Unlike the other species of this genus, longitudinal keels on caudal spiny tubercles of this species are very well developed. Regenerated tail distinctly thickened in the middle part.

COLOR AND PATTERN The main background of the dorsal surface is ashen gray, rarely

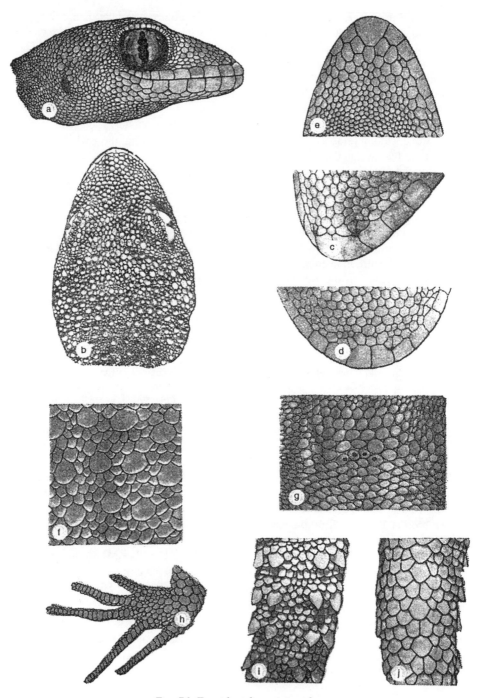

FIG. 76. *Tenuidactylus spinicauda*:
a, head from the side; b, head from above; c-d, nasal scales; e, chin; f, dorsum;
g, preanal pores; h, hindfoot from below; i, tail from above; j, tail from below
(d – ZIL 4047, Shahrud, Iran; others – ZIK SR 364, Danata, Turkmenia).

with ochrish or brownish hue. On the sides of the head, a small brown stripe runs from the nostril through the eye and to the ear. Pileus and supralabials are covered with small brown dots. On the back, from the base of the neck to the tail, there are seven to eight wavy brown small transverse stripes, frequently edged with black. There are transverse bands and spots on the tail and limbs. The ventral surface is grayish white. In juveniles, the tail is pinkish brown (as is the body's main background) above and below, but it is lighter (more ochrish) than the body.

DISTRIBUTION (Fig. 74). In the U.S.S.R. [it is] distributed in Kopetdagh Mountains and Badkhyz (Turkmenia), in Iran – at the type locality and probably throughout all of the Turkmen-Khorasan Mountains massif.

LEGEND to Fig. 74.
U.S.S.R.: Turkmenia: 1 – Danata Settlement; Kazandjik; Eyshem Well; Oboy (ZIK); 2 – Mondzhukly locality (ZIK); 3 – 20 km S Kara-Kala Settlement (ZIK); Kara-Kala, Mondzhukly Ridge [107]; 4 – Kopetdagh: Tuzly-Tepe (ZIK); Aydere [107]; 5 – Karanky locality (ZIK); 6 – Gaudan Pass (ZIK); 7 – near Kalininsky Settlement (ZIK); Meymily Well [13]; 8 – W Badkhyz [13]; 9 – Chaacha locality (ZIK).
Iran: 1 – Shahrud (ZIL).

SEXUAL DIMORPHISM AND AGE VARIATION. Juvenile specimens are shorter tailed. Females lack preanal pores.

SYSTEMATICS AND GEOGRAPHICAL VARIATION. Originally, it was assigned to the genus *Alsophylax*, but the presence in this species of angularly bent digits and some other characters allowed us to reassign it to the group of thin-toed geckos [150]; the correctness of this was later supported by karyological analysis [73]. The difference in the number of nasal scales between the holotype and the other 40 specimens is probably due to individual variation.

HABITAT AND QUANTITATIVE DATA. It inhabits argillaceous {clay – MG} hills in the foothills with arid climate vegetation (sagebrush, fescue), in the vicinity of Kazandjik, Iskander Station on hill slopes, covered with crushed roofing slate {most likely, this was meant to be "crushed schist" – MG} with stones, rare *Ephedra* bushes, sagebrush, grasses, *Lamiaceae* {MG} (estimated cover of 10-100%), or without them. Most often, the slope exposure is northern, less frequently, eastern or western. The incline is to 45°, although these lizards are also encountered on talus slopes and rock valleys (vicinity of Danata, Aydere, Mondzhukly settlements), as well as on rocky slopes with limestone and schist slab outcrops (upper Chulinka River, Bolshiye Karanki Gorge in central Kopetdagh, Chaacha locality in eastern Kopetdagh). The spiny-tailed gecko's habitat elevations are 500-1800 m. Individuals are found singly. The numbers of these geckos do not fluctuate as a result of unfavorable conditions in a particular year: in the vicinity of Danata village, they were recorded as stable from 1963 to 1982, although the numbers of many [other] common species would drop sharply after harsh winters. In corresponding habitats within an area of 300 × 200 m (south of Iskander Station, 4 May 1983), three specimens were caught, that is one individual per 2 hectares. In individual cases, they were encountered with greater frequency: in the vicinity of Danata village on 24 April 1982, in a 150 m^2 area, two lizards were found; and on 6 April 1982, south of Kara-Kala, in an area of 2000 m^2, one gecko was found. Usually, in order to catch a spiny-tailed gecko, a collector had to turn rocks over from 0500 until 1000 hrs.

{p. 178} RESPONSE TO TEMPERATURE. Active lizards were obtained at air temperatures of 22-24°. On 20 April 1977, an active gecko was found under a rock when the air temperature was 8°. The disturbed lizard tried to escape. Eggs were incubated at temperatures of 20-27°.

DAILY ACTIVITY CYCLE. This lizard has a dusk and nocturnal mode of life. We observed the first active lizard on 27 April 1975 at 2100 hrs. In the daytime, they are encountered only in their shelters.

SEASONAL ACTIVITY CYCLE. In the spring, the first specimens are encountered in mid-March [17]. In our collection, the earliest find is dated 15 April 1963 (vicinity of Danata Village). The last individuals were also obtained there on 29 October 1981, and on 30 October 1981, their squeaks were heard at dusk.

SHEDDING. Shedding has not been observed in adult individuals. Hatchlings shed six days after hatching, on 15-16 August 1983, and while hatching (Fig. 77).

FEEDING. Among the 10 lizard stomachs obtained in the spring, only one contained food remnants (a cicada). According to published data [17], spiny-tailed geckos eat small

FIG. 77. Hatchling of *Tenuidactylus spinicauda*.

arthropods, primarily spiders, and heteropterans. In captivity, these lizards would eat aphids, small mealworms, and cockroaches.

REPRODUCTION. In our sample, there are almost twice as many females as males (19 females to 10 males). These lizards reach sexual maturity at a body size of 26-28 mm. Females with 5 × 7 mm eggs were caught in the western Kopetdagh on 1 and 19 April. We caught gravid females on 27 April 1975 and 20 April 1977. We obtained gecko egg clutches (one, more often two eggs) on 10 May 1983 (30 km south of Iskander Station), 14 May 1975 (from Danata village), 30 May 1983 (Bolshiye Karanki Gorge, central Kopetdagh), 10 June 1983 (same place). The eggs are 6.7-8.0 × 8.2-10 mm (n = 8), 7.1 × 9 mm on average. Eggs are oval and white just after laying. Two days later they turn pink if they are developing. A month before complete development they pale, and then dark spots appear on the interior. According to our observations, incubation lasts 71-80 days. Eggs from the same clutch that are incubated under identical conditions can develop with a difference of one day. We got our first hatchlings on 8 August 1983, and in nature, the current year's hatchlings were obtained on 29 October 1981.

{p. 179} GROWTH RATE. Hatchling geckos have body lengths of up to 17.5-18 mm and a tail length of 14 mm. Sexual maturity occurs, probably, after the second hibernation.

ENEMIES are unknown. Once, under a rock, a northern wolf snake (*Lycodon striatus*) was discovered alongside a spiny-tailed thin-toed gecko. It may attack these lizards occasionally.

BEHAVIORAL ATTRIBUTES. Geckos of this species externally resemble the gray thin-toed gecko but are less agile and quick. They do not jump; they are slow-moving and, although they live primarily on horizontal slopes, they are capable of movement on vertical and ceiling surfaces, according to observations in captivity. Daytime shelters are under rocks, in hollows in bush roots, and in cracks in the ground. They hibernate in cracks in rocks, partially covered with soil, [and] in rodent burrows. We observed shells of their eggs under roots of a sagebrush bush, in old talus at a depth of 50 cm, [and] under a rock. Their mating behavior has not been observed. Their voice is similar to the squeaks of the gray thin-toed gecko but quieter.

PRACTICAL IMPORTANCE AND PROTECTION. In order to preserve the gene pool of this rare species that has a limited distribution in our country, it has been included in the *Red*

Books of the U.S.S.R. and T[urkmen].S.S.R. It is protected on the territory of the Kopetdagh Preserve and its Meano-Chaacha Extension.

LEBANESE THIN-TOED GECKO — *TENUIDACTYLUS AMICTOPHOLIS* (HOOFIEN, 1967)

Type locality: Hermon Mountain near Megdel Shams Village, Lebanon, about 2000 m elevation. Karyotype not studied.

1967 — *Cyrtodactylus amictopholis* Hoofien, Isr. J. Zool., Tel Aviv, vol. 16, p. 205.
1981 — *Gymnodactylus (Mediodactylus) amictopholis*, Szczerbak and Golubev, Vestn. Zool., Kiev, no. 5, p. 40.
1984 — *Tenuidactylus amictopholis*, Szczerbak and Golubev, Vestn. Zool., Kiev, no. 2, p. 54.

DIAGNOSIS. No dorsal tubercles; diameter of ear opening smaller than half of eye longitudinal diameter; no more than 90 GVA; scales of tail (except tubercles) and limbs smooth.

HOLOTYPE [230] TAU* 7072, adult female. "Mt. Hermon above Megdel Shams about 2000 m elevation, 20 Aug. 1967, Hoofien et al." L = 36 mm; tail regenerated; two nasal scales; first nasals separated from each other by three scales; eight supralabial scales; mental scale triangular, scales in first pair of postmentals contact each other and are followed by three to four scales, gradually becoming smaller; abdominal scales large; about 70 scales around midbody, only 20 of them are ventrals; no preanal pores; 13 subdigital lamellae.

DEFINITION (Fig. 70) (from 2 specimens in TAU). L = 34.3-35.4 mm; tails regenerated; HHW 49-53; EED 29-30; 8-9 supralabial scales; 6-7 infralabials; 15-17 scales across head; 95-102 GVA; 18-20 venter; 17-18 subdigital lamellae; 2 preanal pores.

Mental scale triangular or pentagonal; three to four pairs of postmentals, scales in first pair contact each other; two nasal scales, the first ones separated from each other by two scales; subcaudal plates on unregenerated part of tail well developed (two on each segment: 1/1); other scales of tail and limbs without keels.

COLOR AND PATTERN In color and pattern, this species differs from all the other ones in the subgenus. In life [230] the body has a golden bronze hue. There are 10 thin dark transverse worm-like bands on the back, intermixed with occasional small spots. On the base of the tail, there are two noticeable yellow transverse bands that share a border with the black edging of the unregenerated base of the tail. After capture or fixation, the bright color disappears and only a brownish gray background remains.

DISTRIBUTION (Fig. 39). Known only from type locality.

{p. 180} LEGEND to Fig. 39.
Lebanon: SW foothills of Hermon Mountain: Megdel Shams Village [230].

ECOLOGY is almost unknown. All data derives from the author of the description [230].

HABITAT. It lives in rocky areas, probably on a mountain plateau, with sparse vegetation, at an altitude of 2000 m.

RESPONSE TO TEMPERATURE. It lives in the harsh conditions of high mountain altitudes, where, in some areas, snow remains throughout the summer. The air temperature does not exceed 22° during the daytime and there is frost at night.

DAILY ACTIVITY CYCLE. This is a diurnal species (!). It is active during the warmest time of the year [in the] afternoon, when the substrate and the air are sufficiently warmed.

SEASONAL ACTIVITY CYCLE, SHEDDING, FEEDING and other issues of ecology have not been studied.

BEHAVIORAL ATTRIBUTES. It is very cautious and agile. The authors noticed that catching these geckos is very difficult.

---

* {TAU – the collection of the Tel-Aviv University Zoological Museum. We were able to examine two specimens of this unique species owing exclusively to the courtesy of Dr. H. Mendelssohn. Unfortunately, we could not acknowledge him in our book, because at that time (1986) any positive mention of the state of Israel could have led to the banning of the book. For the same reason, Dr. Y. Werner (Hebrew University, Jerusalem, Israel), who helped by providing his gecko publications, was not mentioned either. The authors would like to take this opportunity to express their belated but sincere apologies and thanks to Drs. Mendelssohn and Werner. – MG.}

BAMPUR THIN-TOED GECKO — *TENUIDACTYLUS SAGITTIFER* (NIKOLSKY, 1899)

Type locality: Bampur in eastern Kerman, eastern Iran.
Karyotype: not studied {error in original – MG}
1899 — *Gymnodactylus sagittifer* Nikolsky, Ann. zool. Mus. Imp. Acad. Sci., vol. 4, no. 4, p. 379.
1974 — *Cyrtodactylus sagittifer* (ex errore), S. Anderson, Field.: Zool., Chicago, vol. 65, no. 4, p. 33.
1984 — *Tenuidactylus sagittifer*, Szczerbak and Golubev, Vestn. Zool., Kiev, no. 2, p. 54.

DIAGNOSIS. Dorsal tubercles oval, keeled; diameter of ear opening smaller than half of longitudinal diameter of eye ; no more than 90 GVA; scales and plates of tail as well as scales of upper front surfaces of limbs with distinct keels.

LECTOTYPE. ZIL 9331, adult male. "Bampur River in E Kerman, 30 June 1898, N. Zarudny." L = 29.7 mm, tail damaged; 14 scales across head; 10 supralabial scales; 7 infralabials; 85 GVA; 17 ventral scales across midbody; two pairs of postmentals, scales in first pair in contact with each other; four preanal pores; 18 subdigital lamellae; three nasal scales on each side.

DEFINITION (Fig. 78) (from 5 specimens in ZIL). L general (n = 5) 24.1-31.9 (29.4 ± 1.37) mm; L/LCD 0.95; HHW (n = 5) 64-68 (66.40 ± 0.98); EED (n = 2) 36-38; supralabial scales (n = 10) 9-10 (9.70 ± 0.15); infralabials (n = 10) 7-8 (7.40 ± 0.16); scales across head (n = 5) 14-15 (14.40 ± 0.24); DTL (n = 5) 19-26 (23.40 ± 1.21); scales around dorsal tubercle (n = 5) 12-13 (12.21 ± 0.20); GVA (n = 5) 83-88 (85.60 ± 0.87); ventral scales across midbody (n = 5) 16-19 (16.97 ± 0.99); subdigital lamellae (n = 10) 18-19 (18.55 ± 0.16); preanal pores (n = 3) 4. Mental scale pentagonal; no additional dorsal tubercles; two pairs of subcaudal plates per segment (2/2), they, as well as all other scales of tail and upper front surfaces of limbs, are keeled; one to two pairs of postmental scales, sometimes they are not defined; three nasal scales, usually first nasals separated from each other by a scale, but may be in contact.

COLOR AND PATTERN (as fixed in alcohol). The main color is light yellow. On the sides of the head, from the eye to the ear area and along the border between the parietal and temporal areas, there are two small brown stripes. The pileus has occasional, randomly distributed brown spots. On the upper surface of the body, there are five or six M-shaped transverse bands, which may be broken. The side segments merge into longitudinal stripes on the sides of the dorsum, as in the gray thin-toed gecko. The tail has transverse brown spots that also flow together into longitudinal stripes on the sides of the tail. On the fore- and hindlimbs, there are three or four indistinct transverse bands. The abdomen is light.

DISTRIBUTION. Known only from type locality (Fig. 7). Iran: Bampur; Farra (Purra) (ZIL); 30 km W Bampur Settlement, 27°10'N, 60°10'E (CAS).

{p. 181} SEXUAL DIMORPHISM AND AGE VARIATION. No preanal pores in females. Absence of juvenile specimens does not allow the study of age variation.

SYSTEMATICS AND GEOGRAPHICAL VARIATION. O. Wettstein [315] erroneously synonymized this species with *T. heterocercus* [307]. Variation was not found on the basis of the specimens examined.

ECOLOGY has not been studied. N. A. Zarudny [60] reported that this species was found "in the Bampur Hollow and on the heights that surround it." Of the three specimens, two were caught on tree trunks in an acacia forest and one on a wall of a subterranean home. During the day, it does not move much and is easy to catch.

{p. 182} ASIA MINOR THIN-TOED GECKO — *TENUIDACTYLUS HETEROCERCUS* (BLANFORD, 1874)

Type locality: Hamadan, western Iran.
Karyotype: not studied.
1874 — *Gymnodactylus heterocercus* Blanford, Ann. Mag. Nat. Hist., ser. 4, vol. 13 {error in original – MG}, p. 453.
1974 — *Cyrtodactylus heterocercus*, S. Anderson, Field.: Zool., Chicago, vol. 65, no. 4, p. 33.
1984 — *Tenuidactylus heterocercus*, Szczerbak and Golubev, Vestn. Zool., Kiev, no. 2, p. 54.

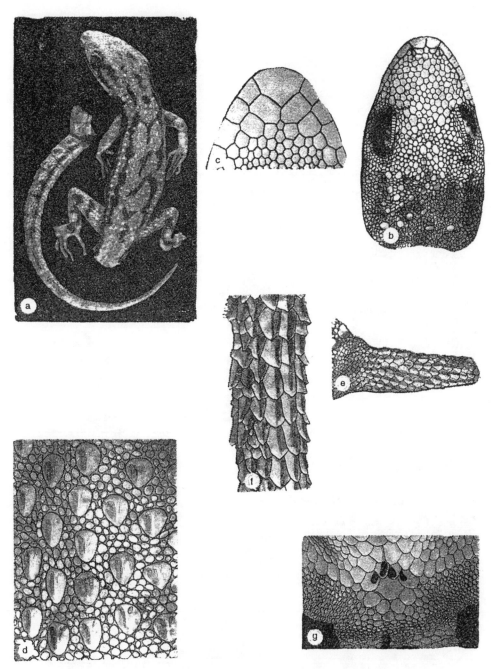

FIG. 78. *Tenuidactylus sagittifer*:
a, general view from above; b, head from above; c, chin; d, dorsum; e, shoulder from front;
f, tail from below; g, preanal pores (ZIL 9331, Bampur, Iran).

{p. 183} DIAGNOSIS. Dorsal tubercles oval, keeled; diameter of ear opening less than half of longitudinal diameter of eye; no less than 90 GVA; scales and plates of tail as well as scales of upper front surfaces of limbs with distinct keels.

LECTOTYPE. TZM no. R2532 (Fig. 79), adult female "Persia dono Doria." L = 47.3 mm; tail damaged; 14 scales across head; 9 supralabial scales; 8 infralabials; 118 GVA; 30 ventral scales across midbody; 12 longitudinal rows of dorsal tubercles; three pairs of postmentals, scales in first pair contact each other; no preanal pores; 21 subdigital lamellae; first nasal

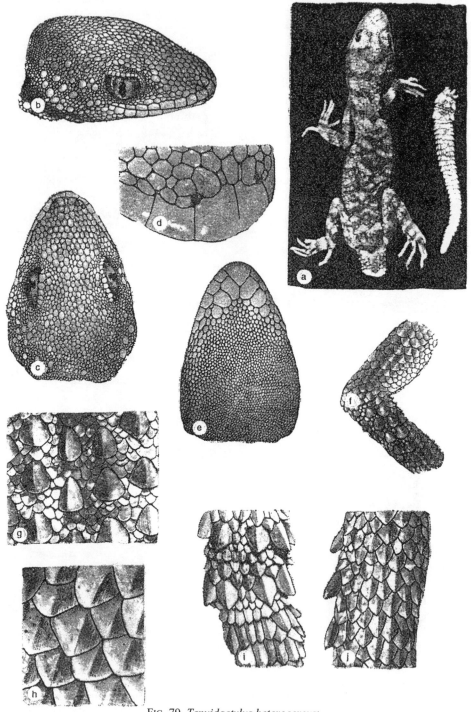

FIG. 79. *Tenuidactylus heterocercus*:
a, general view from above; b, head from the side; c, head from above; d, nasal scales; e, head from below; f, forelimb from above; g, dorsum; h, scales on side of abdomen; i, tail from above; j, tail from below (d, f, h – NMW 7286, Hamadan, Iran {error in original – MG}; others – NMW 19725, Persepolis, Iran {error in original – MG}).

scales are pressed rearward from nostrils by second and third nasal scales; abdominal scales keeled.

DEFINITION (Fig. 79) (from 17 specimens in TZM, SMF, NMW, FMNH). L general (n

= 17) 36.2-49.9 (43.92 ± 1.06) mm; L/LCD 0.86; HHW (n = 12) 48-59 (53.42 ± 0.91); EED (n = 10) 28-48 (38.50 ± 1.97); supralabial scales (n = 34) 9-12 (9.41 ± 0.17); infralabials (n = 34) 7-9 (7.50 ± 0.09); scales across head (n = 15) 12-16 (13.80 ± 0.28); DTL (n = 16 {error in original – MG}) 15-24 (20.25 ± 0.54); scales around dorsal tubercle (n = 17) 10-13 (11.41 ± 0.19); GVA (n = 15) 98-122 (109.07 ± 1.63); ventral scales across midbody (n = 16) 24-30 (26.00 ± 0.41); subdigital lamellae (n = 34) 18-23 (20.56 ± 0.23); preanal pores (n = 4) 2.

Mental scale triangular or pentagonal; three (82.3%) or four (17.7%) pairs of postmentals, scales in first pair usually (94.1%) contact each other; no additional dorsal tubercles; two to three nasal scales, {the first nasals – MG} usually separated from each other by one scale (82.3%), sometimes there are two to three, or they are absent (5.9% each); lower caudal scales homogeneous, they are distinctly keeled, as are scales of upper front surfaces of arms; spiny caudal scales are so large in this species that they can, in contrast with other species of the subgenus, contact each other on one segment.

COLOR AND PATTERN (as fixed in alcohol). The main background is gray. Many indistinct brown transverse bands create a marbled pattern on the upper surface. The pattern on the head and tail is not clear, because the specimens examined have been fixed for a long time.

DISTRIBUTION (Fig. 80). Anatolia (southern Turkey) and Hamadan (western Iran); the discovery of this species in Persepolis (Fars Prov., western Iran – NMW and FMNH) requires confirmation. Also, the record for El-Basra [269] requires confirmation; probably the authors mistook *T. scaber* for this species.

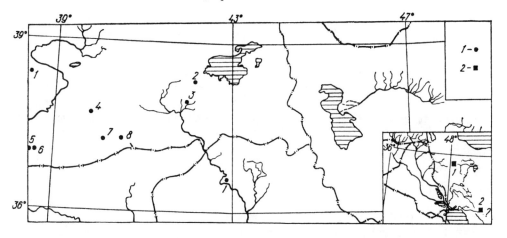

FIG. 80. Distribution of *T[enuidactylus] heterocercus mardinensis* (1) and *T. h. heterocercus* (2).

LEGEND to Fig. 80.
Iran: 1 – Hamadan (NMW); 2 – ruins of Persepolis (the inset).

SEXUAL DIMORPHISM AND AGE VARIATION. Females lack preanal pores. Age variation not studied.

SYSTEMATICS AND GEOGRAPHICAL VARIATION. R. Mertens [254] separated geckos from Anatolia into a separate subspecies. Later, he expressed doubts that this form belongs to *G. heterocercus*, rather than to *G. kotschyi*. Indeed, these species are very closely related but transitional forms between them are not known. The presence of more than a 600 km gap between the two populations seems most doubtful. Probably it reflects poor knowledge of this region's herpetofauna. There is no doubt that this species will be found in the northern parts of Syria and Iraq; it is probably distributed more widely in western Iran, also.

Key to the Subspecies of T[enuidactylus] heterocercus

1 (2). 12 to 14 longitudinal rows of dorsal tubercles . . . .
. . . . *T. heterocercus heterocercus* (Blanford)

2 (1). 10 longitudinal rows of dorsal tubercles . . *T. heterocercus mardinensis* (Mertens)

{p. 184} ASIA MINOR THIN-TOED GECKO — *TENUIDACTYLUS HETEROCERCUS HETEROCERCUS* (BLANFORD, 1874)

Type locality: Hamadan, western Iran.

1924 — *Gymnodactylus hetrocercus heterocercus*, Mertens, Senckenbergiana, Frankfurt am Main, vol. 6, p. 84.

DIAGNOSIS. 12 – 14 longitudinal rows of dorsal tubercles; abdominal scales often with distinct keels, especially on the flanks; first nasal scale separated from contact with nostril by second and third ones.

Other characters are contained in the description of the species.

TYPE OF SPECIES.

DISTRIBUTION. Known from type locality, Hamadan (western Iran); the records of this gecko in Persepolis require confirmation.

MARDIN ASIA MINOR THIN-TOED GECKO — *TENUIDACTYLUS HETEROCERCUS MARDINENSIS* (MERTENS, 1924)

Type locality: Mardin, southern Turkey.

1924 — *Gymnodactylus heterocercus mardinensis* Mertens, Senckenbergiana, Frankfurt am Main, vol. 6, p. 84 (Terra typica designata (Mertens, 1952): Mardin, Kurdistan, Turkey).

DIAGNOSIS. 10 longitudinal rows of dorsal tubercles; abdominal scales smooth; nostril contacts three nasal scales.

Other characters are contained in the description of the species.

NEOTYPE. SMF 8198, adult male. "Mardin, SO-Anatolien. O. Wolter, 1918." The specimen is dehydrated; measurements were not determined. 15 scales across head; 9 supralabial scales; 7 infralabials; 102 GVA; 25 ventral scales across midbody; 10 longitudinal rows of dorsal tubercles; three pairs of postmentals, scales in first pair in contact with each other; two preanal pores; 21/20 subdigital lamellae; three nasal scales; abdominal scales smooth.

DISTRIBUTION. Anatolia; possible finds in northern parts of Syria and Iraq (Fig. 80).

LEGEND to Fig. 80.

Turkey: 1 – Malatia [181]; 2 – Bitlie [181]; 3 – Siirt [181]; 4 – Diarbakir [181]; 5 – Gaziantep [181]; 6 – Nizip (SMF); 7- Derik [181]; 8 – Mardin [181].

Iraq: 1 – Nineveh Prov.: Al Naravan, 30 km N Al Mawzil [269].

ECOLOGY has not been studied. According to the data from M. Basoglu and I. Baran [181 {error in original – MG}], they are found on the walls of residences, in cracks. They emit sounds that can be expressed as "chek--chek," which can be heard after sunset. They catch insects near electric lights, are highly mobile, and have been observed from the end of spring until the end of July.

{p. 185} Subgenus *Cyrtopodion* Fitzinger, 1843, stat. n[ov].

Type species: *Stenodactylus scaber* Heyden, 1827.

1843 — *Cyrtopodion* Fitzinger, Syst. Rept. pp. 18, 93 {error in original – MG}.
1984 — *Mesodactylus* Szczerbak and Golubev, Vestn. Zool., no. 2, p. 54.

DIAGNOSIS. No more than 10 preanal pores in males; length of suture between scales of first pair of postmentals larger than half of height of mental scale; shape of dorsal tubercles trihedral to triangular-roundish; caudal tubercles broadly contact each other on lateral edges; one to two smaller tubercles, larger than surrounding scales, on the frontal part of the base of each such tubercle; a row of subcaudal scales narrower than tail width; height of one scale less than half its width.

ADDITIONAL CHARACTERS. Height of first supralabial scale from nostril to edge of mouth distinctly larger or insignificantly smaller than its width along edge of mouth; poorly developed, flattened, triangular protuberances may be present in upper posterior part of orbit; usually two to 10 small trihedral or conical tubercles among small granules on the posterior surface of thigh; tail gradually thins toward end, row of subcaudal scales may be replaced by two rows or just regular scales. Main color and pattern usually light and dark gray.

Includes six species. Known from Southwest Asia; one species penetrates Central Asia.

Key to Species of the Subgenus Cyrtopodion[1]

1 (2). Two nasal scales . . . . *T. elongatus* (Blanford)
2 (1). Three nasal scales 3
3 (4). Scales with preanal pores significantly larger than surrounding scales; tail segment covered below with three rings of homogeneous scales . *T. agamuroides* (Nikolsky)
4 (3). Scales with four to nine preanal pores do not differ from surrounding ones in size; tail segment covered below with one to two longitudinal rows of large plates . . . 5
5 (8). No more than 25 scales across mid-belly 8
6 (7). No more than nine well developed preanal pores . . . . *T. scaber* (Heyden)
7 (6). Ten to 12 very poorly developed pores from each side of thigh join eight well-developed preanal pores on each side . . . . *T. montiumsalsorum* (Annandale)
8 (5). No fewer than 25 scales across mid-belly 9
9 (10). 100-128 GVA . . . . *T. kacchensis* (Stoliczka)
10 (9). 124-142 GVA . . . . *T. watsoni* (Murray)

KACHHI THIN-TOED GECKO — *TENUIDACTYLUS KACHHENSIS* (STOLICZKA, 1872)

Type locality: Kachh Prov., southwest India.
Karyotype: not studied {error in original – MG}
1872 — *Gymnodactylus kachhensis* Stoliczka, Proc. Asiat. Soc. Bengal, Calcutta, (1), p. 79.
1874 — *Gymnodactylus brevipes*, Blanford, Ann. Mag. Nat. Hist., ser. 4, vol. 13, p. 453. {missing in original – MG}
1884 — *Gymnodactylus petrensis* Murray, Vertebr. Zool. Sind, p. 362.
1935 — *Gymnodactylus kachhensis kachhensis*, M. A. Smith, Fauna Brit. Ind., Rept., Amph., vol. 2, p. 43.
1954 — *Cyrtodactylus kachhensis*, Underwood, Proc. Zool. Soc. London, 1954, vol. 124, no. 3, p. 475.
1984 — *Tenuidactylus kachhensis*, Szczerbak and Golubev, Vestn. Zool., Kiev, no. 2, p. 54.

DIAGNOSIS. Three nasal scales; scales with preanal pores are equal to surrounding ones; large plates are present on lower surface of tail segments; 100-128 GVA; 28-35 ventral scales across midbody; four to eight well developed preanal pores.

{p. 186} LECTOTYPE. NMW 17383:5, adult male. "Katch, Stoliczka, 1874; 11.98." L = 42.3 mm; LCD = 50.3 mm; L/LCD = 0.84; 13 scales across head; 10 supralabial scales; 8 infralabials; 100 GVA; 27 ventral scales across midbody; seven preanal pores; 20/21 subdigital lamellae.

DEFINITION (Fig. 81) (from 86 specimens in ZIK, NMW, FMNH, MNHP [= MNHN], UMMZ, AMNH). {187} L adult males (n = 29) 35.3-45.8 (38.91 ± 0.58) mm; L adult females (n = 24) 35.3-43.8 (39.24 ± 0.51) mm; L/LCD (n = 15) 0.76-0.98 (0.83 ± 0.01); HHW (n = 32) 55-65 (59.31 ± 0.56); EED (n = 11) 38-48 (42.01 ± 0.01); supralabial scales (n = 172) 10-12 (10.85 ± 0.05); infralabials (n = 172) 7-10 (8.30 ± 0.05); GVA (n = 85) 13-18 (14.92 ± 0.12); DTL (n = 62) 17-25 (21.40 ± 0.83); scales around dorsal tubercle (n = 83) 11-15 (12.69 ± 0.11); GVA (n = 79) 100-128 (113.04 ± 0.62); ventral scales across

---

[1] Study of photographs of *G. brevipes* sent from Calcutta by Dr. B. K. Tikader, the small number of preanal pores (4) and absence of large dorsal tubercles makes this specimen, which belongs to this subgenus, [suggests it is] closely related to *T. kachhensis*, [but] the presence of large abdominal scales (20 to 22) – [is closer] to *T. scaber*. The absence of material does not allow us to determine its taxonomic allocation.

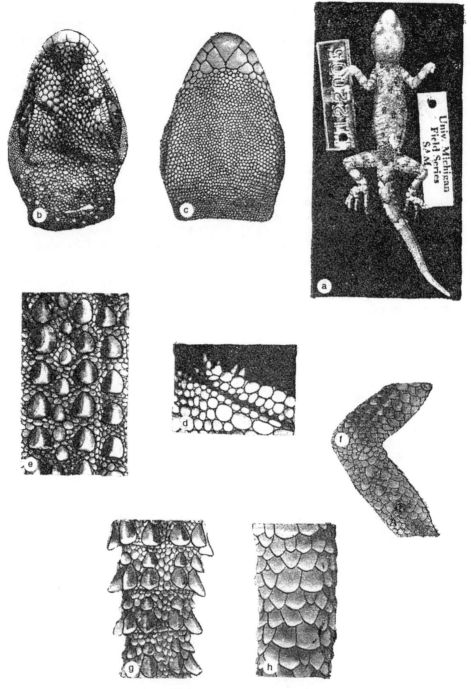

FIG. 81. *Tenuidactylus kachhensis*:
a, general view from above; b, head from above; c, head from below; d, upper edge of orbit; e, dorsum; f, forelimb from above; g, tail from above; h, tail from below (ZIK SR 1446, Haleji, Pakistan).

midbody (n = 76) 28-35 (30.47 ± 0.21); subdigital lamellae (n = 162) 18-23 (21.42 ± 0.10); preanal pores (n = 37) 4-8 (5.57 ± 0.19).

Mental scale triangular or pentagonal; two (23.4%) or three (76.6%) pairs of postmentals, 98.8% of scales of first pair contact each other; no additional dorsal tubercles (90.6%) or sometimes one (7.8%) **{p. 188}** or two (1.6%); always three nasals; first nasal scales separated by one (86.1%) or two (12.3%) scales, or they contact each other in 1.6%); subcaudal scales make combinations on one segment of 1/2/2, 2/2/2, 0/0/2, 1/1/2, and on the end of tail usually 1/1 or 2/2.

COLOR AND PATTERN are very similar to those of the closely related *T. scaber*. From the neck to the lumbar area, there are rows of four to five transverse bands in the form of spots. Three to five spots, sometimes connected with thin lines, correspond to each band. There are seven to ten transverse bands along the tail. The limbs have an indistinct transverse pattern. The underside is white.

DISTRIBUTION (Fig. 82). Southern Pakistan, extreme southwestern part of India: Kachh Peninsula, Sind, coastal zone of Las-Bela; to the north penetrates via river channels, but is not known with certainty above 29°N; R. Mertens' [258] and M. Khan's [235] records of this species in the northern parts of Pakistan (Swat and Northwest Frontier Province) are doubtful and require confirmation.

LEGEND to Fig. 82.
Pakistan: 1 – Mirpur (MNHP [= MNHN]); 2 – Dadu Dist.: Unapur (AMNH); 3 – 26 km N Bela [263]; 4 – Dadu Dist.: 8 km S Sehwan (AMNH); 5 – Las Bela, 13 km S Diwana (AMNH); 6 – 67 km SW Liari [263]; 7 – Tatta Dist.: Jerruck (AMNH); 8 – Khan Chowki (UMMZ), Bhawani [259]; 5 km SW Bund Murad Han [263]; 10 – Tatta Dist.: Haleji (AMNH); 11 – 12 km NW Karachi (UMMZ); Karachi (AMNH); Landhi; Khorangi [258]; 12 – 5 km E Landhi {error in original – MG}; Buleydji [263]; 13 – 1.6 km S Gujjo [263]; 14 – 3 km W Gharo [263]; 15 – Tatta Dist.: near Pir-Patho (AMNH); 16 – Kachh Peninsula, India (NMW); 17 – Sonda [258]; 18 – Attock [258] (needs re-examination); 19 – Ziarat [258]; 20 – near Lalian [235] {misidentified, *fide* M. S. Khan, Pakistan, pers. commun. – MG}.

Not located: Rabwah [235] {misidentified, *fide* M. S. Khan, pers. commun. – MG}; Ladha {error, see 11 and 12 above – MG} [232] {error in original – MG}.

SEXUAL DIMORPHISM AND AGE VARIATION. Males differ from females by presence of preanal pores; age variation not studied.

GEOGRAPHICAL VARIATION AND SYSTEMATICS are not discernable in available material.

ECOLOGY is practically unstudied. There are scant data from S. Minton [283].

HABITAT AND QUANTITATIVE DATA. It inhabits dry, rocky places. During the day, it hides under rocks and euphorb bushes. In [human] residences it is encountered rarely. This is an abundant species.

SEASONAL ACTIVITY CYCLE. Judging from collections (the extreme dates of occurrence are 24-25 January 1959; 20 February 1960 and 6 December 1959), the species is active year-round.

REPRODUCTION. It lays two 7 × 9.5 mm eggs.

PAKISTANI THIN-TOED GECKO — *TENUIDACTYLUS WATSONI* (MURRAY, 1892)

Type locality: Quetta, West Pakistan.
Karyotype: not studied.
1892 — *Gymnodactylus watsoni* Murray, Zool. Baloochist. S. Afghanist., p. 68.
1923 — *Gymnodactylus ingoldbyi* Procter, J. Bombay nat. Hist. Soc., vol. 29, p. 121.
1935 — *Gymnodactylus kachhensis watsoni*, M. A. Smith, Fauna Brit. Ind., Rept. Amph., vol. 2, p. 44.
1966 — *Cyrtodactylus watsoni*, Minton, Bull. Amer. Mus. Nat. Hist., (N.Y.), vol. 134, art. 2, p. 79.
1984 — *Tenuidactylus watsoni*, Szczerbak and Golubev, Vestn. Zool., Kiev, no. 2, p. 84.

DIAGNOSIS. Three nasal scales; scales that contain preanal pores are equal in size to surrounding ones; large scales present on underside of caudal segment; 124-142 GVA, 28-35 ventral scales across midbody; five to nine well-developed preanal pores.

TYPE SPECIMEN. Location is unknown.

DEFINITION (Fig. 83) (from 20 specimens in ZIK, AMNH, SAM, SMF). L adult males (n = 14) 42.0-53.1 (48.09 ± 1.03) mm; L adult females (n = 5) 35.0-54.0 (48.58 ± 3.48) mm; L/LCD (n = 5) 0.75-0.87 (0.805 ± 0.019); HHW (n = 18) 48-60 (56.4 ± 0.66); EED (n = 18) 37-52 (44.2 ± 0.011); supralabial scales (n = 40) 10-13 (11.40 ± 0.14); lower {p. 189} labials (n = 40) 8-10 (8.57 ± 0.11); scales across head (n = 20) 12-16 (13.80 ± 0.29); DTL

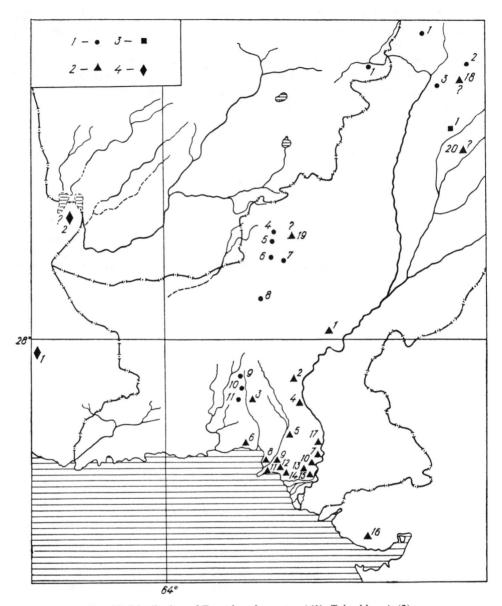

FIG. 82. Distribution of *Tenuidactylus watsoni* (1), *T. kachhensis* (2), *T. montiumsalsorum* (3), *T. agamuroides* (4).

(n = 19) 17-26 (22.16 ± 0.42); scales around tubercle (n = 19) 10-15 (12.15 ± 0.29); GVA (n = 17) 124-142 (132.11 ± 1.39); ventral scales across midbody (n = 18) 30-40 (35.55 ± 0.57); subdigital lamellae (n = {p. 190} = 38) 19-27 (22.76 ± 0.31); preanal pores (n = 14) 5-9 (7.36 ± 0.29).

Mental scale triangular or pentagonal; two (10%) or three (90%) pairs of postmentals, scales of first pair always contact each other; 90% of specimens do not have additional dorsal tubercles, or have one to two such tubercles (5% each); always three nasal scales; first nasals separated from each other by one (85%) or two (15%) scales; usual combination of subcaudal scales on one segment is 1/1, or 2/2 on the first one, but sometimes there are 1/2/1 and 2/2/2. Dorsal tubercles trihedral.

COLOR AND PATTERN The background is gray. There are five to eight transverse M-shaped bands on the back and 10-12 dark brown bands on the tail. There are five to six

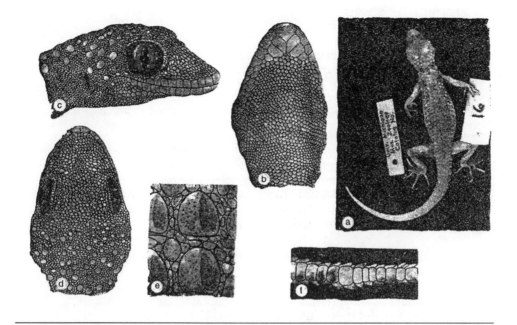
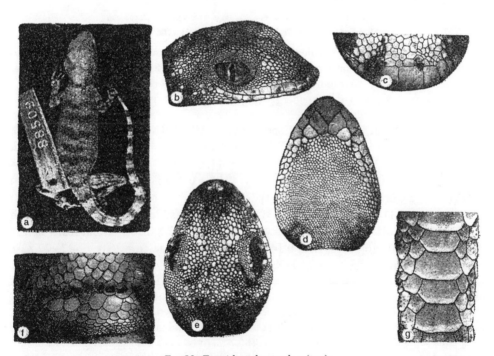

FIG 83. *Tenuidactylus scaber* (top):
a, general view from above; b, head from below; c, head from the side; d, head from above; e, dorsum; f, toe from below (ZIK SR 1451, Sanghar, Pakistan).
Tenuidactylus watsoni (bottom):
a, general view from above; b, head from the side; c, snout in front; d, head from below; e, head from above; f, preanal pores; g, tail from below (ZIK SR 1447, Wad, Pakistan).

indistinct small bands on the limbs. There is an indistinct pattern on the head. The lower surfaces are white.

DISTRIBUTION (Fig. 82). Northeast Afghanistan, northern and central Pakistan, almost does not reach below 28°N.

LEGEND to Fig. 82.
Afghanistan: 1 – Jalalabad [245].
Pakistan: 1 – NW Frontier Prov.: slopes near Udigram (AMNH); 2 – Manshera [258]; 3 – 13 km S Campbellpore (AMNH); 4 – Sibi Dist.: Kachh (AMNH); 5 – Quetta; 16 km S Quetta (AMNH); 6 – Quetta Dist.: 5 and 11 km S Kolpur (AMNH); 7 – Sibi Dist.: near Mach, Bolan River (AMNH); 8 – 13 km S Kalat (AMNH); 9 – 3 km S Wad (AMNH); 10 – 53 km S Wad (AMNH); 11 – Las Bela: 32 and 37 km N Bela (AMNH).
Not located: Ladha [232] {error in original – MG} [but see comment p. 183 under "Legend. Not located . . ." Eds.]

SEXUAL DIMORPHISM AND AGE VARIATION. Females lack preanal pores. Age variation was not studied.

GEOGRAPHICAL VARIATION was not found.

ECOLOGY has not been studied. There are records of this species at 1500-6000 ft. elevations (corresponding to 461-1884 m [263]). Based on collections, the earliest and latest finds are 18 March 1963 and 24 October 1961 [respectively].

ROUGH THIN-TOED GECKO — *TENUIDACTYLUS SCABER* (HEYDEN, 1827)

Type locality: vicinity of Tor, Sinai.
Karyotype: not studied.
1827 — *Stenodactylus scaber* Heyden in Rüppell {error in original – MG}, Atl. Reise nördl. Afr. Rept., p. 15.
1836 — *Gymnodactylus scaber*, Dumeril and Bibron (part.), Erpetol. gen., vol. 3, p. 421.
1954 — *Cyrtodactylus* {error in original – MG} *scaber*, Underwood, Proc. Zool. Soc. London, vol. 124, no. 3, p. 475.
1982 — *Cyrtodactylus basoglui* Baran and Gruber, Spixiana, vol. 5, 2, Munich, p. 121.
1984 — *Tenuidactylus scaber*, Szczerbak and Golubev, Vestn. Zool., Kiev, no. 2, p. 54.

DIAGNOSIS. Three nasal scales; scales that contain preanal pores equal [in size] to surrounding ones; large plates on underside of tail segment; 85-106 GVA; 16-32 ventral scales across midbody; four to nine well-developed preanal pores.

LECTOTYPE. Kept in Senckenberg Museum, Frankfurt am Main, FRG [Germany] [245].

DEFINITION (Fig. 83) (from 137 specimens in ZIK, ZIL, ZMMSU, UMMZ, SAM, AMNH, NMW). L adult males (n = 62); 34.4-50.0 (43.65 ± 0.49) mm; L adult females (n = 52) 34.3-53.2 (45.32 ± 0.73) mm; L/LCD (n = 45) 0.65-0.97 (0.76 ± 0.059); HHW (n = 47) 45-62 (53.00 ± 0.53); EED (n = 55) 30-60 (49.03 ± 0.009); supralabial scales (n = 206) 10-14 (11.38 ± 0.06); infralabials (n = 264) 7-10 (8.31 ± 0.04); scales across head (n = 137) 10-16 (12.72 ± 0.11); DTL (n = 117) 20-31 (24.74 ± 0.21); scales around dorsal tubercle (n = 132) 12-18 (15.05 ± 0.11); GVA (n = 119) 85-106 (90.87 ± 0.36); ventral scales across midbody (n = 120) 16-23 (19.47 ± 0.14); subdigital lamellae (n = 260) 19-27 (22.82 ± 0.09); preanal pores (n = 68) 4-9 (5.78 ± 0.09).

Mental scale triangular or pentagonal; two (21.1%), three (76.9%) or four (2.0%) pairs of postmentals, scales in first pair always contact each other; no additional dorsal tubercles in 93.7% specimens, sometimes it can be one (4.7%) or two (1.6%); three nasal scales, displaced first nasal scale occurs in [only] one specimen [and only] on one side; first na- **{p. 191}** -sals separated from each other by one (85.5%) or two (15.3%) scales, these scales are in contact in 2.2%; usual combination of subcaudal plates on one segment is 1/1. Dorsal tubercles trihedral, large, almost contact each other in longitudinal rows.

COLOR AND PATTERN (Fig. 83). Similar to *T. kachhensis* (see above).

DISTRIBUTION (Fig. 84). Distributed from the African coast of the Red Sea throughout Arabian Peninsula and Southwest Asia [including] Syria, southern Turkey, Iraq, Iran, Afghanistan, Pakistan, to western border of Thar Desert; records from western India are probable {inhabits Radjastan Desert [Biswas and Sanyal, 1977] – MG}.

LEGEND to Fig. 84.
Turkey: 1 – Suruc [46]; 2 – Urfa [46]; 3 – Kiziltepe [46]; 4 – Mardin [46].

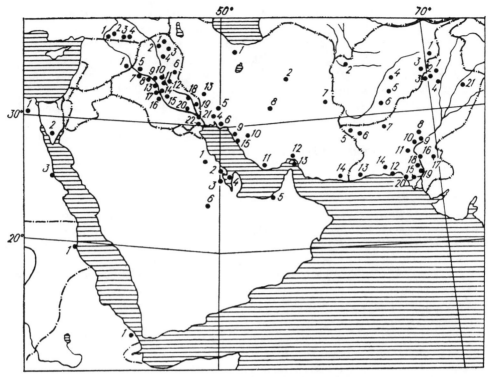

FIG. 84. Distribution of *Tenuidactylus scaber*.

Egypt: 1 – near Cairo [249]; 2 – Et-Tor (Tor, Sinai – terra typica restricta [307]); 3 – Quseir [249].
Sudan: 1 – Port-Sudan [249].
Ethiopia: 1 – Eritrea [249].
Syria: 1 – Abu-Kamal [173].
Iraq: 1 – Aqrah [269]; 2 – Tall Kayf, Al Mawsil [269]; 3 – Irbil Prov.: Koysanjak [269]; 4 – Halfaya [269]; 5 – Rawa [269]; 6 – Khanakin (ZIL); 7 – Anah [269]; 8 – Haditha [269]; 9 – Ath Tharthar [269]; 10 – Balad [269]; 11 – Diyala [269]; 12 – Al Khalis [269]; 13 – Ar Ramadi, Al Habbaniyah [269]; 14 – Al Hillah [269]; 15 – Ad Diwaniyah [269]; 16 – Al Kufah, An Najaf [269]; 17 – Karbala, Khan al Hammad [269]; 18 – Ali Al Gharbi [269]; 19 – Amarah [269]; 20 – An Nasiriyah [269]; 21 – Al Basrah [269]; 22 – Safwan [269].
Iran: 1 – Tehran (ZIL); 2 – Boshruye [285]; 3 – Shush, 32°11'N, 48°15'E (CAS); 4 – Khuzestan Prov.: Kurait, E of Ahvaz {error in original – MG}, 31°17'N, 48°49'E [168]; 5 – Mashjid-Soleyman (CAS); 6 – Agha Jari (CAS); 7 – Birjand (ZIL); 8 – Yazd (NMW); 9 – Kazerun (CAS); 10 – Shiraz (CAS); 11 – 1 km W Bandar-Langeh (CAS); Bandar-Langeh [223]; 12 – Bandar-Abbas [223]; 13 – Hormoz Island (NMW); 14 – Chah-Bahar [303]; 15 – Ahram (CAS).
Arabian Peninsula: 1 – El-Haba (CAS); 2 – Dhahran; Ras Tannura; Al Hubar [226]; 3 – Abqaiq (CAS); 4 – Hawar Island, Bahrain (ZMMSU); 5 – Tarif (CAS); 6 – Er-Ryad (CAS).
Afghanistan: 1 – Jalalabad [240]; 2 – Herat to Islam Qala, 34°22'N,62°10'E [209]; 3 – Khost (Matun) (CAS); 4 – 32 km S Mukur, 32°28'N, 67°30'E [209]; 5 – Shahi Safah, 60 km NE Kandahar (CAS); 6 – Kandahar, 31°30'N, 65°47'E [172].
{p. 192} *Pakistan:* 1 – Ladha [232]; 2 – Kirghi; Tank [232]; 3 – Wana [232]; 4 – Dera Ismail Khan [232]; 5 – Nok-Kundi [258]; 6 – Dalbandin [258]; 7 – Killi Jamaldini [258]; 8 – Shahdadkhoi [258 {this locality is attributed to this author by mistake, actual reference has not been found – MG}; 9 – Sukkur [202]; 10 – Mohenjo Daro [258]; Larkana Dist. (AMNH); 11 – Bubak, 15 km W Sehwan [258]; 12 – Las Bela: 65 km SW Liari (AMNH); 13 – Zarain Mauntains [258]; 14 – Gandrani Caves (AMNH); 15 – Mir Pur Sakro (MNHP); 16 – Hala [258]; 17 – 3 km NW Sanghar; Jamrao (AMNH); 8 km S Sanghar [263]; 18 – 9 km S Sehwan (AMNH); 4 km NW Tirth Lakhi [263]; Tirth (SAM); 19 – Hyderabad Dist.: 1.6 km W Kotri [263]; 20 – Bhawana [235]; 21 – Chiniot [235].
Not located on maps: Abu Guraib, Ar Rashdiyah, Al Kadhaniyah, Al Adhamiyah {error in original

– MG}, Babylon Prov. all from Khalaf, Monsaib [269]; Shah Baran, Mansuabad [284]; Rabwah, Lalian [235]; Kaur Bridge, Sarwekai [232] {error in original – MG}; Khasawiayah (CAS).

SEXUAL DIMORPHISM AND AGE VARIATION. As in other representatives of the genus, females of this species lack preanal pores. Juvenile specimens have shorter tails than adults.

SYSTEMATICS AND GEOGRAPHICAL VARIATION. Recently [181], a new species, *Cyrtodactylus basoglui*, was described from Turkey. As we were able to determine [46], it corresponds to *T. scaber* in all characters. Thus, I. Baran and U. Gruber discovered a new species for the Turkish fauna, rather than for science as a whole. We compared seven samples from all parts of the range. It was found that the rough thin-toed gecko lacks variation in scalation, but it does exhibit [geographic] variation in body size: the smallest geckos of this species inhabit the eastern part of the range, Pakistan, the largest occur in the western part, Egypt and Iran.

ECOLOGY has been studied insufficiently.

HABITAT AND QUANTITATIVE DATA. It inhabits rocky hill slopes with dry grassy cover or sands with xerophytic vegetation [263]. It is the most common species in buildings at the bases of mountains.

RESPONSE TO TEMPERATURE. It was found on 5 December in fairly cold weather [263].

DAILY ACTIVITY CYCLE. It is active at dusk and at night [168].

SEASONAL ACTIVITY CYCLE. It can be encountered from the end of July until November [168]. Based on collection materials, the first individual was obtained on 17 February 1962, in the vicinity of Liari in Pakistan, and the last, on 5 December 1959 in the vicinity of Dadu Village in Pakistan.

REPRODUCTION. They begin laying eggs in mid-August. Eggs are 7×10.5 mm [168].

BEHAVIORAL ATTRIBUTES. They frequently hunt insects attracted to light [168]. This is probably a synanthropic species.

SALT RANGE THIN-TOED GECKO — *TENUIDACTYLUS MONTIUMSALSORUM* (ANNANDALE, 1913)

Type locality: Salt Range, western Punjab, Pakistan.
Karyotype: not studied.
1913 — *Gymnodactylus montiumsalsorum* Annandale, Rec. Ind. Mus., Calcutta, vol. 9, part 5, p. 313.
1966 — *Cyrtodactylus montiumsalsorum*, Minton, Bull. Amer. Mus. Nat. Hist., N. Y., vol. 134, art. 2, p. 78.
1984 — *Tenuidactylus montiumsalsorum*, Szczerbak and Golubev, Vestn. Zool., Kiev, no. 2, p. 54.

DIAGNOSIS. Three nasal scales; scales, which contain preanal pores, are equal [in size] to surrounding ones; large scales on underside caudal segment are present; 86 GVA; 19 ventral scales across midbody; 10 to 12 very poorly developed pores, penetrate thighs, adjoin eight well developed preanal pores from the each side.

DEFINITION[1] (Fig. 85) (from holotype BMNH 1904.11.19.1, adult male. Salt Range, Punjab). L = 37.2 mm; tail damaged; HHW 58; EED 44; 10 supralabial scales; 7/8 infralabials; 11 scales across head; DTL 30; 15 scales around dorsal tubercle; 86 GVA; 19 ventral scales across midbody; {p. 193} 21/20 subdigital lamellae; eight preanal pores, 10-12 very poorly developed pores on each side of thigh join the 8 preanal.

Mental scale pentagonal; two pairs of postmentals, scales of first pair contact each other; additional dorsal tubercles do not contact the regular ones; three nasal scales, first nasals separated from each other by a scale; dorsal tubercles trihedral.

COLOR AND PATTERN. The specimen is faded.

DISTRIBUTION. Known only from the type locality (Fig. 82).

{p. 194} SYSTEMATICS. As it was noted previously [153], the presence of additional poorly developed preanal pores in both specimens is probably an anomaly; it fits within the limits of *T. scaber* diagnosis in all other characters. However, the final identification of *T. montiumsalsorum* as *T. scaber* will be possible only after additional comparative material is obtained.*

---

[1] A photograph of a paralectotype, which is kept in Zoological Survey of India, was also used.

* {The distinctness of this species was recently demonstrated by M. S. Khan (1989) – MG.}

FIG. 85. *Tenuidactylus montiumsalsorum*:
a, general view from above; b, head from the side; c, snout in front; d, head from below; e, head from above; f, dorsum; g, preanal pores (BMNH 1904.11.19.1, Punjab, Pakistan).

ECOLOGY has not been studied.

KASHGHAR THIN-TOED GECKO — *TENUIDACTYLUS ELONGATUS* (BLANFORD, 1875)

Type locality: Yangihissar, western China.
Karyotype (possibly): 2n = 42; 38 acrocentrics, 2 subtelocentrics, 2 telocentrics, NF = 42 {error in original – MG}.

1875 — *Gymnodactylus elongatus* Blanford, J. Asiat. Soc. Bengal, Calcutta, vol. 44, part 2, p. 193.
1954 — *Cyrtodactylus elongatus,* Underwood, Proc. Zool. Soc. London, vol. 124, no. 3, p. 475.
1984 — *Tenuidactylus elongatus,* Szczerbak and Golubev, Vestn. Zool., Kiev, no. 2, p. 54.

DIAGNOSIS. Two distinctly bulbous nasal scales; scales that contain preanal pores are more or less equal to surrounding ones; large plates are present on underside of caudal segment; 102-132 GVA; 26-30 ventral scales across midbody; four to seven well-developed preanal pores.

TYPES are kept in the Indian Museum, Calcutta.

DEFINITION (Fig. 86) (from 33 specimens in ZIK, ZIL, and ZMMSU). L adult males (n = 10) 43.9-55.6 (49.27 ± 1.05) mm; L adult females (n = 18) 42.2-56.8 (50.03 ± 0.94) mm; L/LCD (n = 12) 0.7-0.8 (0.76 ± 0.01); HHW (n = 29) 54-67 (59.03 ± 0.006); EED (n = 12) 39-44 (41.25 ± 0.46); supralabial scales (n = 66) 10-13 (11.80 ± 0.10); infralabials (n = 65) 8-11 (9.72 ± 0.11); scales across head (n = 33) 15-19 (16.71 ± 0.18); DTL (n = 27) 14-19 (16.41 ± 0.26); scales around dorsal tubercle (n = 33) 8-12 (10.45 ± 0.16); GVA (n = 32) 102-132 (119.37 ± 1.13); ventral scales across midbody (n = 29) 26-30 (27.41 ± 0.23); subdigital lamellae (n = 64) 19-23 (21.23 ± 0.13); preanal pores (n = 14) 4-7 (5.80 ± 0.24).

Mental scale pentagonal; one (3.7%), two (87%), or three (9.3%) pairs of postmentals, scales in first pair contact each other; additional dorsal tubercles contact regular tubercles only in 14.7%, others lack the contact or the additional tubercles; always two nasal scales, they are very bulbous, first nasal scales separated from each other by two (41.2%), three (53%), or four (5.8%) scales; combinations of subcaudal plates on one segment are: 1/1/1 (on first segments up to II – VIII), then up to the end 1/1, rarely 1/2. Dorsal tubercles roundish-triangular.

COLOR AND PATTERN. Dorsally, the background is light gray. On the sides of the head, there is a weakly defined small stripe from the eye to the neck and in the corner of the mouth. There are five to six brown transverse bands on the body from the neck to the lumbar area. The tail has 7-10 bands of the same color. The interspaces equal the width of the bands. The under surface is white.

DISTRIBUTION. (Fig. 15) P. R. C. and P. R. M. {China and Mongolia – MG}: Kashgharia Gobi Desert up to Khara-Khoto ruins and adjacent, from northwest, parts of Transaltai Gobi)*.

LEGEND to Fig. 15.
China: 1 – Yangihissar [187]; 2 – ruins of Khara-Khoto, C Gobi (ZIL).
Mongolia: 1 – Nogontsav [78]; 2 – Shara-Khulsny-Bulak [78]; 3 – Tsugalbar-Us [78]; 4 – Tsuvra-Khara-Ula [78]; 5 – Khutsin-Khara-Ula [78].

SEXUAL DIMORPHISM AND AGE VARIATION. Females lack preanal {error in original – MG} pores. Age variation not studied.

SYSTEMATICS AND GEOGRAPHICAL VARIATION. Samples from the Gobi were available. Geographical variation was not found there. One should take into account that this species is described from Kashgharia, 2,000 km to the west. The possibility that the western samples will differ from the eastern ones should not be excluded.

ECOLOGY has not been studied sufficiently.

{p. 195} HABITAT AND QUANTITATIVE DATA. It is found in rocky terrain, in canyons, and on outliers. It is a rare species [31].

RESPONSE TO TEMPERATURE. A gecko of this species that was available to us was studied in a thermal gradient apparatus (n = 42). It was observed in the temperature range of 28.5°-40°. The preferred temperature was 34.1° ± 0.5°. It is apparently a heat-loving species.

DAILY ACTIVITY CYCLE. Based on observations in the terrarium, this is a nocturnal species.

SEASONAL ACTIVITY CYCLE has not been studied. The latest known record is from the vicinity of Khara-Khoto from 10-12 December 1926 (from the ZIN collection).

---

* {See also Borkin *et al.*, 1990 and Zhao and Adler, 1993 – MG.}

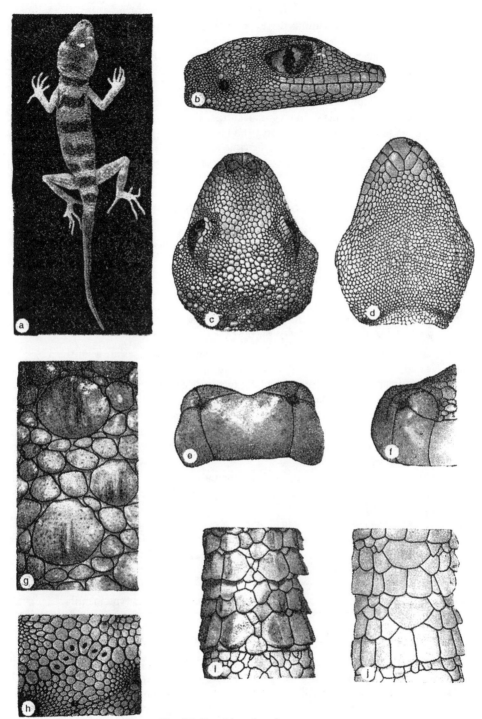

FIG. 86. *Tenuidactylus elongatus*:
[N.B. In the original publication, figures 86 and 87 are reversed. This is figure 87 in the original, but it should have been 86 and is correctly positioned here. Eds.]
a, general view from above; b, head from the side; c, head from above; d, head from below; e, rostral and nasal scales in front; f, nasal scales from the side; g, dorsum; h, preanal pores; i, tail from above; j, tail from below (ZIK SR 2642, southern Gobi, Mongolia).

FEEDING has not been studied. In captivity, it willingly ate mealworms.

REPRODUCTION. In our sample, there are nearly twice as many females {p. 196} as males (18 to 10). Detailed data on reproduction are not available.

ENEMIES. There is a known instance of an attack on this gecko by the steppe ribbon snake (*Psammophis lineolatum*) [31].

BEHAVIORAL ATTRIBUTES. According to our observations in the terrarium, the Kashghar geckos, unlike the other species of the genus, burrow in the sand well and behave like a psammophilous species and not like vertical surface dwellers. At the same time, in the Gobi they have been found in rock cracks, caves and under rocks [31].

PRACTICAL IMPORTANCE AND PROTECTION. This is a rare and unstudied species. It is included in the *Red Book of the Peoples' Republic of Mongolia* and is protected in the Gobi Biosphere Preserve.

AGAMUROID THIN-TOED GECKO — *TENUIDACTYLUS AGAMUROIDES* (NIKOLSKY, 1900)

Type locality: Pensareh (Pendzhsara), eastern Kirman, Iran.
Karyotype: not studied.

1900 — *Gymnodactylus agamuroides* Nikolsky, Ann. Zool. Mus. for 1899 {error in original – MG}, SPb [St. Petersburg], vol. 4, no. 4, p. 384.
1963 — *Cyrtodactylus agamuroides*, S. Anderson, Proc. Calif. Acad. Sci., San Francisco, ser. 4, vol. 31, p. 438 (err. det.).
1966 — *Agamura agamuroides*, Minton, Bull. Amer. Mus. Nat. Hist., N.Y., vol. 134, art. 2, p. 80 (err. det.).
1984 — *Tenuidactylus agamuroides*, Szczerbak and Golubev, Vestn. Zool., Kiev, no. 2, p. 54.

DIAGNOSIS. Three nasal scales; scales containing preanal pores are significantly larger than surrounding ones; three rings of scales are present on lower surface of each tail segment; 120 GVA; 28 ventral scales across midbody; two well-developed preanal pores.

DEFINITION (Fig. 87) (from lectotype ZIL 9327, adult male. "Pendzh-Sara, E Kirman, 10 Aug. 1898, N. Zarudny"). L = 38.7 mm; LCD = 49.1 mm; L/LCD = 0.79; HHW 63; EED 40; 15/14 supralabial scales, 10 infralabials; 16 scales across head; DTL 18; 10 scales around dorsal tubercle; 120 GVA; 28 ventral scales across midbody; 20/21 subdigital lamellae; two preanal pores, no additional pores.

Mental scale pentagonal; two pairs of postmentals, scales of first pair contact each other; no additional dorsal tubercles; first nasal scales separated from each other by a scale; dorsal tubercles roundish-triangular with weak keels.

COLOR AND PATTERN (as fixed in alcohol). It resembles *T. kachhensis* and *T. kirmanensis*: there are seven transverse bands between the neck and the lumbar area, each of which is formed by three to five spots. There are 13 solid transverse bands on the tail and four to five transverse bands on the limbs. The head has an indistinct pattern. The lower surface is white.

DISTRIBUTION (Fig. 82). Known with certainty only from the type locality.

LEGEND to Fig. 82.
Iran: 1 – Pensareh [89]; Neizar in Sistan ([89] – this record requires confirmation).

SYSTEMATICS. As was shown [47], S. Anderson [168] and S. Minton [263] incorrectly determined their specimens, attributing this species to *Agamura* (cf. *persica*). This can be explained not only by the fact that the diagnosis of the agamuroid thin-toed gecko was constructed on the basis of specimens from different species, but also by their close relationship.

ECOLOGY has not been studied. According to N. A. Zarudny's data [60], they are found in abundance on the ruins of earthen buildings, in shady areas of clay, and [on] rock cliffs. Early in the morning and in the evenings they can be seen in flat areas within a few hundred paces of their diurnal cover. They are also active during the daytime, but only in the shade.

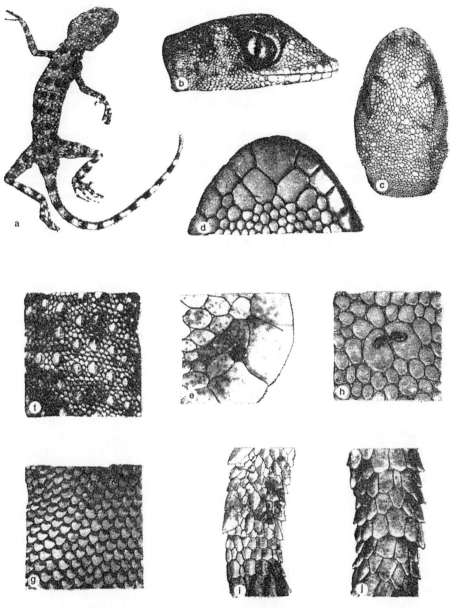

FIG. 87. *Tenuidactylus agamuroides*:
[N.B. See note, Fig. 86; figs. 86 and 87 reversed in original; but legends were correctly placed. Eds.]
a, general view from above; b, head from the side; c, head from above; d, chin; e, dorsum; f, belly; g, nasal scales; h, preanal pores; i, tail from above; j, tail from below (ZIL 9327, Pensareh, Iran).

{p. 197} Tibeto-Himalayan group of *Tenuidactylus*

This group, in many characters, is transitional between East Asian naked-toed geckos of the genus *Cyrtodactylus* and the Palearctic thin-toed geckos of the genus *Tenuidactylus*. However, despite the presence of some ancestral characters, this group is generally closely related to the latter genus and is **{p. 198}** considered as being within its limits. Almost all the species are represented by a single or a few specimens in museum collections, including *T. fasciolatus*, which may also belong to this group. Type specimens of four species, *T. lawderanus*,[1] *T. himalayanus* {error in original – MG},[2] *T. fasciolatus*[3] {error in original – MG}, and *T. dattensis*[4] {correct spelling is *dattanensis*, see Khan, 1980:11 – MG} were inaccessible to us, and their published descriptions do not take into account many of the principal characters used in the present work. This forced us to not include the diagnoses of the species noted above. It is very difficult to determine not only the subgeneric allocation but also the species independence on the basis of incomplete descriptions. On the other hand, the high level of variation of this entire group should be noted, as we did previously [153]. This phenomenon has already led to the appearance of numerous synonyms: sometimes, their number approaches the number of known specimens (*T. lawderanus, T. stoliczkai*). Examination of the species types that inhabit the Himalayas proper will probably increase their number.

Thus, the key to the species and the brief group characterization are just preliminary.

BRIEF CHARACTERISTICS OF TIBET-HIMALAYAN THIN-TOED GECKOS GROUP. Tail is distinctly depressed at the base, thick, abruptly thinner in the last third, equal to or slightly longer or shorter than the body length; in some species, tail segments are not defined, tail tubercles and subcaudal plates are very poorly developed; three to four nasal scales, first one larger than others; usually height of first supralabial scale distinctly less than its width; commonly head scales small, homogeneous; frontal depression is often better defined than in other congeners (but not so strong as in *Cyrtodactylus*); postmentals are always defined well, scales of first pair always in contact with each other; dorsal tubercles small, usually roundish, sometimes oval-triangular with indistinct keels, no additional tubercles; other scales always smooth; 120-160 GVA; four to six preanal pores; limbs sometimes slightly shortened; basal phalanges somewhat widened in some species. Pattern and color vary greatly: some species retain a marbled pattern on head and fragments of neck crossband and body transverse crossbars broken in the middle; this pattern in some species turns into M-shaped bands (sometimes preserved only as partial spots), common in Palearctic geckos; these bands appear as thin transverse vermicular lines.

Key to the Species of the Tibeto-Himalayan Group of Thin-Toed Geckos
[of the Genus Tenuidactylus]

1 (6). Caudal segments not defined 6
2 (3). No more than 130 GVA . . . . *T. tibetanus* (Boulenger)
3 (2). No fewer than 130 GVA 4
4 (5). Dorsal transverse bands very thin, vermiculate . *T. mintoni* (Golubev et Szczerbak)
5 (4). Dorsal transverse bands wide, often interrupted in the middle . . . .
 . . . . *T. chitralensis* (M. A. Smith)
6 (1). Caudal segments well developed 7
7 (8). Wide transverse continuous dorsal bars; L/LCD no more than 0.8; no subcaudal plates . . . . *T. stoliczkai* (Steindachner)
8 (7). Spot-shaped fragments of transverse dorsal bars; L/LCD no more than 0.8; sub-caudal plates well developed on less than last third of tail . . . .
 . . . . *T. kirmanensis* (Nikolsky)

{p. 199} TIBETAN THIN-TOED GECKO — *TENUIDACTYLUS TIBETANUS* (BOULENGER, 1905)

Type locality: Chaksam Ferry, Tsangpo Valley (Brahmaputra), Tibet, China.
Karyotype not studied.

1905 — *Alsophylax tibetanus* Boulenger, Ann. Mag. Nat. Hist., London, ser. 7, vol. 15, p. 378.
1981 — *Gymnodactylus tibetanus*, Golubev, Vopr. Gerpetologii, Leningrad, p. 39.
1984 — *Tenuidactylus tibetanus*, Szczerbak and Golubev, Vestn. Zool., Kiev, no. 2, p. 55.

{p. 200} DIAGNOSIS. No more than 130 GVA; dorsal crossbars wide, interrupted in the

---

[1] Distribution [according to 291] in India (Fig. 89): 1 – Simla, Himalaya; 2 – Ambala; 3 – Tehri Garhwal; 4 – Almora; 5 – W Nepal.

[2] Distribution (Fig. 89): 1 – India: Jammu and Kashmir: Doda, Kishtwar [210].

[3] Distribution (Fig. 89): 2 – India: Simla; 3 – Kumaon; Subatu; Almora, 1500 m elevation [291].

[4] Distribution (Fig. 89) in Pakistan: 1 – NW Frontier Prov.: Datta [330] (possibly, specimen SMF 63548, see Fig. 90, in part, is this form).

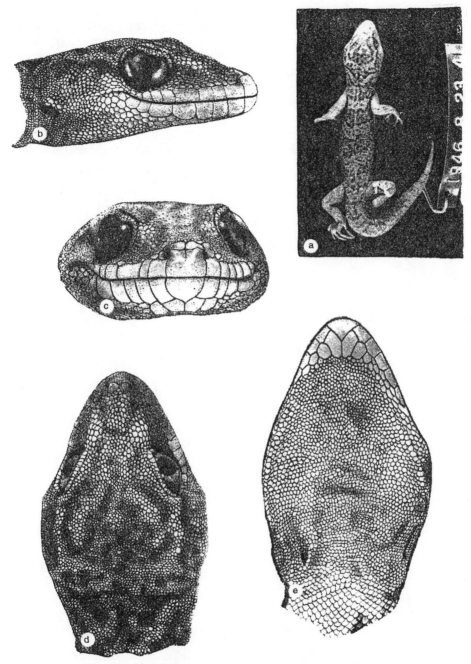

FIG. 88. *Tenuidactylus tibetanus*:
a, general view from above; b, head from the side; c, head in front; d, head from above;
e, head from below (BMNH 1946.8.23.41, Changpo Valley, Tibet).

middle; caudal segments not defined; two-thirds of lower tail surface covered by plates; L/LCD more than 1; {error in original – MG} 25 scales across head.

DEFINITION (Fig. 88) (from holotype BMNH 1905.2.8.5 RR 1946.8.23.41, adult female. "Tsangpo Valley, Tibet. Pres: Lt Col. L A. Waddell" {error in original – MG}). L = 52.3 mm; LCD = 45 mm; L/LCD = 1.16; HHW 57; EED 60; 9/10 supralabial scales; 7/8 infralabials; 25 scales across head; DTL 26; 8-9 scales around dorsal tubercle; 126 GVA; near 30 ventral scales across midbody; 21/21 subdigital lamellae; no preanal pores.

{p. 201} Mental scale triangular; three to four pairs of postmentals, scales of first pair contact each other; first nasal scales also contact each other; three nasal scales; subcaudal plates defined in last two thirds (in 1/1 combination per segment); caudal tubercles defined only on tail base; dorsal tubercles smooth, roundish, arranged in more or less regular rows; limbs somewhat shortened; adpressed forelimb extends slightly beyond anterior edge of orbit with finger tips; basal phalanges of digits not widened.

COLOR AND PATTERN. The main background is coffee-brown. The head has a marbled pattern, which is formed by small dark spots and vermiculate small lines. There are fragments of a neck stripe, characteristic of the species from the genus *Cyrtodactylus*, visible on the neck. Between the shoulder and the lumbar area, there are four wide transverse bands, with straight edges, and darker than the main background color. Each band is broken in the middle

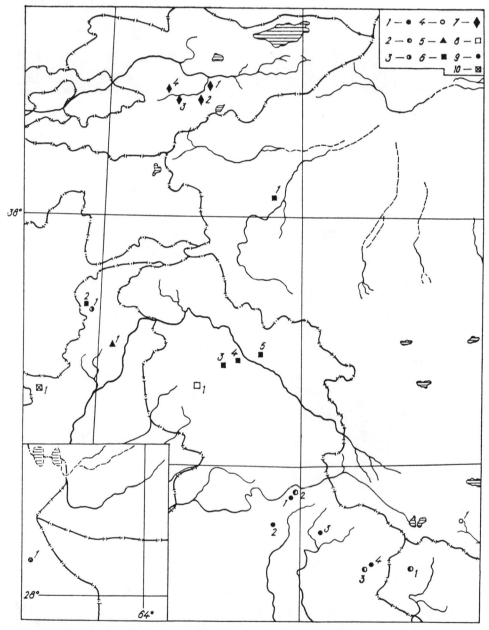

FIG. 89. Distribution of *Tenuidactylus lawderanus* (1), *T. himalayanus* (2), *T. chitralensis* (3), *T. tibetanus* (4), *T. mintoni* (5), *T. stoliczkai* (6), *Alsophylax tokobajevi* (7), *T. fasciolatus* (8), *T. kirmanensis* (9), *T. dattanensis* (10) {error in original – MG}.

and has a dark brown border along front and rear edges. The tail and limbs lack a pattern. The lower surface is white.

DISTRIBUTION (Fig. 89). Known only from type locality.

SYSTEMATICS. This species has been transferred to the thin-toed gecko genus from *Alsophylax* [37]. It is very closely related to *T. chitralensis* and, probably, to *T. lawderanus*, *T. himalayanus*, and *T. fasciolatus*.

ECOLOGY not studied. It is known that the type specimen was caught at 12,000 ft elevation [205].

CHITRAL THIN-TOED GECKO — *TENUIDACTYLUS CHITRALENSIS* (M. A. SMITH, 1935)*

Type locality: Karakal in the Bumhoet Valley, Chitral, northern Pakistan.
Karyotype: not studied.
1935 — *Gymnodactylus chitralensis* M. A. Smith, Fauna Brit. Ind., Rept. Amph., vol. 2, p. 46.
1969 — *Gymnodactylus (Cyrtodactylus) chitralensis*, Mertens, Stutt. Beitr. Naturk., Stuttgart, no. 197, p. 24.
1984 — *Tenuidactylus chitralensis*, Szczerbak and Golubev, Vestn. Zool., Kiev, no. 2, p. 55.

DIAGNOSIS. No more than 130 GVA; dorsal crossbars wide, interrupted in the middle; caudal segments not defined; a row of enlarged plates may be present on underside of last third of tail; L/LCD = 0.98; 26-32 scales across head.

TYPES are kept in the Indian (Calcutta) and British (London) museums.

DEFINITION (Fig. 90) (from 3 specimens in SMF and MNHP [= MNHN]). L = 36.3-59.5 mm; L/LCD = 0.80; HHW 49-56; EED 63; supralabial scales (n = 6) 10-12 (11.31 ± 0.28); infralabials (n = 6) 8-9 (8.89 ± 0.01); scales across head 26-32; DTL 8-9; GVA 142-163; ventral scales across midbody 30-34; subdigital lamellae (n = 6) 15-22 (17.80 ± 1.31); eight preanal pores.

Mental scale triangular or pentagonal; two to three pairs of postmentals, scales of first pair contact each other; first nasal scales separated from each other by a scale; three to four nasal scales; subcaudal plates may be defined on last third of tail (combinations 1/2 and 1/1 on a segment); caudal tubercles present only on the tail base; dorsal tubercles roundish or slightly oval, smooth, arranged in relatively regular rows; limbs somewhat shortened, adpressed forelimb does not reach tip of snout with finger tips; basal digit phalanges not widened.

COLOR. Similar to color of preceding species or differs by narrower transverse slightly zigzaging bands.

DISTRIBUTION (Fig. 89). Inhabits Chitral in northern Pakistan.

ECOLOGY not studied. Earliest record from Abbotabad on 12 May 1958 (from collection specimens).

{P. 202} MINTON'S THIN-TOED GECKO — *TENUIDACTYLUS MINTONI* (GOLUBEV ET SZCZERBAK, 1981)

Type locality: Udigram, Swat, Pakistan.
Karyotype: not studied.
1981 — *Gymnodactylus mintoni* Golubev and Szczerbak, Vestn. Zool., Kiev, no. 3, p. 40.
1984 — *Tenuidactylus mintoni*, Szczerbak and Golubev, Vestn. Zool., Kiev, no. 2, p. 55.

{p. 203} DIAGNOSIS. No less than 130 GVA; dorsal transverse bands thin, vermiculate; caudal segments not defined; no subcaudal scales along entire tail; L/LCD = 1.18; near 30 scales across head.

DEFINITION (Fig. 91) (from the holotype SAM 683, adult female (?) "Udigram in Swat, W. Pakistan, 30 March, S. Minton"). L = 38.4 mm; L/LCD = 1.18; HHW 60; EED {p. 204}

---

* {M. S. Khan (1992) synonymized this name with *Gymnodactylus walli* Ingoldby, 1922. Also, he noted that one of the specimens, which is illustrated in Fig. 90 under the name *T. chitralensis* (SMF 63548), is one of the paratypes of *Gymnodactylus dattanensis* (Khan, 1980) (see also footnote 4, p. 193 [p. 198 in original]); another specimen, MNHN 1916.63, was mistakenly determined by us as *T. chitralensis*, Upper Indus Valley, Pakistan; it belongs to *Cyrtodactylus nebulosus* (central India) – MG.}

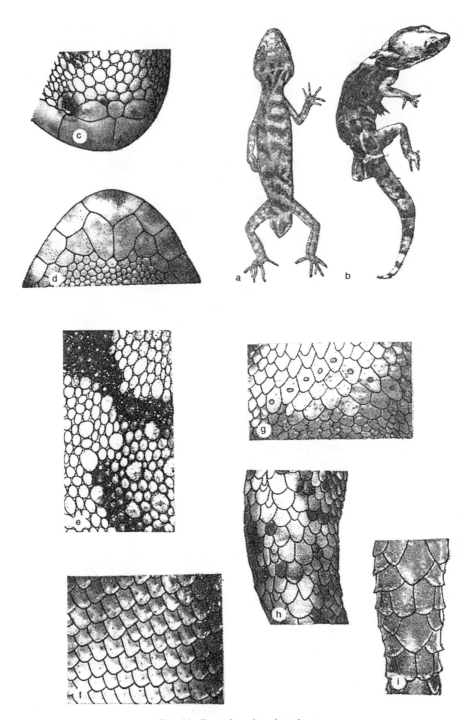

FIG. 90. *Tenuidactylus chitralensis*:
a-b, general view from above; c, snout in front; d, chin; e, dorsum; f, belly; g, preanal pores; h, tail from above; i, tail from below (b, e – MNHP [= MNHN] 1916.63, Upper Indus, Pakistan; others – SMF 63548, Abbotabad, Pakistan).

32; 11 supralabial scales; 9/8 infralabials; near 30 scales across head; DTL 8; 150 GVA; 36 ventral scales across midbody; 17/17 subdigital lamellae; no preanal pores.

Mental scale pentagonal; two to four pairs of postmentals, scales of first pair contact each other, one or two last pairs {p. 205} are poorly differentiated from gular scales; first nasal scales separated from each other by two scales; three to four nasal scales, the last is

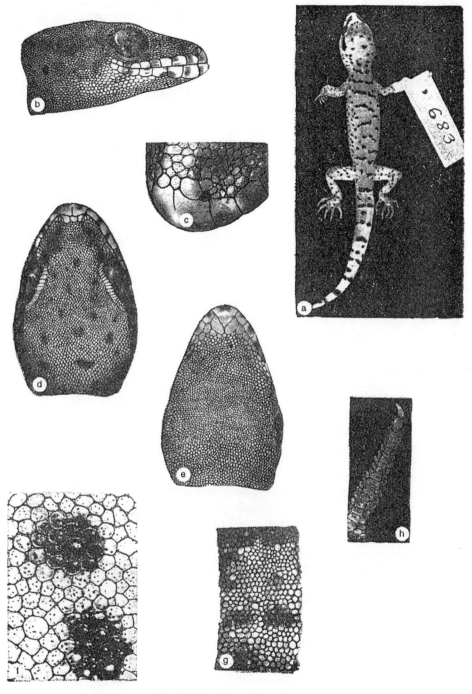

FIG 91. *Tenuidactylus mintoni*:
a, general view from above; b, head from the side; c, nasal scales; d, head from above; e, head from below; f, dorsum; g, tail from above; h, toe from below (SAM 683, Swat, Pakistan).

poorly differentiated; no subcaudal plates; dorsal and caudal tubercles almost not defined; limbs somewhat shortened, adpressed forelimb reaches anterior edge of orbit with finger tips; basal digital phalanges slightly widened.

COLOR AND PATTERN [263]. The main background color on the back is amber, becoming lemon on the tail. On the back, there is a series of eight uneven, broken bands that are dark, very thin and, in the majority, have a white rear border. Small dark dots and spots are scattered on the sides and on the dorsal surfaces of the head, tail and limbs. The abdomen is pale yellow. The labial scales have alternating black and white bars. A brown stripe, also edged with white on top, runs from the nostril, through the eye and to the ear.

DISTRIBUTION (Fig. 89). Known only from type locality.

ECOLOGY has not been studied.

STOLICZKA'S THIN-TOED GECKO — *TENUIDACTYLUS STOLICZKAI* (STEINDACHNER, 1868)

Type locality: Karoo, north of Dras, Kashmir.

Karyotype: not studied.

1868 — *Gymnodactylus stoliczkai* Steindachner, Reise Novara, Zool., 1. Rept., p. 15.
1872 — *Cyrtodactylus yarkandensis* J. Anderson, Proc. Zool. Soc. London, p. 381.
1922 — *Gymnodactylus walli* Ingoldby, J. Bombay Nat. Hist. Soc., vol. 28, p.1051.
1954 — *Cyrtodactylus stoliczkai*, Underwood, Proc. Zool. Soc. London, vol. 124, no. 3, p. 475.
1984 — *Tenuidactylus stoliczkai*, Szczerbak and Golubev, Vestn. Zool., Kiev, no. 2, p. 55.

DIAGNOSIS. Fewer than 150 GVA; dorsal transverse crossbars wide, M-shaped; caudal segments well defined; no enlarged subcaudal plates along entire tail; L/LCD = 0.85-0.90; 17-20 scales across head.

DEFINITION (Fig. 92) (from holotype NMW 16756, adult female, "Karoo bei Dras, Kashmir, 1866, Stoliczka," type *Gymnodactylus walli* BMNH 1910.7.12.1, and *Cyrtodactylus yarkandensis* BMNH 72.3.22.4). L = 27-48 mm; L/LCD = 0.85-0.98; HHW 52-59; EED 28-36; supralabial scales (n = 6) 10-11 (10.87 ± 0.02); infralabials (n = 6) 7-8 (7.88 ± 0.02); scales across head 17-20; DTL 8-10; GVA 120-149; ventral scales across midbody 27-32; subdigital lamellae (n = 6) 23-24 (23.5 ± 0.01); no preanal pores (all specimens examined were females).

Mental scale triangular or pentagonal; three to four pairs of postmentals, scales in first pair contact each other; first nasal scales separated from each other by one to two scales; three nasal scales; extrimities moderate in size; adpressed forelimb reaches tip of snout with finger tips; basal digit phalanges not widened; tail flattened and segments very distinct (8-10 segments); three to four conical tubercles on middle part of each side of a segment, they do not contact each other; subcaudal plates not defined. Dorsal tubercles roundish, smooth, appear almost in disorder; small tail scales smooth.

COLOR AND PATTERN. On light background, from the neck to the sacrum, there are seven to eight wide, M-shaped transverse dark bands. A neck stripe runs to the nostrils through the eyes on both sides of the head. Each of these stripes becomes darker from the front (closer to the head) to the rear, where it is bordered by a thin brown line. The tail has about 12 transverse bands of the same color. On top of the head the pattern is indistinct. The limbs are covered with thin, small transverse bands. The lower surfaces are white.

DISTRIBUTION (Fig. 89). Known from Chitral (northern Pakistan) and western Kashgharia* (China); inhabits the mountain systems of northern India between the aforementioned areas.

{p. 206} LEGEND to Fig. 89.
China: 1 – Yarkand [164 {error in original – MG}]*.
Pakistan: 2 – Chitral [258].
India: 3 – Kashmir: Karoo N of Dras (NMW); 4 – Kashmir: Kargil [263]; 5 – Ladakh: Saspool, Ghompa; Hemis [324].

SYSTEMATICS AND GEOGRAPHICAL VARIATION not studied.**

ECOLOGY not studied. An early record is from Chitral on 26 April 1962 (from collection specimens).

---

* This record is likely in error (see Zhao and Adler, *Herpetology of China*, 1993, p. 179, top). This species is probably not native to China; at least no specimens with *bona fide* collecting data have been secured in the country.

** {M. S. Khan (1992) and N. N. Szczerbak (1992) described two new species, *Tenuidactylus baturensis* and *Alsophylax boehmei* respectively, which are very close to (if not the same as) this species – MG}

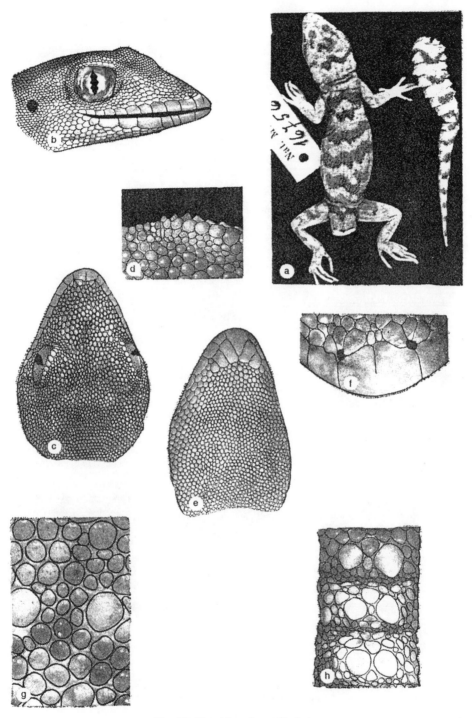

FIG. 92. *Tenuidactylus stoliczkai*:
a, general view from above; b, head from the side; c, head from above; d, upper edge of orbit; e, head from below; f, snout in front; g, dorsum; h, tail from the side (NMW 16756, Kashmir, Pakistan).

{p. 207} KIRMAN THIN-TOED GECKO — *TENUIDACTYLUS KIRMANENSIS* (NIKOLSKY, 1900)

Type locality: Kuh-e-Taftan, eastern Kirman, Iran.

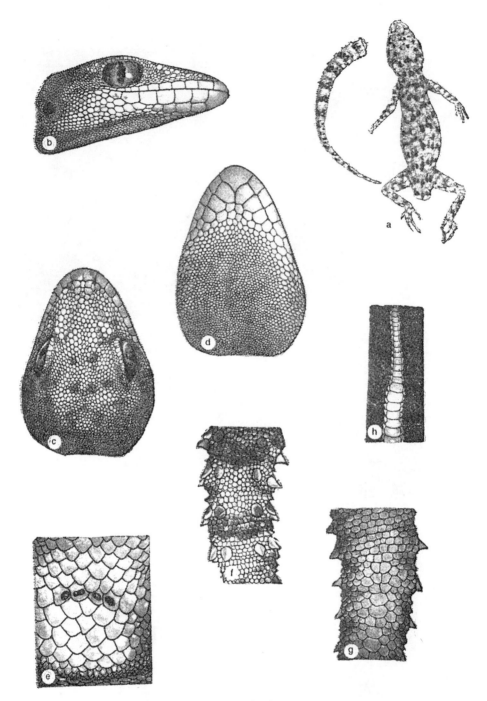

FIG. 93. *Tenuidactylus kirmanensis*:
a, general view from above; b, head from the side; c, head from above; d, head from below; e, preanal pores; f, tail from above; g, tail from below; h, toe from below (ZIL 9330, eastern Kirman, Iran)

Karyotype: not studied.
1900 — *Gymnodactylus kirmanensis* Nikolsky, Ann. Zool. Mus. for 1889, vol. 4, p. 381.
1963 — *Cyrtodactylus kirmanensis*, S. Anderson, Proc. Calif. Acad. Sci., San Francisco, ser. 4, vol. 31, no. 16, p. 474.
1984 — *Tenuidactylus kirmanensis*, Szczerbak and Golubev, Vestn. Zool., Kiev, no. 2, p. 55.

DIAGNOSIS. Less than 140 GVA; dorsal transverse bars defined in fragments as spots; caudal segments well developed; subcaudal plates defined along entire tail; L/LCD = 0.71–0.78; 17–20 scales across head.

LECTOTYPE. ZIL 9330, adult male. "E. Kirman, 17 June 1898, N. Zarudny." L = 50.4 mm; tail damaged; 17 scales across head; 11 supralabial scales; 8 infralabials; 131 GVA; near 30 ventral scales across midbody; 22 subdigital lamellae; 4 preanal pores.

DEFINITION (Fig. 93) (from 8 specimens in ZIL and NMW). L general (n = 8) 37.4-50.7 mm; L/LCD (n = 2) 0.71-0.78; HHW (n = 1) 53; EED (n = 7) 32-42 (39.14 ± 1.69); supralabial scales (n = 15) 11-12 (11.47 ± 0.13); infralabials (n = 16) 7-9 (8.0 ± 0.16); scales across head (n = 7) 15-20 (17.43 ± 0.65); DTL (n = 8) 15-16 (15.62 ± 0.18); GVA (n = 7) 120-132 (128.0 ± 1.59); ventral scales across midbody (n = 3) 26-30; subdigital lamellae (n = 14) 20-24 (22.51 ± 0.34); 4 preanal pores.

Mental scale triangular or pentagonal; three to four pairs of postmentals, scales of first pair contact each other; first nasal scales separated by a scale; three nasal scales; limbs of moderate length; adpressed forelimb reaches tip of snout with finger tips; basal digit phalanges slightly widened; caudal segments well-developed, tubercles situated as in preceding species; subcaudal tubercles arranged in combinations 1/1/2, 1/1/1, 2/1/1, 2/2/2 on a single segment; dorsal tubercles roundish and smooth on neck and shoulders, become oval and oval-triangular toward flanks and rear part of body, weak keels appear on them; small tail scales slightly keeled.

COLOR AND PATTERN (as fixed in alcohol). On the light background of the back, there are about eight dark transverse bands formed from separate spots, in places forming a chessboard-like pattern. The tail has eight to ten bands of the same color. There are small dark spots on the head and small transverse bands on the limbs. The lower surfaces are white.

DISTRIBUTION. Known only from type locality (Fig. 89).

SEXUAL DIMORPHISM AND AGE VARIATION. Females lack preanal pores; no juveniles among examined specimens.

SYSTEMATICS. Previously, this species was placed in the subgenus *Mediodactylus* [46], but then [153] it was relocated into the group of Tibet-Himalayan species. The reason was that this species is closely related to both of these groups and, additionally (to a lesser degree), to the species of *Cyrtopodion* subgenus. The majority of major scale characters links this species with the Tibet-Himalayan group.

ECOLOGY has not been studied. Some data are presented from N. A. Zarudny's observations [60].

HABITAT AND QUANTITATIVE DATA. This species is encountered in abundance in the mountainous terrain between the Sistan and Bampur Depressions. It has been found on the plains and in the mountains. It inhabits sheer rocky cliffs in the mountains; river banks and dry channels; shady terraces, cracks, niches; occasionally, on loose fragments of rock boulders. It is most frequently found on granites and, less often, on conglomerates and other rocks.

{p. 208} DAILY ACTIVITY CYCLE. It is encountered in the morning and evening more often than at night.

BEHAVIORAL ATTRIBUTES. It is sociable with other individuals. It has excellent day vision and, consequently, is a difficult species to catch. Every individual has its own observation point to which it returns after being frightened or chased away. The authors have caught them in May.

Genus Spider Geckos — *Agamura* Blanford, 1874

Type species: *Agamura cruralis* Blanf.
1874 — *Agamura* Blanford, Ann. Mag. Nat. Hist., London, ser. 4 {error in original – MG}, vol. 13, p. 455.
1973 — *Rhinogekko* de Witte, Bull. Inst. Roy. Sci. Nat. Belg., vol. 49, no. 1, p. 1.
1984 — *Agamura*, Szczerbak and Golubev, Vestn. Zool., Kiev, no. 2, p. 55.

DIAGNOSIS. Extremities very thin and long, adpressed forelimb reaches tip of snout with wrist joint[1]; length of foot with toes distinctly longer than crus; tail becomes abruptly thicker

---

[1] Except *Agamura femoralis*, which has shorter limbs [290].

at the base, not fragile, shorter or slightly longer than body; preanal pores developed only in the males; digits slightly angularly bent, not widened, sometimes slightly compressed, clawed, covered below by a longitudinal row of transversely widened smooth lamellae; pupil vertical with serrated edges; all scales and plates of body smooth (except tail tubercles, weak longitudinal keels may be discerned).

ADDITIONAL CHARACTERS. Rostral scale quadrangular or (slightly developed) pentagonal; three nasal scales; snout scales roundish-polygonal, become slightly smaller on neck, where they are intermixed with conical tubercles; height of first supralabial scale is equal to or slightly smaller than its width along edge of mouth; ear opening small, oval; postnasal cavities developed, frontal cavity not present. Mental scale pentagonal or trapezoid; one to three pairs of postmentals (scales of first pair contact each other) may be replaced by a row of gular scales, much larger than others. Small conical dorsal tubercles, arranged in poorly developed longitudinal (about 26) and transverse (8-10) rows, intermixed with small roundish scales, which gradually become enlarged on flanks; ventral scales hexagonal-roundish, often with serrated free (rear) edge; three to four poorly developed caudal tubercles widely contact each other by lateral edges on each side of segment; on underside of tail longitudinal row of large plates or somewhat enlarged scales, two to three per segment. Base of claw covered by upper and lower plates, which are contacted by one (above) and two (on the sides) rows of scales. Roundish flat tubercles intermixed with small scales on upper surfaces of hindlimbs (they are absent on forelimbs).

SEXUAL DIMORPHISM AND AGE VARIATION are probably developed similarly in all species of the genus; no preanal pores in females; young specimens have agamuroid habitus (long, slender limbs, relatively large head) expressed less distinctly than in adults (which repeatedly led to their erroneous identification as thin-toed geckos).

The genus includes four species distributed in Southwest Asia (Afghanistan, Iran, Pakistan).

Key to Species of the Genus Agamura

1 (4). Homogeneous scales on the lower surfaces of thighs 4
2 (3). More than 30 ventral scales across midbody . . . . *A. persica* (Duméril)
3 (2). Less than 20 ventral scales across midbody . . . . *A. gastropholis* (Werner)
4 (1). A row of enlarged scales on the lower surfaces of thighs 5
{p. 209} 5 (6). Nasal scales distinctly bulbous, 17 to 22 ventral scales
  across midbody . . . . *A. femoralis* (Smith)
6 (5). Nasal scales sharply bulbous and extended into small tubes;
  26 to 28 ventral scales across midbody . . . . *A. misonnei* (de Witte)

PERSIAN SPIDER GECKO — *AGAMURA PERSICA* (DUMÉRIL, 1856)

Type locality: "Persia."
Karyotype not studied.
1856 — *Gymnodactylus persicus* Duméril, Arch. Mus. Hist. Nat. Paris, vol. 8, p. 481.
{p. 210} 1874 — *Agamura cruralis* Blanford, Ann. Mag. Nat. Hist., London, ser. 4, vol. 13, no. 78, p. 455.
1876 — *Agamura persica*, Blanford, East. Persia, Zool., vol. 2, p. 358.

DIAGNOSIS. 30 to 50 ventral scales across midbody; nasal scales slightly bulbous; no row of enlarged scales on lower surfaces of thighs; two to four feebly defined preanal pores; tail shorter than body {including head – MG}.

TYPES are kept in the Paris Museum [245].

DESCRIPTION (Figs. 94, 96) (from 10 specimens in ZIL, CAS, and ZML) L general (n = 9) 41.1-72.0 (52.21 ± 3.77) mm; L/LCD (n = 8) 1.05-1.42 (1.22 ± 0.05); HHW (n = 7) 53-60 (56.86 ± 0.96); EED (n = 7) 20-40 (30.57 ± 2.84); number of supralabial scales (n = 20) 9-12 (11.19 ± 0.21); infralabials (n = 20) 12-16 (14.62 ± 0.30); scales across head (n = 8) 21-27 (23.62 ± 0.75); DTL (n = 7) 11-12 (11.57 ± 0.20); scales around tubercle (n = 8) 8-11 (10.25 ± 0.37); GVA (n = 8) 130-158 (148.86 ± 3.67); ventral scales across midbody

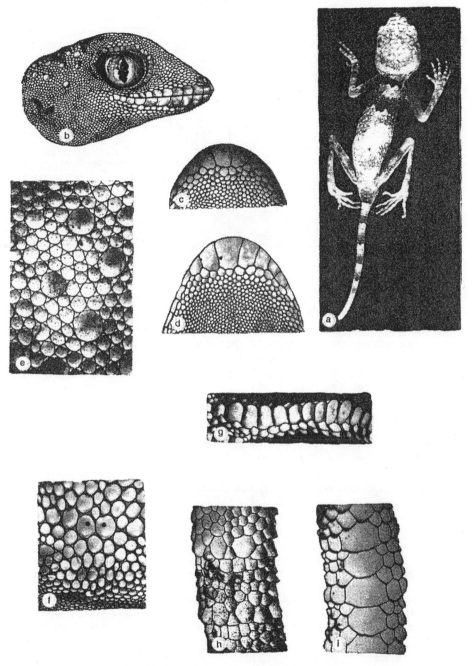

FIG. 94. *Agamura persica* (eastern population):
a, general view from above; b, head from the side; c-d, chin; e, dorsum; f, preanal pores;
g, toe from below; h, tail from above; i, tail from below (a-b, g – ZML 57/3791, Hassan,
Guilan, Pakistan; others – CAS 120282, Girishk, Afghanistan).

(n = 9) 37-50 (40.86 ± 1.62); preanal pores (n = 4) 2-4; subdigital lamellae (n = 14) 16-24 (20.57 ± 0.71).

Mental scale trapezoid (80%) or pentagonal (triangular) (20%); postmentals present (20%) or absent (80%), in the latter case, mental scale is bordered by three (25%), four (25%), five (37.5%), or six (12.5%) gular scales; first nasal scales separated from each other by two (30%), three (50%), four or five (10% each) scales.

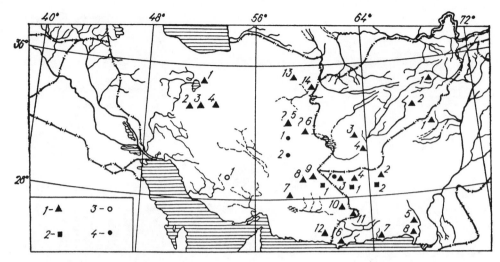

FIG. 95. Distribution of *Agamura persica* (1), *A. femoralis* (2), *A. gastropholis* (3), *A. misonnei* (4).

COLOR AND PATTERN (as fixed in alcohol). The main body background is ochrish. There is an indistinct pattern on the head. Between the neck and the lumbar area, there are five to seven wide dark M-shaped bands, of which the neck band and the following one may be much better developed than the others. On the tail and each of the limbs, there are several small transverse bands of the same color. The lower body surfaces are light.

DISTRIBUTION (Fig. 95). Central and eastern Iran, southern and southeastern Afghanistan, central and western Pakistan.

LEGEND to Fig. 95.
Iran: 1 – Tehran Prov.: Shah-Abbas, 34°44′N, 52°10′E (CAS); 2 – Pir Bakran near Esfahan [315]; 3 – Galatappeh, Esfahan Prov. [315]; 4 – Siah-Kuh, Prov. Tehran (Prov. Esfahan, on maps) [216]; 5 – between Faizabad and Chahar-Farsangh in Sarr-Chah Hollow [59]; 6 – Basiran [87]; 7 – Kirman Prov.: Jebal Barez [315]; 8 – 32 km W Zahedan, 29°28′N, 60°41′E [216]; 9 – E Kirman: Duz-Ab (ZIL; determined as *G. agamuroides*); 10 – Terra of Sib (ZIL); 11 – Eskan, Baluchistan [291]; 12 – Bahu Kalat [291 {error in original – MG}]; 13 – Torbat-e Heydariyeh [303]; 14 – 6 km W Ayubi, 34°18′N, 60°35E′ [303].

{p. 211} *Afghanistan:* 1 – Paghman, 34°36′N, 68°56′E [172]; 2 – Chahardeh (Mazanderan) [96]; 3 – Hassan Guilan (Djilan), SW Girishk [317]; 4 – 50 km W Girishk (CAS).

Pakistan: 1 – Wana [232]; 2 – Kalat [258]; 3 – N Nok Kundi [258]; 4 – Merui [258]; 5 – 30 to 37 km N Bela [263]; 6 – Jiwani [258]; 7 – Ormara [258]; 8 – Cape Monze near Karachi, Waziristan [245].

Not located: Zaudi Achmed-Wüste [310]; Kidsh, Mekranwüste (ZFMK).

SEXUAL DIMORPHISM AND AGE VARIATION. Besides the presence of preanal pores in males, noted previously, they have a swelling in the cloacal area, which is absent in females. Age variation, see above.

SYSTEMATICS AND GEOGRAPHICAL VARIATION. M. Smith [291] synonymized *A. persica* and *A. cruralis* Blanf., and this point of view now has become generally accepted. However, the appearance of enlarged postmental scales is recorded in specimens of this species from Iranian Baluchistan (type locality of Blanford's species) and to the north, to the Sistan Depression and Desht-i-Lut Desert. These specimens also have deviations of some other traits. Lack of comparative material makes it impossible to reach a final taxonomic conclusion yet. However, considering the presence of high endemism within the genus *Agamura* in this region (*A. femoralis, A. misonnei*), the existence of a different subspecies, *A. persica cruralis* Blanf., can be assumed (Fig. 96).

ECOLOGY has not been studied adequately.

HABITAT. According to I. S. Darevsky's observations in Iran (in litt.), the Persian agamura inhabits rocky semi-desert, covered with black rock waste, on the border of sand semi-desert, and on the slopes of hills. N. A. Zarudny [60] caught these lizards among large rock fragments "at the foot of a rock slide" and on solid elevated steppe with abundant rock waste and thin, undersized sagebrush. S. Minton [263] encountered them on rocky terraces at 100-3000 ft elevations (30-942 m).

DAILY ACTIVITY CYCLE. This is a nocturnal species [49].

SEASONAL ACTIVITY CYCLE. From collection materials, there are records from 22 December 1962 in the Iranian province of Esfahan and on 9 February 1901 in the country of Sib (Baluchistan).

REPRODUCTION. It lays eggs in June and young appear at the end of September [263].

BEHAVIORAL ATTRIBUTES. It moves on elevated limbs with movements resembling those of a dog (from observations in captivity). The tail is almost unbreakable [49].

FARSIAN SPIDER GECKO — *AGAMURA GASTROPHOLIS* (WERNER, 1917) COMB. N.

Type locality: Iran: Prov. Fars.
Karyotype not studied.
1917 — *Gymnodactylus gastropholis* Werner, Verh. zool.-botan. Ges. Wien, vol. 67, p. 194.
1951 — *Gymnodactylus agamuroides*, Wettstein, Sber. Osterr. Acad. Wiss. Wien, math.-naturwiss. Kl., Abt. 1, vol. 160, p. 432.

DIAGNOSIS. 14-18 ventral scales across midbody; nasal scales slightly bulbous; homogeneous scales on lower surfaces of thighs; four well developed preanal pores; tail not longer than body.

LECTOTYPE. ZFMK 27095, adult male, Prov. Fars, Iran [308]. L = 49.4 mm; tail damaged; 20 scales across head; 12/13 supralabial scales; 9/10 infralabials; 90 GVA; 16 ventral scales across midbody; 8 scales around dorsal tubercle; four preanal pores; 22/22 subdigital lamellae.

DEFINITION (Fig. 96) (from four specimens of ZFMK type series). L general (n = 3) 44.6-50.0 (48.0 ± 1.71) mm; tails are damaged in all specimens; HHW (n = 3) 58-63 (60.0 ± 1.53); EED {p.212} (n = 3) 20-37 (31.0 ± 5.51); number of supralabial scales (n = 8) 12-14 (12.87 ± 0.23); infralabials (n = 8) 9-10 (9.25 ± 0.16); scales across head (n = 4) 18-20 (18.75 ± 0.48); DTL (n = 3) 11-14 (12.33 ± 0.88); scales around dorsal tubercle (n = 3) 8-9 (8.25 ± 0.25); GVA (n = 4) 84-94 (91.0 ± 1.08); ventral scales across midbody (n = 4) 14-18 (16.0 ± 0.82); preanal pores (n = 4) 4; subdigital lamellae (n = 8) 22-24 (23.0 ± 0.33).

{p. 213} Mental scale pentagonal or triangular (100%); three pairs (100%) of postmentals, scales of first pair contact each other by wide suture (100%); first nasal scales separated from each other by three scales (100%).

COLOR AND PATTERN (as fixed in alcohol). The geckos have largely lost their color and pattern, but from the remaining elements it can be concluded that they do not differ from those of the related species: five to seven dark transverse bands, each on the back and tail over the lighter background, indistinct dark stripes on the limbs and white lower surface.

DISTRIBUTION. Known only from the type locality.

SYSTEMATICS. This species, which was originally described as {*Gymnodactylus* – MG}, was identified as *G. agamuroides* by O. Wettstein [315]. However, it is apparent through comparison of these and closely related species that the form "*gastropholis*" is not a {*Gymnodactylus* – MG} but a typical agamuroid gecko; it is distinguished from the species of the latter genus by very large abdominal scales and may be considered as an independent species. In all probability, both young specimens "*G. agamuroides*" from Kirman, which were used for comparison with *G. gastropholis* by O. Wettstein, belong to this species.

ECOLOGY not studied.

MISONNE'S SPIDER GECKO — *AGAMURA MISONNEI* (DE WITTE, 1973), COMB. N.

Type locality: Iran: Dast-i-Lut Desert, 31°13′N, 58°47′E.
Karyotype not studied.
1973 — *Rhinogekko misonnei* de Witte, Bull. Inst. R[oy]. Sci. Nat. Belg., Bruxelles, vol. 49, no. 1, p. 1.

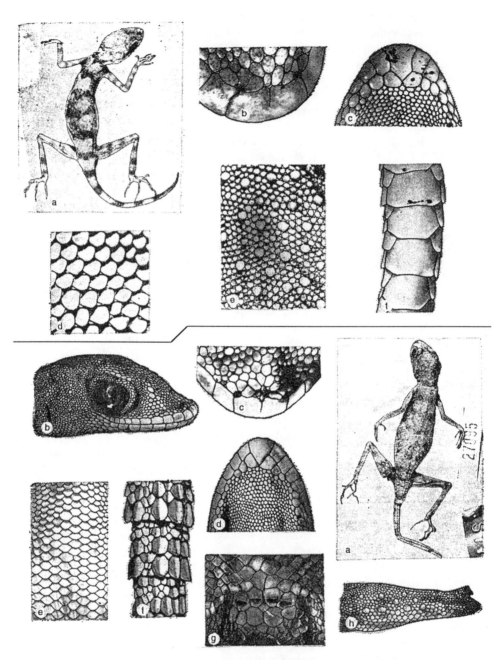

FIG. 96. *Agamura persica* (top; western population, *"cruralis"*):
a, general view from above; b, nasal scales; c, chin; d, belly; e, dorsum;
f, tail from below (ZIL 9328, terra Sib, Iran).
Agamura gastropholis (bottom):
a, general view from above; b, head from the side; c, snout from above; d, head from below;
e, belly; f, tail from above; g, preanal pores; h, thigh (ZFMK 27095, Fars Prov., Iran).

DIAGNOSIS. 26-28 ventral scales across midbody; nasal scales strongly bulbous and extended into tubes; a row of 9-12 enlarged scales on the lower surfaces of thighs; 4-8 very poorly developed {preanal – MG} pores; tail somewhat longer than body.

HOLOTYPE is kept in the Instute of Royal Sciences of Nature in Belgium (Brussels),

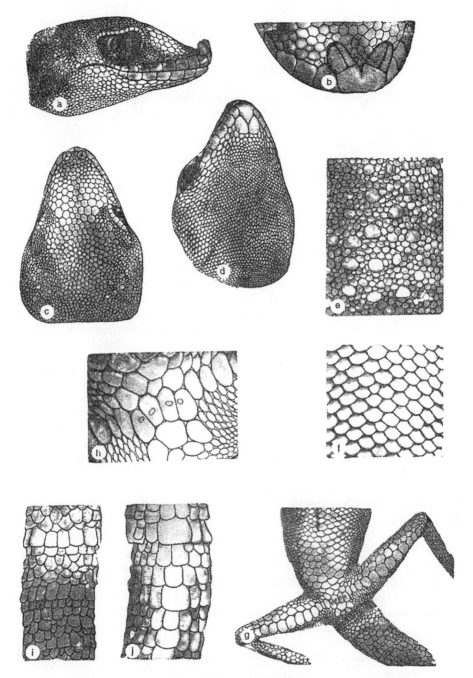

FIG. 97. *Agamura misonnei*:
a, head from side; b, snout in front; c, head from above; d, head from below; e, dorsum; f, belly; g, thigh from below; h, preanal pores; i, tail from above; j, tail from below (BZ 24.703 Reg. 25/6, Kuh-i-Bakhtu, Iran).

collected in "désert du Dasht-I-Lut, Iran, 31°13′N et 58°47′E"; Paratype – "Kuh-I-Bakhtu, désert du Lut, Iran oriental 31°40′N, 58°21′E," 12 Apr 1972, leg. X. Misonne coll.

DEFINITION (Fig. 97) (from paratype in the same collection and [320, 321]. L = 56.9-61.0 mm; LCD = 58.0-73.0 mm; L/LCD = 0.84-0.96; HHW 56; EED 53; 9-12 supralabial scales; 8-11 infralabials; 16 scales across head; DTL 13; 8-9 scales around dorsal tubercle; 120 GVA; 26-28 ventral scales across midbody; 4-8 preanal pores.

Mental scale pentagonal, single pair of large postmentals contact each other by wide suture; at the beginning of tail on underside on a single segment, there is a longitudinal row of three scales, larger than others, and further down tail – large quadrangular plates (two on a segment).

COLOR AND PATTERN. The main background of the body's upper surface is coffee-gray. There are five wide transverse bands on the body and seven on the tail. The limbs also have small dark bands that are lighter than the others. The lips and the rostral scale are covered with dark brown spots. The lower surfaces are whitish.

DISTRIBUTION (Fig. 95). At present it is known only from the Dasht-i-Lut Desert and southern foothills of Chagai Ridge (Iran and Pakistan).

LEGEND to Fig. 95.
Iran: 1 – Kuh-i-Bakhtu, 31°40'N, 58°21'E (BZM); 2 – Dast-i-Lut, 31°13'N, 58°47'E [320].
Pakistan: 1 – N Nok Kundi [320].

SYSTEMATICS. Generic independence of this unique gecko was defended repeatedly by the author of the description [321]. However, precisely {**p. 214**} from this work, as well as from other sources [290, 263, 172] it follows, that the *"misonnei"* form is so closely related to *A. femoralis* that its separation into an independent genus does not seem advisable.

ECOLOGY not studied. The authors of the description [320] noted that the holotype was collected in a desert without vegetation, and the paratype – in a gravel desert with very scant vegetation.

{P. 215} KHARAN SPIDER GECKO — *AGAMURA FEMORALIS* SMITH, 1933

Type locality: Kharan, Baluchistan in Pakistan.
Karyotype: not studied.
1933 — *Agamura femoralis* M. A. Smith, Rec. Ind. Mus., Calcutta, vol. 35, p. 17.

DIAGNOSIS. 17-21 ventral scales across midbody; nasal scales distinctly bulbous; a row of enlarged scales on lower surfaces of thighs; five to six well-developed {preanal – MG} pores; tail slightly longer than body.

TYPE is kept in the British Museum (Natural History).

DEFINITION [290 and 263] (Fig. 98). L = adult 52-60 mm; L/LCD = 0.91; ear opening diameter less than half of eye diameter; 12 supralabial scales; 11 infralabials; a pair of postmentals that is followed by another, much smaller, pair of scales; 17-21 ventral scales across midbody; abdominal scales become larger toward posterior part of body; six (rarely, five) preanal pores; adpressed hindlimb {error in original – MG} only reaches neck; a row of 9-12 enlarged scales stretches along lower surface of each thigh; along {**p. 216**} lower surface of each {tail – MG} segment, there are several scales (two in each of three longitudinal rows) that are larger than others.

COLOR AND PATTERN. There are five dark brown transverse bands on the body and 8-10 on the tail over the lighter background. The lower surface is white.

DISTRIBUTION (Fig. 95). Known only from Kharan Desert and, adjacent to the north, Chagai Mountains on the Afghan-Pakistan border.

LEGEND to Fig. 95.
Iran: 1 – Koh-i-Taftan [263]; Mirjawe [321].
Pakistan: 1 – N Nok Kundi [258]; 2 – Kharan [258].

ECOLOGY not studied. According to S. Minton's report [263], this species inhabits rock outcroppings among dunes; large eggs were seen in females on 7 June.

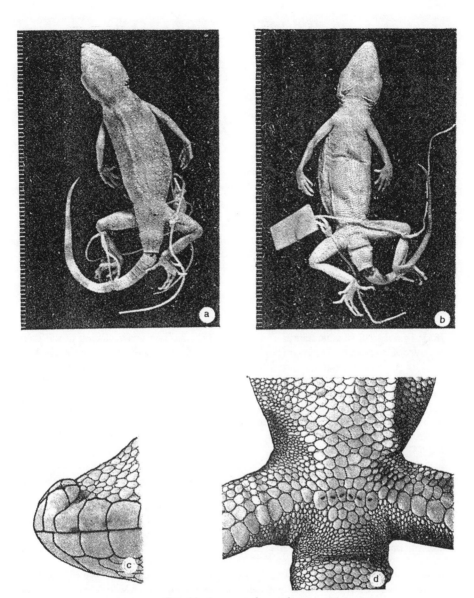

FIG. 98. *Agamura femoralis*:
a, general view from above; b, general view from below; c, snout from the side; d, preanal pores (Holotype, BMNH 1912.3.26.12; photo by Dr. E. N. Arnold).

{p. 217} LITERATURE CITED

1. Alekperov, A. M. 1978. Zemnovodnie i presmikayushchiesya Azerbaydzhana [The Amphibians and reptiles of Azerbajan]. Baku: Elm, 264 pages, errata (in Russian).
2. Alenitsyn, V. 1876. Gadi ostrovov i beregov Aralskogo morya [Reptiles of the islands and shores of the Aral Sea]. Trudi Aralo-Kasp. ekspeditsii [Transactions of the Aralo-Caspian Expedition]. Pril. k Tr. St.-Peterburg. Ob-va estestvoispitatelej [Additions Trans. St.-Petersb. Soc. Nat.], 1876(3):1-64 (in Russian).
3. Alpheraki, S. 1871. Kuldzha i Tyan-Shan: Putevie zametki [Kuldja and Tien-Shan travel notes]. St. Petersburg: IRGO [Imper. Russ. Geogr. Soc.], 193 pages (in Russian).
4. Andrushko, A. M. 1949. Materiali po biologii *Alsophylax pipiens* (Pallas) [Biological data on *Alsophylax pipiens* (Pallas)]. Nauch. Byull. Leningr. Un-ta, [Sci. Bull. Leningrad Univ.], 1949(23):25-39 (in Russian).
5. Andrushko, A. M. 1955. Presmikayushchiesya Kazakhskogo Nagorya i ih khozyajstvennoe znachenie [Reptiles of the Kazakh Upland and their economic significance]. Uchen. zap. Leningr. Un-ta [Sci. Mem. Leningrad Univ.], ser. Biol. Nauk, 1955(181), pt. 38:19-33 (in Russian).
6. Andrushko, A. M. 1968. Novij vid gekkonchika *Alsophylax kashkarovi* sp. n. (Reptilia, Sauria, Gekkonidae) iz Srednej Azii [A new species of a pygmy gecko, *Alsophylax kashkarovi* sp. n. (Reptilia, Sauria, Gekkonidae) from Middle Asia]. Zool. Zhur. [Zool. Jour.], 47(2):308-311 (in Russian).
7. Andrushko, A. M. 1977. Sovremennoe ponimanie statusa i strukturi rodov *Alsophylax* (Fitzinger, 1843) i *Bunopus* (Reptilia, Sauria, Gekkonidae) [The modern understanding of the status and structure of the genera *Alsophylax* (Fitzinger, 1843) and *Bunopus* (Reptilia, Sauria, Gekkonidae), p. 10-12. In: Voprosi gerpetologii. Avtoref. dokl. IV-j Vsesoyuz. gerpetol. konf. [Problems of Herpetology. Fourth All-Union Herpetol. Conf. Abstracts]. Leningrad: Nauka Press (in Russian).
8. Andrushko, A. M., N. O. Lange, and E. N. Yemelyanova. 1939. Ekologicheskie nablyudeniya nad reptiliyami v rayone Kyzyl-Arvat, stantsii Iskander i v rayone g. Krasnovodska, Turkmeniya [Ecological observations on reptiles in the Kyzyl-Arvat District, Iskander Station, and the vicinity of Krasnovodsk City]. Vopr. ekologii i biotsenologii [Problems of ecology and biocenology], 1939(4):207-252 (in Russian).
9. Andrushko, A. M., and N. E. Mikkau. 1964. Rasprostranenie i obraz zhizni afganskogo litorinkha (*Lytorhynchus ridgewayi* Boulenger, 1887) s ekologo-geograficheskim obzorom poda *Lytorhynchus* Peters, 1862 [Distribution and life history of *Lytorhynchus ridgewayi* Boulenger, 1887 with an ecological and geographical review of genus *Lytorhynchus* Peters, 1862]. Vestn. Leningr. Un-ta [Herald Leningrad Univ.], ser. Biol., 1964(9), pt. 2:15-19 (in Russian).
10. Atayev Ch. 1974. Proniknovenie presmikayushchihsya gor v rechnie dolini Turkmenii [Penetration of mountain reptiles into the river valleys of Turkmenia]. Izv. AN TSSR [Proc. Acad. Sci. Turkmenian S.S.R.], ser. Biol. Nauk, 1974(3):38-43 (in Russian).
11. Atayev, Ch. 1975. O zimnej aktivnosti presmikayushchihsya na Kopetdage i v Badkhize [On winter activity of reptiles in Kopetdagh and Badkhyz]. Izv. AN TSSR [Proc. Acad. Sci. Turkmenian S.S.R.], ser. Biol. Nauk, 1975(4):63-67 (in Russian).
12. Atayev, Ch. 1977. K rasprostraneniyu i ekologii nekotorih vidov presmikayushchihsya Turkmenistana [On the distribution and ecology of some species of reptiles of Turkmenistan]. Izv. AN TSSR [Proc. Acad. Sci. Turkmenian S.S.R.], ser. Biol. Nauk, 1977(1):80-82 (in Russian).
13. Atayev, Ch., Yu. K. Gorelov, and S. Shammakov. 1978. Materiali po redkim i ischezayushchim vidam presmikayushchihsya fauni Turkmenistana [Data on rare and endangered species of reptiles of the Turkmenistan fauna]. Izv. AN TSSR [Proc. Acad. Sci. Turkmenian S.S.R.], ser. Biol. Nauk, 1978(4):81-83 (in Russian).
14. Akhmedov, M. I., and N. N. Szczerbak. 1978. *Gymnodactylus caspius insularis* ssp. n. (Reptilia, Sauria) DD novij podvid kaspijskogo gekkona s ostrova Vulf v Kaspijskom more [*Gymnodactylus caspius insularis* ssp. n. (Reptilia, Sauria), a new subspecies of *Gymnodactylus caspius* Eichw. from Vulf Island in the Caspian Sea]. Vestn. Zool. [Herald Zool.], 1978(2):80-82 (in Russian).
15. Bannikov, A. G. 1958. Materiali po faune amfibij i reptilij Mongolii [Data on the amphibian and reptile fauna of Mongolia]. Byull. Mosk. ob-va ispitatelej prirodi [Bull. Moscow Soc. Nat.], Otd. Biol., 58(2):71-91 (in Russian).
16. Bannikov, A. G. 1974. Po zapovednikam Sovetskogo Soyuza [Nature reserves of the Soviet Union]. Moscow: Mysl Press, 235 pages (in Russian).

17. Bannikov, A. G., I. S. Darevskij, V. G. Ishchenko, A. K. Rustamov, and N. N. Szczerbak. 1977. Opredelitel zemnovodnih i presmikayushchihsya fauni SSSR [Guide to the amphibians and reptiles of the U.S.S.R.]. Moscow: Prosveshchenie Press, 414 pages (in Russian).
18. Bedriaga, J. von. 1907-1909. Nauchnye rezultati puteshestviya N. M. Przevalskogo po Tsentralnoj Azii [Scientific results of the travels of N. M. Przewalsky to Central Asia]. Otd. zool., vol. 3, sect. 1. Zemnovodnie i presmikayushchiesya [Amphibians and reptiles], pts. 2-3, Lacertilia. St. Petersburg: Imper. Acad. Sci., 71-502 pages (in Russian and German).
19. Bogdanov, O. P. 1955. O vidovoj samostoyatelnosti *Alsophylax laevis* i ego rasprostranenii v Uzbekistane [On species independence of *Alsophylax laevis* and its distribution in Uzbekistan]. Dokl. AN SSSR [Rep. Acad. Sci. U.S.S.R.], 101(5):959-960 (in Russian).
20. Bogdanov, O. P. 1955. O rasprostranenii *Alsophylax laevis* Nik. v Uzbekistane [On the distribution of *Alsophylax laevis* Nik. in Uzbekistan]. Dokl. AN UzSSR [Rep. Acad. Sci. Uzbekistan S.S.R.], 1955(6):43-54 (in Russian).
21. Bogdanov, O. P. 1956. Neskolko popravok k statye V. P. Kostina o zemnovodnih i presmikayushchihsya drevnej delti Amu-Daryi i Ustyurta [Several corrections to V. P. Kostin's article on the amphibians and reptiles of the ancient delta of Amu-Darya and Ustyurt]. Tr. In-ta zoologii i parazitologii AN UzSSR [Trans. Inst. Zool. Parasitol. Acad. Sci. Uzbek S.S.R.], 1956(8):194-195 (in Russian).
22. Bogdanov, O. P. 1960. Fauna Uzbekskoj SSR. Ch. 1. Zemnovodnie i presmikayushchiesya. [Fauna of Uzbek S.S.R., pt. 1. Amphibians and reptiles]. Tashkent: Izd-vo AN UzSSR [Acad. Sci. Uzbek S.S.R. Press], 254 pages (in Russian).
23. Bogdanov, O. P. 1962. Presmikayushchiesya Turkmenii [Reptiles of Turkmenia]. Ashkhabad: Izd-vo AN TSSR [Acad. Sci. Turkmenian S.S.R. Press], 235 pages (in Russian).
24. Bogdanov, O. P. 1965. Ekologiya presmikayushchihsya Srednej Azii [Ecology of reptiles of Middle Asia]. Tashkent: Nauka Press, 259 pages (in Russian).
25. Bogdabov, O. P. 1971. Ekologiya gladkogo gekkonchika v vesennij period [Ecology of the smooth pygmy gecko during the spring season]. Ekologiya [Ecology], 1971(1):105-107 (in Russian).
26. Bogdanov, O. P. 1972. Novie dannie po rasprostraneniyu pantsirnogo gekkonchika v Turkmenii [New data on the distribution of the armored pygmy gecko in Turkmenia]. Izv. AN TSSR [Proc. Acad. Sci. Turkmenian S.S.R.], ser. Biol. Nauk [ser. Biol. Sci.], 1972(1):91 (in Russian).
{p. 218} 27. Bogdanov, O. P. 1977. O sokrashchenii chislennosti nekotorih vidov bespozvonochnih i reptilij Ustyurta v svyazi s usihaniem Arala [On the reduction of quantity of some species of invertebrates and reptiles in connection with dessication of the Aral Sea]. Vestn. Karakalp. fil. AN UzSSR [Herald Karakalpak Branch Acad. Sci. Uzbek S.S.R.], 1977(3):57 (in Russian).
28. Bogdanov, O. P., and O. A. Petrusenko. 1973. Pro zhivlennya gekkona sirogo D *Gymnodactylus russowi* Strauch, 1887 [Feeding in the gray rock gecko, *Gymnodactylus russowi* Strauch, 1887]. Zb. prats zoomuzeyu [Coll. Art. Zool. Mus.], 1973(35):66-67 (in Ukrainian).
29. Bondarenko, D. A., and G. S. Antonova. 1977. Landshaftnoe raspredelenie reptilij na plato Ustyurt [A landscape distribution of reptiles on the Ustyurt Plateau], p. 41-42. In: Voprosi gerpetologii: Avtoref. dokl. IV-j Vsesoyuz. gerpetol. konf. [Problems of herpetology. Fourth All-Union Herp. Conf. Abstracts], Leningrad: Nauka (in Russian).
30. Borisov, A. A. 1967. Klimat SSSR [Climate of the U.S.S.R.]. Moscow: Prosveshchenie Press, 296 pages (in Russian).
31. Borkin, L. Ya., Kh. Munkhbayar, and D. V. Semenov. 1983. Amfibii i reptilii Zaaltayskoj Gobi [Amphibians and Reptiles of Transaltay Gobi]. Priroda [Nature], 1983(10):68-75 (in Russian).
32. Vavilov, N. I. 1967. Zakon gomologicheskih ryadov v nasledstvennoj izmenchivosti. Linneevskij vid kak sistema [The law of homologous series in variation. Linnean species as a system]. Leningrad: Nauka Press, 92 pages (in Russian).
33. Vashetko, E. V., and Z. Ya. Kamalova. 1974. Cherepahi, yashcheritsi [Tortoises, lizards], p. 60-75. In: Pozvonochnie zhivotnie Ferganskoj dolini [Vertebrate animals of Ferghan Valley]. Tashkent: FAN [Uzbek Branch Acad. Sci.] (in Russian).
34. Velikanov, V. P. 1977. O novih nahodkah pisklivogo gekkonchika i obiknovennogo shchitomordnika v Turkmenii [On the findings of squeaky pygmy gecko and Halys viper in Turkmenia]. Izv. AN TSSR. [Proc. Acad. Sci. Turkmenian S.S.R.], ser. Biol. Nauk, 1977(5):81-82 (in Russian).
35. Velikanov, V. P. 1977. O gerpetofaune Sarikamishskoj kotlovini [On the herpetofauna of Sarykamysh Hollow], p. 56-57. In: Voprosi gerpetologii: Avtoref. dokl. IV-j Vsesoyuz. gerpetol.

konf. [Problems of Herpetol. Fourth All-Union Herp. Conf. Abstracts]. Leningrad: Nauka Press (in Russian).

36. Golubev, M. L. 1979. O geograficheskoj izmenchivosti i taksonomii gladkogo gekkonchika *Alsophylax laevis* Nikolsky, 1905 (Sauria, Gekkonidae) [Geographic variability and taxonomy of the smooth pygmy gecko *Alsophylax laevis* Nikolsky, 1905 (Sauria, Gekkonidae), p. 55-64. In: Ekologiya i sistematika amfibij i reptilij [Ecology and systematics of amphibians and reptiles]. Leningrad: Zool. In-ta AN SSSR (Tr. Zool. In-ta AN SSSR, T. 89) [Proc. Zool. Inst. Acad. Sci. U.S.S.R., vol. 89] (in Russian).

37. Golubev, M. L. 1981. Ob obyeme roda *Alsophylax* Fitzinger, 1843 (Reptilia, Gekkonidae) [Content of the genus *Alsophylax* Fitzinger, 1843 (Reptilia, Gekkonidae)], p. 39. In: Voprosi gerpetologii: Avtoref. dokl. V-j Vsesoyuz. gerpetol. konf. [Problems of Herpetology. Fifth All-Union Herp. Conf., Abstracts]. Leningrad: Nauka Press (in Russian).

38. Golubev, M. L. 1984. *Alsophylax tadjikiensis* Golubev, stat. n. (*Alsophylax laevis tadjikiensis* Golubev, 1979:62). Vestn. Zool. [Herald Zool.], 1984(2):73 (in Russian).

39. Golubev, M. L. 1984. *Alsophylax przewalskii* (Reptilia, Gekkonidae): nekotorie nomenklaturnie trudnosti [*Alsophylax przewalskii* (Reptilia, Gekkonidae): some nomenclatural difficulties]. Vestn. Zool. [Herald Zool.], 1984(4):18-22 (in Russian).

40. Golubev, M. L. 1984. Struktura roda *Tropiocolotes* (Reptilia, Gekkonidae) [Structure of the genus *Tropiocolotes* (Reptilia, Gekkonidae)]. Vestn. Zool. [Herald Zool.], 1984(6):12 (in Russian).

41. Golubev, M. L. 1985. Novie tochki nahodok *Alsophylax loricatus szczerbaki* Golubev et Sattorov (Reptilia, Gekkonidae) [New localities for *Alsophylax loricatus szczerbaki* Golubev et Sattorov (Reptilia, Gekkonidae)]. Vestn. Zool. [Herald Zool.], 1985(2):87 (in Russian).

42. Golubev, M. L., and T. S. Sattarov. 1979. O podvidah u pantsirnogo gekkonchika *Alsophylax loricatus* Strauch, 1887 (Reptilia, Sauria, Gekkonidae) [On subspecies of the armored pygmy gecko, *Alsophylax loricatus* Strauch, 1887 (Reptilia, Sauria, Gekkonidae)]. Vestn. Zool. [Herald Zool.], 1979(5):18-24 (in Russian).

43. Golubev, M. L., and N. N. Szczerbak. 1979. Novij vid roda *Tropiocolotes* Peters, 1880 (Reptilia, Sauria, Gekkonidae) iz Afganistana [A new species of the genus *Tropiocolotes* Peters, 1880 (Reptilia, Sauria, Gekkonidae) from Afghanistan]. Dokl. AN UkrSSR [Rep. Acad. Sci. Ukrainian U.S.S.R.], ser. B, 1979(4):309-312 (in Russian).

44. Golubev, M. L., and N. N. Szczerbak. 1981. Novij vid roda *Gymnodactylus* (Reptilia, Gekkonidae) iz Pakistana (A new species of the genus *Gymnodactylus* (Reptilia, Gekkonidae) from Pakistan]. Vestn. Zool. [Herald Zool.], 1981(3):40-45 (in Russian).

45. Golubev, M. L., and N. N. Szczerbak. 1981. *Carinatogecko* gen. n. (Reptilia, Gekkonidae) novij rod gekkonovih yashcherits iz Yugo-Zapadnoj Azii [*Carinatogecko* gen. n. (Reptilia, Gekkonidae), a new genus of gecko lizards from Southwest Asia]. Vestn. Zool. [Herald Zool.], 1981(5):34-41 (in Russian).

46. Golubev, M. L., and N. N. Szczerbak. 1983. *Gymnodactylus scaber* (Heyden), 1827 = *Cyrtodactylus basoglui* Baran et Gruber, 1982, syn. n. Vestn. Zool. [Herald Zool.], 1983(2):54 (in Russian).

47. Golubev, M. L., and N. N. Szczerbak. 1985. O vzaimootnosheniyah dvuh rodov palearcticheskih gekkonov: *Tenuidactylus* i *Agamura* (Reptilia, Gekkonidae) [On the relationships of two palearctic gecko genera, *Tenuidactylus* and *Agamura* (Reptilia, Gekkonidae)], p. 59-60. In: Voprosi gerpetologii: Avtoref. dokl. VI-j Vsesoyuz. gerpetol. konf. {original text contains a typo - MG} [Problems of Herpetology. Sixth All-Union Herp. Conf. Abstracts], Tashkent. Leningrad: Nauka Press (in Russian).

48. Gorelov, Yu. K., I. S. Darevskij, and N. N. Szczerbak. 1974. Dva novih dlya fauni SSSR vida yashcherits iz semeystva gekkonov [Two lizard species of the family Gekkonidae new for the U.S.S.R. fauna]. Vestn. Zool. [Herald Zool.], 1974(4):33-39 (in Russian).

49. Darevskij, I. S. 1969. Yashcheritsi [Lizards], p. 202-316. In: Zhizn' zhivotnih [Animal Life], vol. 4, ch. 2. Moscow: Prosveshchenie Press (in Russian).

50. Darevskij, I. S. 1978. Kakoj vid eublefara (Sauria, Gekkonidae) vstrechaetsya v Srednej Azii [Which eublepharid species inhabit Middle Asia?]. Novie vidi zhivotnih [New species of animals], Tr. Zool. in-ta AN SSSR [Proc. Zool. Inst., Acad. Sci. U.S.S.R.], 61:204-209 (in Russian).

51. Darevskij, I. S. 1979. Izuchenie funktsionalnih osobennostej priznakov cheshujchatogo pokrova u presmikayushchihsya [Study of functional peculiarities of pholidosis characters in reptiles], p. 182-185. In: Sostoyanie i perspektivi razvitiya morfologii: (Materiali k Vsesoyuz. soveshch.) [Conditions and perspectives of morphology development (Materials for the All-Union Meeting)]. Moscow: Science Press (in Russian).

52. Dementyev, G. P., N. N. Kartashev, and A. N. Soldatova. 1953. Pitanie i prakticheskoe znachenie nekotorih hishchnih ptits v Yugo-Zapadnoj Turkmenii [Feeding and practical significance of some birds of prey in Southwest Turkmenia]. Zool. Zhur. [Zool. Jour.], 32(3):361-375 (in Russian).
53. Dementyev, G. P., A. K. Rustamov, and E. P. Spangenberg. 1955. Materiali po faune nazemnih pozvonochnih Yugo-Vostochnoj Turkmenii [Matherials on the terrestrial vertebrate fauna of Southeast Turkmenia]. Tr. Turkm. s.-kh. in-ta [Proc. Turkmen Agricultural Inst.], 7:125-183 (in Russian).
54. Dubinin, V. B. 1954. Ekologo-faunisticheskij ocherk zemnovodnih i presmikayushchihsya Khavastskogo rayona Tashkentskoj oblasti UzSSR [Ecological and faunistic sketch of amphibians and reptiles of Khavast District of Tashkent Region of Uzbek S.S.R.]. Tr. In-ta zoologii i parazitologii AN UzSSR [Transact. Inst. Zool. Parasitol. Acad. Sci. Uzbek S.S.R.], 3:159-170 (in Russian).
55. Dubrovskij, Yu. A. 1967. Novie nahodki reptilij v stepyah Kazakhstana [New findings of reptiles in the steppes of Kazakhstan]. Byul. Mosk. ob-va ispitatelej prirodi [Bull. Moscow Soc. Nat.], 72(1):146-147 (in Russian).
56. Eremchenko, V. K., and N. N. Szczerbak. 1984. Novij vid gekkona - *Alsophylax tokobayevi* sp. nov. iz Tyan-Shanya (Reptilia, Sauria, Gekkonidae) [A new species, *Alsophylax tokobayevi* sp. nov. from Tien-Shan (Reptilia, Sauria, Gekkonidae)]. Vestn. Zool. [Herald Zool.], 1984(2):46-50 (in Russian).
57. Eremchenko, V. K., and N. N. Szczerbak. 1985. Rasprostranenie i ekologiya gekkonchika Tokobayeva *Alsophylax tokobayevi* Jeriomtschenko et Szczerbak, 1984 (Reptilia, Gekkonidae) [Distribution and ecology of the pygmy gecko, *Alsophylax tokobayevi* Jeriomtschenko et Szczerbak, 1984 (Reptilia, Gekkonidae)]. Izv. AN KirgSSR [Proc. Acad. Sci. Kirghizian S.S.R.], 1985(3):60-63 (in Russian).
58. Zarudnij, N. 1896. Materiali dlya fauni amfibij i reptilij Orenburgskogo kraya [Materials on the amphibian and reptile fauna of the Orenburg Region]. Byul. Mosk. ob-va ispitatelej prirodi [Bull. Moscow Soc. Nat.], 1896(3):361-370 (in Russian).
59. Zarudnij, N. 1897. Zametka o cheshujchatih i golih gadah iz severo-vostochnoj Persii [A note on reptiles and amphibians from Northeast Persia]. Ezhegodnik Zool. muzeya AN [Ann. Zool. Mus. Imper. Acad. Sci.], 2(3):349-361 (in Russian).
60. Zarudnij, N. 1904. O gadah i ribah vostochnoj Persii [Reptiles and fishes of East Persia]. Zap. IRGO [Mem. Imper. Russ. Geogr. Sci.], 36(3):1- 42 (in Russian).
{p. 219} 61. Zarudnij, N. 1915. Poezdka po Aralskomu moryu letom 1914 goda [The Aral Sea field trip in the summer of 1914]. II. Gadi Arala [The reptiles of Aral]. Izv. Turkest. otd. RGO [Proc. Turkestan Branch Russ. Geogr. Sci.], 11(1):113-115 (in Russian).
62. Zakhidov, T. 1938. Biologiya reptilij yuzhnih Kzyl-Kumov i hrebta Nura-Tau [Biology of the reptiles of south Kyzylkum and Nuratau Ridge]. Tashkent, 1938. 52 pages (Tr. Sredneaz. Un-ta [Transact. Middle Asian Univ.], ser. 8, Zoologiya, no. 54) (in Russian).
63. Kaluzhina, M. V. 1951. Morfologiya i biologiya otryada yashcherits Zeravshanskoj dolini [Morphology and biology of the lizards of Zeravshan Valley]. Tr. biol.-pochvennogo fakult. Uzb. Un-ta [Sci. Pap. Biol. Soil Faculty Uzbek Univ.], Zoologiya, N. s. [Zoology, new ser.], 1951(46):75 (in Russian).
64. Kartashev, N. N. 1955. Materiali po amphibiyam i reptiliyam Yu.-Z. Turkmenii [Materials on the amphibians and reptiles of Southwest Turkmenia]. Uchen. zap. Mosk. Un-ta [Sci. Notes Moscow Univ.], 1955(171):173-202 (in Russian).
65. Kireyev, V. A. 1981. Zemnovodnie i presmikayushchiesya hrebta Zheltau [Amphibians and reptiles of Zheltau Ridge], p. 64-65. In: Voprosi gerpetologii: Avtoref. dokl. V-j Vsesoyuz. gerpetol. konf. [Problems of Herpetology. Fifth All-Union Herp. Conf. Abstracts]. Leningrad: Nauka Press (in Russian).
66. Kovachev, V. 1905. Prinos za izuchvane zemnovodnite i vlechuchite v Bulgariya [Contributions to the study of amphibians and reptiles in Bulgaria]. Sb. za nar. umotvoreniya, 1905(21):1-13 (in Bulgarian).
67. Kostin, V. P. 1956. Zametki po rasprostraneniyu i ekologii zemnovodnih i presmikayushchihsya drevnej delti Amu-Darji i Kara-Kalpakskogo Ustyurta [Distribution and ecological notes of reptiles of ancient delta of Amudarya and Karakalpak Ustyurt]. Tr. In-ta zoologii i parazitologii AN UzSSR [Trans. Inst. Zool. Parasitol. Acad. Sci. Uzbek S.S.R.], 1956(8):47-65 (in Russian).
68. Krasnaya kniga SSSR / Gl. redkol.: A. M. Borodin i dr. [The Red Data Book of the U.S.S.R. / Eds.: A. M. Borodin, etc.]. 1978. Moscow: Lesnaya Promishlennost Press, 460 pages (in Russian).

69. Krasnaya kniga SSSR / Gl. redkol.: A. M. Borodin i dr. [The Red Data Book of the U.S.S.R. / Eds.: A. M. Borodin, etc.]. 1984. Moscow: Lesnaya Promishlennost Press, 390 pages (in Russian).
70. Kubykin, R. A. 1975. Ekologo-faunisticheskij obzor reptilij ostrovov oz. Alakol (Vost. Kazakhstan) [Ecological-faunistic review of reptiles of the Alakol Lake islands]. Izv. AN KazSSR [Proc. Acad. Sci. Kazakh S.S.R.], ser. Biol., 1975(3):10-16 (in Russian).
71. Lazdin, V. V. 1915. Marshrut poezdok s zoologicheskoj tsel'yu v Vostochnuyu Bukharu i na zapadnij Pamir letom 1915 goda [The route of the zoological trips to East Bukhara and West Pamir in summer 1915]. Ezhegodnik Zool. muzeya AN [Ann. Zool. Mus. Imper. Acad. Sci.], 20:LIV-LVIII (in Russian).
72. Mayr, E. 1971. Printsipi zoologicheskoj sistematiki [Principles of systematic zoology]. Moscow: Mir Press, 454 pages (in Russian).
73. Manylo, V. V., and N. N. Szczerbak. 1984. Kariotipi gekkonov podroda *Mediodactylus* (Reptilia, Gekkonidae) fauni SSSR [Karyotypes of the gecko subgenus *Mediodactylus* (Reptilia, Gekkonidae) from the U.S.S.R. fauna]. Vestn. Zool. [Herald Zool.], 1984(3):81-83 (in Russian).
74. Mezhdunarodnij kodeks zoologicheskoj nomenklaturi / Pod red. A. N. Svetovidova, Ya. I. Starobogatova [International code of zoological nomenclature / A. N. Svetovidov and Ya. I. Starobogatov, Eds]. 1966. Moscow: Nauka Press, 100 pages (in Russian).
75. Mednikov, B. M. 1981. Sovremennoe sostoyaniye i razvitie zakona gomologicheskih ryadov v nasledstvennoj izmenchivosti [Modern conditions and development of the law of homologous series in hereditary variation], p. 127-135. In: Problemi novejshej istorii evolyutsionnogo ucheniya [Problems of the latest history of the doctrine of the evolution]. Leningrad: Nauka Press (in Russian).
76. Mikhailovskij, M. K. 1904. K gerpetofaune Zakaspijskoj oblasti [On the herpetofauna of the Transcaspian Region]. Ezhegodnik Zool. muzeya AN [Ann. Zool. Mus. Imper. Acad. Sci.], 9:39-44 (in Russian).
77. Munkhbayar, Khorlooghiyn. 1978. Zemnovodnie i presmikayushchiesya Mongolskoj Narodnoj Respubliki: Avtoref. diss. . . . kand. biol. nauk [The amphibians and reptiles of the People's Republic of Mongolia: an abstract of Ph.D. thesis]. Tashkent: Tashkent State Univ., 38 pages (in Russian).
78. Munkhbayar, Khorlooghiyn. 1981. Novie dannie o rasprostranenii nekotorih amfibij i reptilij Mongolskoj Narodnoj Respubliki [New data on the distribution of some amphibians and reptiles in the People's Republic of Mongolia], p. 52-56. In: Fauna i ekologiya amfibij i reptilij palearkticheskoj Azii [The fauna and ecology of amphibians and reptiles of Palearctic Asia]. Leningrad: Zool. in-ta AN SSSR, 1981 (Tr. Zool. in-ta AN SSSR, T. 101) [Proc. Zool. Inst. Acad. Sci. U.S.S.R., vol. 101] (in Russian).
79. Morits, L. D. 1922. Spisok presmikayushchihsya, sobrannih v 1921 g. v Zakaspijskoj oblasti, Turkmenistane i Severnoj Persii [A list of reptiles, collected in 1921 in the Transcaspian Region and North Persia]. Tr. Stavrop. s.-kh. in-ta [Trans. Stavropol Agricult. Inst.], 1(11):35-47 (in Russian).
80. Morits, L. D. 1929. Presmikayushchiesya Turkmenistana i sopredelnoj Persii [Reptiles of Turkmenistan and contiguous Persia]. Turkmenovedenie [Turkmenian Studies], 4(6-7):17-19 (in Russian).
81. Muratov, Sh. Kh., and R. Sh. Muratov. 1977. O gladkom gekkonchike Yuzhnogo Tadjikistana [The smooth pygmy gecko of South Tadjikistan], p. 150. In: Voprosi gerpetologii: Avtoref. dokl. IV-j Vsesoyuz. gerpetol. konf. [Problems of Herpetology. Fourth All-Union Herp. Conf.]. Leningrad: Nauka Press (in Russian).
82. Muskhelishvili, T. A. 1964. O faune yashcherits (Sauria, Reptilia) okrestnostej Tbilisi [On the lizard fauna (Sauria, Reptilia) in the vicinity of Tbilisi]. Soobshch. AN GSSR [Rep. Acad. Sci. Georgian S.S.R.], 35(1):199-205 (in Russian).
83. Kireev, V. A. 1984. O severnoj granitse areala serogo golopalogo gekkona [On the northern border of the range of the gray thin-toed gecko], p. 141-142. In: Ekologiya i faunistika amfibij i reptilij SSSR i sopredelnih stran [Ecology and faunistics of amphibians and reptiles of the U.S.S.R. and contiguous countries]. Leningrad: Zool. In-ta AN SSSR [Zool. Inst. Acad. Sci. U.S.S.R.] (in Russian).
84. Neruchev, V. V., and N. F. Vasil'yev. 1978. Fauna reptilij severo-vostochnogo Prikaspiya [The reptile fauna of the northeastern part of the Precaspian region]. Vestn. Zool. [Herald Zool.], 1978(6):36-41 (in Russian).
85. Nikol'skij, A. M. 1887. O faune pozvonochnih zhivotnih dna Balkhashskoj kotlovini [On the

fauna of vertebrate animals of Balkhash Hollow]. Tr. Sib. ob-va estestvoispitatelej [Trans. Siberian Soc. Nat.], Otd. Zool., 18:59-190 (in Russian).

86. Nikol'skij, A. M. 1896. Diagnoses reptilium et amphibiorum novorum in Persia orientali a N. Zarudny collectorum [Diagnoses of new reptiles and amphibians collected by N. Zarudny in East Persia]. Ezhegodnik Zool. muzeya AN [Ann. Zool. Mus. Imper. Acad. Sci.], 1(4):369-372 (in Latin).

87. Nikol'skij, A. M. 1897. Presmikayushchiesya, amfibii i ribi, sobrannie N. A. Zarudnim v Vostochnoj Persii [Reptiles, amphibians, and fishes collected by N. A. Zarudny in East Persia]. Ezhegodnik Zool. muzeya AN [Ann. Zool. Mus. Imper. Acad. Sci.], 2(3):306-348 (in Russian).

88. Nikol'skij, A. M. 1899. Dva novih vida *Teratoscincus* iz Vostochnoj Persii [Two new species of *Teratoscincus* from East Persia]. Ezhegodnik Zool. muzeya AN [Ann. Zool. Mus. Imper. Acad. Sci.], 4(2):145-147 (in Russian).

89. Nikol'skij, A. M. 1900. Presmikayushchiesya, amfibii i ribi vtorogo puteshestviya N. A. Zarudnogo v Persiyu v 1898 g [Reptiles, amphibians and fishes of the second field trip of N. A. Zarudny to Persia in 1898]. Ezhegodnik Zool. muzeya AN [Ann. Zool. Mus. Imper. Acad. Sci.], 4:375-417 (in Russian).

90. Nikol'skij, A. M. 1902. *Gymnodactylus danilewskii* Str. et *Gymnodactylus colchicus* n. sp. (Lacertilia, Gekkonidae). Ezhegodnik Zool. muzeya AN [Ann. Zool. Mus. Imper. Acad. Sci.], 7:3-6 (in Latin).

91. Nikol'skij, A. M. 1903. Novie vidi gadov iz Vostochnoj Persii, privezennie N. A. Zarudnim v 1901 g. [New species of reptiles from East Persia, brought by N. A. Zarudny in 1901]. Ezhegodnik Zool. muzeya AN [Ann. Zool. Mus. Imper. Acad. Sci.], 8:95-98 (in Russian).

92. Nikol'skij, A. M. 1907. Presmikayushchiesya i zemnovodnie, sobrannie N. A. Zarudnim v Persii v 1903-1904 gg. [Reptiles and amphibians collected by N. A. Zarudny in Persia in 1903-1904]. Ezhegodnik Zool. muzeya AN [Ann. Zool. Mus. Imper. Acad. Sci.], 10(3-4):260-301 (in Russian).

93. Nikol'skij, A. M. 1907. *Alsophylax laevis* sp. nov. (Geckonidarum). Ezhegodnik Zool. muzeya AN [Ann. Zool. Mus. Imper. Acad. Sci.], 10(3-4):333-335 (in Latin).

94. Nikol'skij, A. M. 1907. OpredelitelÕ presmikayushchihsya i zemnovodnih Rossijskoj Imperii [Field guide to reptiles and amphibians of the Russian Empire]. Kharkov: Russ. tipogr. i litografiya [Russian Print. Lithog.], 182 pages (in Russian).

{p. 220} 95. Nikol'skij, A. M. 1908. Materiali po gerpetologii Russkogo Turkestana [Materials on herpetology of the Russian Turkestan]. Ezhegodnik Zool. Muz. AN [Ann. Zool. Mus. Imper. Acad. Sci.], 14(3):336-344 (in Russian).

96. Nikol'skij, A. M. 1915. Fauna Rossii i sopredelnih stran [Fauna of Russia and contiguous countries]. Presmikayushchiesya (Reptilia), T. 2. Chelonia i Sauria. St. Petersburg: Izd-vo Imp. Ak. nauk [Imper. Acad. Sci. Press], 532 pages (in Russian).

97. Orlov, N. L. 1981. O vostochnoj granitse dlinnonogogo golopalogo gekkona *Gymnodactylus longipes* Nikolsky, 1896 [About the eastern range limit of the gecko *Gymnodactylus longipes* Nikolsky, 1896], p. 89-91. In: Fauna i ekologiya amfibij i reptilij palearkticheskoj Azii [The fauna and ecology of amphibians and reptiles of the Palearctic Asia]. Leningrad: Zool. In-ta AN SSSR (Tr. Zool. in-ta AN SSSR [Proc. Zool. Inst. Acad. Sci. U.S.S.R.], t. 101) (in Russian).

98. Paraskiv, K. P. 1956. Presmikayushchiesya Kazakhstana [Reptiles of Kazakhstan]. Alma-Ata: Izd-vo AN KazSSR [Acad. Sci. Kazakh S.S.R. Press], 228 pages (in Russian).

99. Pashchenko, Yu. M. 1964. Novie dannie o rasprostranenii nekotorih presmikayushchihsya v Turkmenii [New data on distribution of some reptiles in Turkmenia], p. 52-53. In: Voprosi gerpetologii: (Materiali gerpetol. konf., 12-14 okt. 1964 g.) [Problems of Herpetology: (Materials Gerpetol. Conf., 12-14 Oct., 1964)]. Leningrad: Izd-vo Leningr. Un-ta [Leningrad Univ. Press], (in Russian).

100. Plohinskij, N. A. 1961. Biometriya [Biometry]. Novosibirsk: SO [Siberian Branch] Acad. Sci. U.S.S.R. Press, 366 pages (in Russian).

101. Rokitskij, P. F. 1961. Osnovi variatsionnoj statistiki dlya biologov [Foundations of the statistics of variations]. Minsk: Izd-vo Belorus Un-ta [Belarussian Univ. Press], 221 pages (in Russian).

102. Rustamov, A. K. 1954. Ptitsi pustini Kara-Kum [Birds of Karakum Desert]. Ashkhabad: Izd-vo AN TSSR [Acad. Sci. Turkmenian S.S.R. Press], 344 pages (in Russian).

103. Rustamov, A. K., and Ch. Atayev. 1976. Novie dannie po gerpetofaune Turkmenistana [New data on the herpetofauna of Turkmenistan]. Izv. AN TSSR [Proc. Acad. Sci. Turkmenian S.S.R.], 1976(5):47-53 (in Russian).

104. Rustamov, A. K., and S. Shammakov. 1979. Redkie i ischezayushchie vidi reptilij Turkmenistana

[The rare and endangered reptile species of Turkmenistan]. Ohrana Prirodi Turkmenistana [Protection of Turkmenistan Nature], 1979(5):139-146 (in Russian).
105. Said-Aliev, S. A. 1962. O novih nahodkah pantsirnogo i pisklivogo gekkonchikov iz roda *Alsophylax* v Tadjikistane [New findings of armored and squeaky pygmy geckos of the genus *Alsophylax* in Tadjikistan]. Tr. In-ta zoologii i parazitologii TadjSSR [Trans. Inst. Zool. Parasitol. Tadjik S.S.R.], 1962(22):103-105 (in Russian).
106. Said-Aliev, S. A. 1963. Materiali k faune presmikayushchihsya i zemnovodnih Sev. Tadjikistana [Materials on the reptile and amphibian fauna of North Tadjikistan]. Izv. AN TadjSSR [Proc. Acad. Sci. Tadjik S.S.R.], Otd. Biol., 1963(3):81-94 (in Russian).
107. Said-Aliev, S. A. 1979. Zemnovodnie i presmikayushchiesya Tadjikistana [Amphibians and reptiles of Tadjikistan]. Dushanbe: Donish Press, 145 pages (in Russian).
108. Sattarov, T. 1976. O novih nahodkah nekotorih presmikayushchihsya v Severnom Tadjikistane [New findings of some reptiles in North Tadjikistan]. Dokl. AN TadjSSR [Rep. Acad. Sci. Tadjik S.S.R.], 19(12):49-51 (in Russian).
109. Skalon, N. V. 1982. Zemnovodnie i presmikayushchiesya Yugo-Zapadnogo Kopetdaga [Amphibians and reptiles of Southwest Kopetdagh], p. 146-157. In: Priroda Zapadnogo Kopetdaga [The nature of West Kopetdagh]. Ashkhabad: Ylym Press (in Russian).
110. Solovkin, N. 1915. Otchet o sbore faunisticheskih kollektsij po beregam Astrabadskogo zaliva, a takzhe v rayone Krasnovodskogo zaliva i na o. Chelekene v 1913 g. [Report on a faunal collection made on the shores of Astrabad Sound, as well as near Krasnovodsk Bay and on Cheleken Island in 1913]. Ezhegodnik Zool. muzeya AN [Ann. Zool. Mus. Imper. Acad. Sci.], 20(54):I-XVIII (in Russian).
111. Terentjev, P. V. 1961. Gerpetologiya [Herpetology]. Moscow: Visshaya Shkola Press, 335 pages (in Russian).
112. Terentjev, P. V., and S. A. Chernov. 1961. Opredelitel presmikayushchihsya i zemnovodnih [Guide to reptiles and amphibians]. Third edition. Moscow and Leningrad: Sov. Nauka Press, 340 pages (in Russian).
113. Utemisov, O. 1973. Yashcheritsi kulturnogo landshafta Karakalpakii [Lizards of the cultural landscape of Karakalpakiya], 184-185. In: Voprosi Gerpetologii: Avtoref. dokl. III-j Vsesoyuz. gerpetol. konf. [Problems of Herpetology. Third All-Union Herp. Conf. Abstracts]. Leningrad: Nauka Press (in Russian).
114. Fedchenko, A. P. 1870. Otchet Turkestanskoj uchenoj ekspeditsii Obshchestva s 16 aprelya 1869 g. po 15 aprelya 1870 g. [Report on the Turkestan scientific field trip of the Society from 16 April 1869 to 15 April 1870]. Izv. ob-va lyubitelej estestvoznaniya, antropologii i etnografii [Proc. Soc. Nat., Anthropol. Ethnogr. Amateurs], 8(1):135-189 (in Russian).
115. Frantsevich, L. I. 1979. Obrabotka rezultatov biologicheskih eksperimentov na mikro-EVM "Elektronika B 3-21" [Processing of the results of biological experiments using micro-computer B 3-21]. Kiev: Nauk. Dumka Press, 91 pages (in Russian).
116. Frolov, V. E. 1981. Razmnozhenie chetireh vidov gekkonov v Moskovskom zooparke [Reproduction of four gecko species in the Moscow Zoo], p. 138-139. In: Voprosi gerpetologii: Avtoref. dokl. V-j Vsesoyuz. gerpetol. konf. [Problems of Herpetology. Fifth All-Union Herp. Conf. Abstracts]. Leningrad: Nauka Press (in Russian).
117. Frolov, V. E. "1985" (1987). Neodnokratnoe razmnozhenie scinkovogo gekkona v moskovskom zooparke [Repeated reproduction of the common skink gecko in the Moscow Zoo]. Vestn. Zool. [Herald Zool.], 1985 [Publication of this article was delayed to 1987: see the list of recent literature - MG].
118. Hu Shu-qin, Hu Bu-qing, Ding Han-bo, and Huang Zhu-jian. 1962. Opisanie zhivotnogo mira Kitaya (presmikayushchiesya) [Atlas of Chinese AnimalsDDReptiles]. Peking: Science Press, 67 pages (in Chinese).
119. Zarevskij [Tsarevsky], S. F. 1922. Presmikayushchiesya i zemnovodnie, sobrannie V. Ya. Lazdinim v Yuzhnoj i Vostochnoj Bukhare i s.-v. chasti Zakaspijskoj oblasti letom 1916 g. [Reptiles and amphibians collected in South Bukhara and the southwestern part of the Transcaspian Region by V. Ya. Lasdin in summer 1916]. Ezhegodnik Zool. muzeya AN [Ann. Zool. Mus. Imper. Acad. Sci.], 22(1-3):79-90 (in Russian).
120. Zellarius [Tsellarius], A. Yu. 1975. Fauna, biotopicheskoe razmeshchenie i chislennost presmikayushchihsya Repetekskogo zapovednika [Fauna, distribution, and quantity of reptiles in the Repetek Preserve]. Izv. AN TSSR [Proc. Acad. Sci. Turkmenian S.S.R.], ser. Biol. Nauk, 1975(6):42-47 (in Russian).
121. Chernov, S. A. 1934. Presmikayushchiesya Turkmenii [Reptiles of Turkmenia]. Tr. soveta po

izuch. proizvodit. sil [Scientific works for study of productive forces], 1934(6):255-290 (in Russian).

122. Chernov, S. A. 1935. Yashcheritsi (Sauria) Tadjikistana [Lizards (Sauria) of Tadjikistan]. Tr. Tadj. bazi AN SSSR [Scientific works of Tadjik source Acad. Sci. U.S.S.R.], 1935(5):469-475 (in Russian).

123. Chernov, S. A. 1947. Materiali k gerpetofaune Kazakhskogo nagor'ya, severnogo poberezh'ya Balhasha i gor Ken-Tau [Materials on the herpetofauna of the Kazakh Upland, northern banks of Balkhash, and Khentau Mountains]. Izv. AN KazSSR [Proc. Acad. Sci. Kazakh S.S.R.], ser. Zool., 1947(6):120-124 (in Russian).

124. Chernov, S. A. 1954. Ekologo-faunisticheskij obzor presmikayushchihsya yuga mezhdurech'ya Volga-Ural [Ecological faunistic review of the reptiles of the southern part of the Volga-Ural Divide]. Tr. Zool. in-ta AN SSSR [Proc. Zool. Inst. Acad. Sci. U.S.S.R.], 16:137-158 (in Russian).

125. Chernov, S. A. 1959. Fauna Tadjikskoj SSR [Fauna of Tadjik S.S.R.]. T. 18. Presmikayushchiesya [Reptiles]. Stalinabad: Izd-vo AN TadjSSR [Acad. Sci. Tadjik S.S.R. Press], 202 pages (in Russian).

126. Chegodayev, A. E. 1975. Scinkovij gekkon [Common skink gecko]. Priroda [Nature], 1975(9): 98-100 (in Russian).

127. Shammakov, S. 1964. O nekotorih redkih vidah presmikayushchihsya Turkmenii [On some rare species of reptiles]. Izv. AN TSSR [Proc. Acad. Sci. Turkmemian S.S.R.], ser. Biol. Nauk, 1964(6):86-87 (in Russian).

128. Shammakov, S. 1968. Faunisticheskie materiali o presmikayushchihsya malih hrebtov (Malij Balkhan, Kyuren-Dag, Kara-Goz) Zap. Turkmenii [Faunistic data on the reptiles of the small ridges (Maly Balkhan, Kyurendagh, Karagoez) of West Turkmenia], p. 10-15. In: Gerpetologiya Srednej Azii [Herpetology of Middle Asia]. Tashkent: In-t zoologii i parazitologii AN UzSSR [Inst. Zool. Parasitol. Acad. Sci. Uzbek S.S.R.] (in Russian).

129. Shammakov, S. 1974. K ekologii pantsirnogo gekkonchika (*Alsophylax loricatus* Str.) v Turkmenii [Data on the ecology of the armored pygmy gecko (*Alsophylax loricatus* Str.) in Turkmenia]. Izv. AN TSSR [Proc. Acad. Sci. Turkmenian S.S.R.], ser. Biol. Nauk, 1974(2):76-77 (in Russian).

{p. 221} 130. Shammakov, S. 1981. Presmikayushchiesya ravninnogo Turkmenistana [Reptiles of the plains of Turkmenistan]. Ashkhabad: Ylym Press, 311 pages (In Russian).

131. Shammakov, S., and Ch. Atayev. 1971. O gladkom gekkonchike (*Alsophylax laevis* Nikolsky) v Turkmenii [On the smooth pygmy gecko (*Alsophylax laevis* Nikolsky) in Turkmenia]. Izv. AN TSSR [Proc. Acad. Sci. Turkmenian S.S.R.], ser. Biol. Nauk, 1971(3):65-69 (in Russian).

132. Shammakov, S., and V. P. Velikanov. 1982. O rasprostranenii i ekologii pisklivogo gekkonchika v Turkmenistane [Distribution and ecology of squeaky pygmy geckos in Turkmenistan]. Izv. AN TSSR [Proc. Acad. Sci. Turkmenian S.S.R.], ser. Biol. Nauk, 1982(2):72-73 (in Russian).

133. Shammakov, S., O. S. Sopiyev, and N. M. Feodorova. 1982. Ekologiya scinkovogo gekkona v Karakumah [Ecology of the common skink gecko in the Karakums]. Izv. AN TSSR [Proc. Acad. Sci. Turkmenian S.S.R.], ser. Biol. Nauk, 1982(3):36-42 (in Russian).

134. Sharpilo, V. P. 1976. Paraziticheskie chervi presmikayushchihsya fauni SSSR [Parasitic worms of the reptile fauna of the U.S.S.R. fauna]. Kiev: Nauk. Dumka Press, 287 pages (in Russian).

135. Shestoperov, E. L. 1935. Predvaritel'noe obsledovanie v zoologicheskom otnoshenii Akhcha-Kujminskogo zapovednika [Preliminary zoological investigation of Akhcha-Kuyma Reserve]. Izv. Turkm. mezhduvedomstv. kom. po ohrane prirodi i razvitiyu prirodnih bogatstv [Scientific Works of the Interdepartmental Committee for the Nature Preserve and Development of Natural Resources], ser. Biol., 1935(2):161-192 (in Russian).

136. Shibanov, N. V. 1931. Yashcheritsi [Lizards], p. 596-698. In: Zhizn zhivotnih po A. E. Bremu [Animal life, after A. E. Brehm]. Moscow: Uchpedgiz Press (in Russian).

137. Shnitnikov, V. 1928. Presmikayushchiesya Semirech'ya [The reptiles of the Semirechye Region]. Tr. o-va izuch. Kazakhstana [Trans. Soc. Study Kazakhstan], Kzyl-Orda [Kyzyl-Orda (city)], 8(3):24-28 (in Russian).

138. Szczerbak, N. N. 1960. Do vivchennya kryms'kogo gekona (*Gymnodactylus kotschyi danilewskii* Strauch) [On study of Danilewski's gecko (*Gymnodactylus kotschyi danilewskii* Strauch)]. Dop. AN UkrRSR [Rep. Acad. Sci. Ukrainian S.S.R.], 7(7):970-973 (in Ukrainian).

139. Szczerbak, N. N. 1960. Novie dannie o krimskom gekkone (*Gymnodactylus kotschyi danilewskii* Strauch) [New data on Danilewski's gecko (*Gymnodactylus kotschyi danilewskii* Strauch)]. Zool. Zhur. [Zool. Jour.], 39(9):1390-1397 (in Russian).

140. Szczerbak, N. N. 1965. Novie dannie o razmnozhenii krimskogo gekkona (*Gymnodactylus*

kotschyi danilewskii Str.) [New data on reproduction of the Danilewski's gecko (*Gymnodactylus kotschyi danilewskii* Str.). Zool. Zhur. [Zool. Jour.], 44(9):1421 (in Russian).

141. Szczerbak, N. N. 1966. Zemnovodnie i presmikayushchiesya Krima [Amphibians and reptiles of Crimea]. Kiev: Nauk. Dumka Press, 240 pages (in Russian).

142. Szczerbak, N. N. 1971. Sistematika roda yashchurka - *Eremias* (Sauria, Reptilia) v svyazi s ochagami razvitiya pustinno-stepnoj fauni Palearktiki [Systematics of the racerunner genus *Eremias*, in connection with development centers of the desert-steppe fauna of the Palearctic]. Vestn. Zool. [Herald Zool.], 1971(2):48-55 (in Russian).

143. Szczerbak, N. N. 1974. Yashchurki Palearktiki [Palearctic racerunners]. Kiev: Nauk. Dumka Press, 294 pages (in Russian).

144. Szczerbak, N. N. 1978. *Gymnodactylus turcmenicus* sp. n. (Reptilia, Sauria) novij vid gekkona iz Yuzhnoj Turkmenii [*Gymnodactylus turcmenicus* sp. n. (Reptilia, Sauria), a new gecko species from South Turkmenia]. Vestn. Zool. [Herald Zool.], 1978(3):39-44 (in Russian).

145. Szczerbak, N. N. 1979. Eublefar turkmenskij [The Turkmen fat-tailed lizard]. Priroda [Nature], 1979(6):83-85 (in Russian).

146. Szczerbak, N. N. 1979. Novij podvid scinkovogo gekkona (*Teratoscincus scincus rustamovi* ssp. n., (Sauria, Reptilia) iz Uzbekistana i sistematika vida [A new subspecies of the common skink gecko (*Teratoscincus scincus rustamovi* ssp. n., (Sauria, Reptilia) from Uzbekistan, with systematics of the species]. Ohrana prirodi Turkmenistana [Protection of Turkmen Nature], 1979(5):129-138 (in Russian).

147. Szczerbak, N. N. 1979. Novie nahodki yashcherits i zmej na territorii Srednej Azii [New findings of lizards and snakes on the territory of Middle Asia]. Vestn. Zool. [Herald Zool.], 1979(1):68-70 (in Russian).

148. Szczerbak, N. N. 1981. Osnovi gerpetologicheskogo rajonirovaniya territorii SSSR [The fundamentals of herpetogeographic division of the U.S.S.R. territory], p. 157-158. In: Voprosi gerpetologii: Avtoref. dokl. V-j Vsesoyuzn. gerpetol. konf. [Problems of Herpetology. Fifth All-Union Herp. Conf. Abstracts]. Leningrad: Nauka Press (in Russian).

149. Szczerbak, N. N., and M. L. Golubev. 1977. Vzaimootnosheniya rodov *Gymnodactylus* i *Alsophylax* i ih vnutrirodovaya struktura [Relationships between the genera *Gymnodactylus* and *Alsophylax* and their intrageneric structures], p. 237-238. In: Voprosi gerpetologii: Avtoref. dokl. IV-j Vsesoyuzn. gerpetol. konf. [Problems of Herpetology. Fourth All-Union Herp. Conf. Abstracts]. Leningrad: Nauka Press (in Russian).

150. Szczerbak, N. N., and M. L. Golubev. 1977. Materiali k sistematike palearkticheskih gekkonov (rodi *Gymnodactylus, Bunopus, Alsophylax*) [Materials to the systematics of Palearctic geckos (genera *Gymnodactylus, Bunopus, Alsophylax*)], p. 120-123. In: Gerpetologicheskij sbornik [Herpetological collected papers]. Leningrad: Zool. in-t AN SSSR (Tr. Zool. in-ta AN SSSR, T. 74) [Proc. Zool. Inst. Acad. Sci. U.S.S.R., vol. 74] (in Russian).

151. Szczerbak, N. N., and M. L. Golubev. 1981. Novie nahodki zemnovodnih i presmikayushchihsya v Srednej Azii i Kazakhstane [New findings of amphibians and reptiles in Middle Asia and Kazakhstan]. Vestn. Zool. [Herald Zool.], 1981(1):70-72 (in Russian).

152. Szczerbak, N. N., and M. L. Golubev. 1983. Geograficheskaya izmenchivost' i taksonomiya serogo golopalogo gekkona (*Gymnodactylus russowi* Str., Gekkonidae, Reptilia) [Geographical variation and taxonomy of the gray thin-toed gecko (*Gymnodactylus russowi* Str., Gekkonidae, Reptilia)], p. 160-162. In: Fiziologicheskaya i populyatsionnaya ekologiya [Physiological and population ecology]. Saratov: Izd-vo SGU [Saratov State Univ. Press] (in Russian).

153. Szczerbak, N. N., and M. L. Golubev. 1984. O rodovoj prinadlezhnosti i vnutrirodovoj strukture palearkticheskih golopalih gekkonov (Reptilia, Gekkonidae, *Tenuidactylus* gen. n.) [On generic assignement of the Palearctic Cyrtodactylus lizard species (Reptilia, Gekkonidae, *Tenuidactylus* gen. n.)]. Vestn. Zool. [Herald Zool.], 1984(2):50-56 (in Russian).

154. Szczerbak, N. N., V. V. Zhukova, and E. M. Pisanets. 1981. Kariotipi gekkonov podroda *Cyrtodactylus* (*Gymnodactylus*, Gekkonidae, Sauria, Reptilia) fauni SSSR [Karyotypes of geckos of the subgenus *Cyrtodactylus* (*Gymnodactylus*, Gekkonidae, Sauria, Reptilia) of the U.S.S.R. fauna]. Dokl. AN UkrSSR [Sci. Rep. Acad. Sci. Ukrainian S.S.R.], ser. B, 1981(8):85-87 (in Russian).

155. Eminov, A. 1974. Ekologiya chajkonosoj krachki v Yuzhnoj Turkmenii [Ecology of a gull {*Gelochelidon nilotica* - MG} in South Turkmenia], p. 149-161. In: Fauna i Ekologiya ptits Turkmenii [Fauna and ecology of the birds of Turkmenia]. Ashkhabad: Ylym Press (in Russian).

156. Yablokov, A. V. {Ed. - MG}. 1976. Pritkaya yashcheritsa [The sand lizard {*Lacerta agilis* - MG}]. Moscow: Nauka Press, 374 pages (in Russian).

157. Yablokov, A. V. 1981. Problema priznaka v morfologii [The problem of a character in morphology], p. 121-122. In: Sovremennie problemi evolyutsionnoj morfologii zhivotnih: Tez. dokl. Mezhdunar. simpos [Contemporary problems of evolutionary animal morphology: Abstracts of International Symp.]. Moscow: Nauka Press (in Russian).
158. Yadgarov, T. 1973. Sostav, raspredelenie i chislennost' presmikayushchihsya o. Aral-Pajgambar (bassejn Surkhandarji) [Composition, distribution and quantity of reptiles of Aral-Payghambar Island (Surkhandarya Basin)], p. 229-230. In: Voprosi gerpetologii: Avtoref. dokl. III-j Vsesoyuz. gerpetol. konf. [Problems of Herpetology. Third All-Union Herp. Conf. Abstracts]. Leningrad: Nauka Press (in Russian).
159. Yadgarov, T. 1959. Gladkij gekkonchik (*Alsophylax laevis*) v pustine Kyzylkum [The smooth pygmy gecko (*Alsophylax laevis*) in the Kyzylkum Desert]. Zool. Zhur. [Zool. Jour.], 54(9):1412-1413 (in Russian).
160. Yadgarov, T. 1981. Novie dannie o rasprostranenii i ekologii pantsirnogo gekkonchika (*Alsophylax loricatus*) v Uzbekistane [New distributional and ecological data on the armored pygmy gecko (*Alsophylax loricatus*) in Uzbekistan], p. 159. In: Voprosi gerpetologii: Avtoref. dokl. V-j Vsesoyuz. konf. [Problems of Herpetology. Fifth All-Union Herp. Conf. Abstracts], Ashkhabad, 1981. Leningrad: Nauka Press (in Russian).
161. Yadgarov, T. 1982. Blizkie vidi golopalih gekkonov roda *Gymnodactylus* na stike ih arealov v Srednej Azii [The closely related thin-toed gecko species of the genus *Gymnodactylus* at the junctions of their ranges]. Uzb. Biol. Zhur. [Uzbek Biol. Jour.], 1982(3):41-43 (in Russian).
162. Yakovleva, I. D. 1964. Presmikayushchiesya Kirgizii [Reptiles of Kirghizia]. Frunze: Ylym Press, 272 pages (in Russian).
163. Anderson, J. 1871. On some Indian reptiles. Proc. Zool. Soc. London, 1871(9):149-211.
164. Anderson, J. 1872. On some Persian, Himalayan and other reptiles. Proc. Zool. Soc. London, 1872(2):371-404.
165. Anderson, J. 1901. A list of reptiles and batrachians obtained by Mr. A. Blayney Percival in southern Arabia. Proc. Zool. Soc. London, 1901(3-4):135-152.
166. Anderson, L. G. 1900. Catalogue of Linnean type-specimens of LinnaeusÕs Reptilia in the Royal Museum in Stockholm. Bihang till K. Sven. Vet.-Akad. Handl., 26(1):1-29.
167. Anderson, S. C. 1961. A note on the synonymy of *Microgecko* Nikolsky with *Tropiocolotes* Peters. Wasmann Jour. Biol., 19(2):287-289.
168. Anderson, S. C. 1963. Amphibians and reptiles from Iran. Proc. California Acad. Sci., 31(16):417-498.
169. Anderson, S. C. 1973. A new species of *Bunopus* from Iran and key to lizards of the genus *Bunopus*. Herpetologica, 29(4):355-358.
170. Anderson, S. C. 1974. Preliminary key to the turtles, lizards and amphisbaenians of Iran. Fieldiana: Zool., 65(4):27-44.
171. Anderson, S. C., and A. E. Leviton. 1966. A new species of *Eublepharis* from southwestern Iran (Reptilia: Gekkonidae). Occas. Pap. California Acad. Sci., 53:1-5.
172. Anderson, S. C., and A. E. Leviton. 1969. Amphibians and reptiles collected by the Street expedition to Afganistan, 1965. Proc. California Acad. Sci., 37(2):25-56.
173. Angel, F. 1936. Reptiles et batrachiens de Syrie et de Mesopotamie recoltes par M. P. Pallary. Bull. Inst. Egypt, 18:107-116.
174. Annandale, N. 1913. The Indian geckos of the genus *Gymnodactylus* sp. n. Rec. Indian Mus., 9(5):313.
175. Arnold, E. N. 1977. Little-known geckoes (Reptilia: Gekkonidae) from Arabia with descriptions of two new species from the Sultanate of Oman, p. 81-110. In: The Scientific Results of the Oman Flora and Fauna Survey 1975 [Jour. Oman Stud. Spec. Rep., no. 1].
176. Arnold, E. N. 1980*a*. A review of the lizard genus *Stenodactylus* (Reptilia: Gekkonidae), p. 368-404. In: Fauna of Saudi Arabia, vol. 2. Basel: Pro Entomologia.
177. Arnold, E. N. 1980*b*. The reptiles and amphibians of Dhofar, southern Arabia. Jour. Oman Stud. Spec. Rep., 3:273-332.
178. Arnold, E. N., and M. D. Gallagher. 1977. Reptiles and amphibians from the mountains of Oman with special reference to the Jebel Akhdar region, p. 59-80. In: The Scientific Results of the Oman Flora and Fauna Survey 1975 [Jour. Oman Stud. Spec. Rep., no. 1].
179. Baran, I., and U. Gruber. 1981. Taxonomische Untersuchungen an trkischen Inselformen von *Cyrtodactylus kotschyi* (Steindachner, 1870). Teil I. Die Populationen der nödlichen Ägais, des Marmarmeeres und des Schwarzen Meers (Reptilia: Gekkonidae). Spixiana, 4(3):255-270.

180. Baran, I., and U. Gruber. 1982. Taxonomische Untersuchungen an türkischen Gekkoniden. Spixiana, 5(2):109-138.
181. Basoglu, M., and I. Baran. 1977. Türkije şürüngenleri. Bornova-Izmir: Ege Univ. Fen Fakül. Kit., 76. 272 pages.
182. Bedriaga, J. von. 1905. Verzeichnis der von der Central-Asiatischen Expedition unter Stabs-Kapitn W. Roborowski in den Jahren 1893-1895 gesammelten Reptilien. Annu. Mus. Zool. Acad. Impér. Sci. St.-Petersbourg, 10(3-4):159-200.
183. Beutler, A. 1981. *Cyrtodactylus kotschyi* (Steindachner, 1870) - Ägäischer Bogenfingergecko, p. 53-74. In: W. Böhme (ed.), Handbuch der Reptilien und Amphibien Europas. Vol. 1. Echsen. Wiesbaden: Akad. Verlags.
184. Beutler, A., and U. Gruber. 1977. Intraspezifische Untersuchungen an *Cyrtodactylus kotschyi* (Steindachner, 1870); Reptilia: Gekkonidae. Beitrag zu einer mathematischen Definition des Begriffs Unterart. Spixana, 1(2):165-202.
185. Beutler, A., and E. Frör. 1980. Die Amphibien und Reptilien der Nordkykladen (Griechenland). Mitt. Zool. Ges. Braunau, 8(10-12):255-290.
186. Blanford, W. T. 1874. Descriptions of new Reptilia and Amphibia from Persia and Baluchistan. Ann. Mag. Nat. Hist., ser. 4, 13(78):453-455.
187. Blanford, W. T. 1875. List of Reptilia and Amphibia collected by the late Dr. Stoliczka in Kashmir, Ladak, eastern Turkestan, and Wakhan, with descriptions of new species. Jour. Asiatic Soc. Bengal, 44:191-196.
188. Blanford, W. T. 1876. On some new lizards from Sind with descriptions of new species of *Ptyodactylus, Stenodactylus* and *Trapelus*. Jour. Oman Stud. Spec. Rep., 45(2):18-26.
189. Blanford, W. T. 1878. Scientific results of the Second Yarkand Mission based upon the collections and notes of the late Ferdinand Stoliczka, Ph.D. Reptilia and Amphibia. Calcutta: Office Superintendent Govt. Print., [1-11], 1-26 pages.
190. Blyth, E. 1854. Proceedings of the society. Report of the curator, zoological department. Jour. Asiatic Soc. Bengal, 23:737-740.
191. Böhme, W. (ed.). 1891. Handbuch der Reptilien und Amphibien Europas. Volume 1. Wiesbaden: Akad. Verlags, 520 pages.
192. Boettger, O. 1888. Die Reptilien und Batrachier Transkaspiens. Zool. Jahrb., Abt. Syst., 1888(3): 871-1014.
193. Boettger, O. 1888. Materialien zur herpetologischen Fauna von China II. Ber. Thätig. Offenbach. Ver. Naturk., 26-28:53-191.
194. Boettger, O. 1893. Katalog der Reptilien-Sammlung im Museum der Senckenbergischen Naturforschenden Gesellschaft in Frankfurt am Main. Teil 1. Frankfurt am Main: Senckenb. Nat. Ges., x, 140 pages.
195. Börner, A.-R. 1974. Ein neuer Lidgecko der Gattung *Eublepharis* Gray, 1827. Misc. Art. Saurol., 4:5-14.
196. Börner, A.-R. 1976. Second contribution to the systematics of the southwest Asian lizards of the geckonid genus *Eublepharis* Gray, 1827: Materials from the Indian Subcontinent. Saurologica, 2:1-15.
197. Börner, A.-R. 1981. Third contribution to the systematics of the southwest Asian lizards of the geckonid genus *Eublepharis* Gray, 1827: Further materials from the Indian Subcontinent. Saurologica, 3:1-7.
198. Bons, J. 1959. Les Lacertiliens du Sud-Ouest Marocain. Trav. Inst. Sci. Chérif., ser. Zool., 18:1-130.
199. Boulenger, G. A. 1883. Description of a new genus of geckos. Ann. Mag. Nat. Hist., 5(11):174-176.
200. Boulenger, G. A. 1885. Catalogue of the lizards in the British Museum. Vol. 1. London: Taylor and Francis, 436 pages.
201. Boulenger, G. A. 1887. Catalogue of the lizards in the British Museum. Vol. 3. London: Taylor and Francis, 575 pages.
202. Boulenger, G. A. 1890. The Fauna of British India, including Ceylon and Burma. Reptilia and Batrachia. London: Taylor and Francis, 541 pages.
203. Boulenger, G. A. 1889 {original text contains incorrect date-MG}. The Zoology of the Afghan Delimitation Commission (Reptiles and Batrachians). Trans. Linn. Soc. London, ser. 2, Zool., 5(3):94-106.
204. Boulenger, G. A. 1891. Catalogue of the reptiles and batrachians of Barbary (Morocco, Algeria, Tunisia). Trans. Zool. Soc. London, 13(3):93-164.

205. Boulenger, G. A. 1905. On some batrachians and reptiles from Tibet. Ann. Mag. Nat. Hist., 15:378-379.
206. Brongersma, L. D. 1934. Contributions to Indo-Australian herpetology. Zool. Meded., Rijksmus. Natuurl. Hist., 17(3-4):161-251.
207. Buresch, I., and J. Zonkow. 1933. Untersuchungen über die Verbreitung der Reptilien und Amphibien in Bulgarien und auf der Balkanhalbinsel. 1. Schildkröten und Eidechsen. Mitt. Königl. Naturwiss. Inst. Sofia, 6:150-207.
208. Cherchi, M. A., and S. Spano'. 1963. Una nuova specie di *Tropiocolotes* del Sud Arabia Spedizione Scortecci nell'Hadramaut (1962). Boll. Mus. Ist. Biol. Univ. Genova, ser. Biol Anim., 32(188):29-34.
209. Clark, J., E. D. Clark, S. C. Anderson, and A. E. Leviton. 1969. Report on a collection of amphibians and reptiles from Afghanistan. Proc. California Acad. Sci., ser. 4, 36(10):279-315.
210. Duda, P. L., and D. N. Sahi. 1978. *Cyrtodactylus himalayanus*: a new gekkonid species from Jammu, India. Jour. Herpetol., 12(3):351-354.
211. Duméril, A. M. 1856. Description des reptiles nouveaux ou imparfaitment connus de la collection du Museum d'Histoire Naturelle. Arch. Mus. Nat. Hist. Nat., 8:437-588.
212. Duméril, A. M., and G. Bibron. 1836. Erptologie gnrale. Vol. 3. Geckos et les varans. Paris: Roret, 517 pages.
213. Eichwald, E. 1831. Zoologia specialis quam expositis animalibus tum vivis. Vilnae: J. Zawadzki, 404 pages.
214. Ernst, G. 1983. *Eublepharis macularius*.- Leopardgecko. Aquaria, 30(10):169-171.
215. Fitzinger, L. 1843. Systema reptilium. Fasciculus primus. Amblyglossae. Vienna: Braumüller und Seidel, 106, x pages.
216. Forcart, L. 1950. Amphibien und Reptilien von Iran. Verh. Naturforsch. Ges. Basel, 61:141-156.
217. Flower, S. 1933. Notes on the Recent reptiles and amphibians of Egypt with a list of the species recorded from that kingdom. Proc. Zool. Soc. London, 1933(3):735-851.
218. Gasperetti, J. 1967. Survey of the reptiles of the Sheikhdom of Abu Dhabi. Proc. California Acad. Sci., ser. 4, 35(8):141-156.
219. Gauthier, R. 1967. Ecologie et ethologie des reptiles du Sahara Nord-Occidental (Region de Beni-Abbes). Ann. Mus. Afr. Centr., ser. in-8, 155:1-83.
220. Golubev, M. L. 1981. The content of the genus *Bunopus* (Reptilia: Gekkonidae), p. 17. In: O. Gy. Dely (ed.), I Conferencia Herpetologica Respublicarum Socialisticarum, Budapest.
221. Gray, J. E. 1827. A synopsis of the genera of the saurian reptiles in which some new genera are indicated, and the other reviewed by actual examination. Philos. Mag., ser. 2, 3:53-56.
222. Gray, J. E. 1845. Catalogue of the specimens of the lizards in the collection of the British Museum. London: British Mus., xxviii, 289 pages.
223. Guibé, J. 1966a. Reptiles et amphibiens recoltes par la mission Franco-Iranienne. Bull. Mus. Nat. Hist. Nat., ser. 2, 38(2):97-98.
224. Guibé, J. 1966b. Contribution a l'étude des genres *Microgecko* Nikolsky et *Tropiocolotes* Peters (Lacertilia, Gekkonidae). Bull. Mus. Nat. Hist. Nat., ser. 2, 38(4):337-346.
225. Günther, A. 1864. Description of a new species of *Eublepharis*. Ann. Mag. Nat. Hist., ser. 3, 14:429-430.
226. Haas, G. 1957. Amphibians and reptiles from Arabia. Proc. California Acad. Sci., ser. 4, 29(3):47-86.
227. Haas, G., and J. C. Battersby. 1959. Amphibians and reptiles from Arabia. Copeia, 1959(3):196-202.
228. Haas, G., and Y. L. Werner. 1969. Lizards and snakes from southwestern Asia, collected by Henry Field. Bull. Mus. Comp. Zool. Harvard Coll., 138(6):327-406.
229. Heyden, C. H. G. 1827. Reptilies, p. 1-24. In: E. Rüppell, Atlas zu der Reise im nördlichen Africa. Frankfurt-am-Main: H. L. Brönner.
230. Hoofien, J. H. 1967. Contributions to the herpetofauna of Mount Hermon no. I. *Cyrtodactylus amictopholis* n. sp. (Sauria, Gekkonidae). Israel Jour. Zool., 16(4):205-210.
231. Ingoldby, C. M. 1922 (1923). A new stone gecko from the Himalaya. Jour. Bombay Nat. Hist. Soc., 28(4):1051.
232. Ignoldby, R., and B. Procter. 1924. Notes on a collection of Reptilia from Waziristan and the adjoining portion of the N.W. Frontier Province. Jour. Bombay Nat. Hist. Soc., 9(1-2):117-130.
233. Karaman, M. S. 1965. Eine neue Unterart der Eidechse *Gymnodactylus kotschyi* aus Mazedonien, *Gymnodactylus kotschyi scopjensis* n. ssp. Zool. Anz., 174(4-5):348-351.

234. Kasimir, M. J. 1971. Zur Herpetofauna der Provinz Badghis (N.W. Afghanistan). Deutsch. Aquar.-Terr. Zeitschr., 24:244-246.
235. Khan, M. S. 1972. Checklist and key to lizards of Jhangh District, West Pakistan. Herpetologica, 28(2):94-98.
236. Klingelhöffer, W. 1957. Terrarienkunde. Volume 3. Echsen. Stuttgart: A. Kernen, 264 pages.
237. Kluge, A. G. 1967. Higher taxonomic categories of gekkonid lizards and their evolution. Bull. Amer. Mus. Nat. Hist., 135(1):1-60.
238. Kluge, A. G. 1969. An interpretation of the status of *Gymnodactylus tenuis* Hallowell (Sauria, Gekkonidae). Copeia, 1969(3):623-624.
239. Kluge, A. G. 1983. Cladistic relationships among gekkonid Lizards. Copeia, 1983(2):465-475.
240. Král, B. 1969. Notes on the herpetofauna of certain provinces of Afghanistan. Zool. Listy, 18(1):55-66.
241. Lantz, L. A. 1918. Reptiles from the River Tajan (Transcaspia). Proc. Zool. Soc. London, 1918(7):11-13.
242. Leviton, A. E. 1959. Report on collection of reptiles from Afghanistan. Proc. California Acad. Sci., ser. 4, 29(12):445-463.
243. Leviton, A. E., and S. C. Anderson. 1963. Third contribution to the herpetology of Afghanistan. Proc. California Acad. Sci., 31(12):329-339.
244. Leviton, A. E., and S. C. Anderson. 1967. Survey of the reptiles of the Sheikhdom of Abu Dhabi, Arabian Peninsula. Pt. 2. Systematic account in the Sheikhdom of Abu Dhabi by John Gasperetti. Proc. California Acad. Sci., 35(9):157-192.
245. Leviton, A. E., and S. C. Anderson. 1970. The amphibians and reptiles of Afghanistan, a checklist and key to the herpetofauna. Proc. California Acad. Sci., 38(10):163-206.
246. Leviton, A. E., and S. G. Anderson. 1972. Description of a new species of *Tropiocolotes* (Reptilia: Gekkonidae) with a revised key to the genus. Occas. Pap. California Acad. Sci., 96:1-7.
247. Lichtenstein, H. 1823. Naturhistorische Anhang, p. 112-147. In: E. Eversmann, Reise von Orenburg nach Buchara. Berlin: E. H. G. Christiani.
248. Lictenstein, H. 1856. Nomenclator reptilium et amphibiorum Musei Zoologici Berolinensis. Berlin: Buchdruck. Königl. Akad. Wiss., 48 pages.
249. Loveridge, A. 1947. Revision of the African lizards of the family Gekkonidae. Bull. Mus. Comp. Zool. Harvard Coll., 98(1):1-469.
250. Marx, H. 1956. Keys to the lizards and snakes of Egypt. NAMRU-3, Cairo, 8 pages.
251. Marx, H. 1968. Checklist of the reptiles and amphibians of Egypt. Cairo: Spec. Publ. U.S. NAMRU-3, 91 pages.
252. Marx, H. 1976. Supplementary catalogue of type specimens of reptiles and amphibians in Field Museum of Natural History. Fieldiana: Zool., 69(2):33-94.
253. Meeuwen, H. M. van. 1977. Enkele herpetologische aantekeningen van een reis naar Afghanistan. Lacerta, 35(8):115-124.
254. Mertens, R. 1924. Ein neuer Gecko aus Mesopotamien. Senckenbergiana, 6:84-85.
255. Mertens, R. 1952. Amphibien und Reptilien aus der Turkei. Istanbul Üniv. Fen Fak. Mecmuasi, ser. B, 17(1):41-75.
256. Mertens, R. 1956. Amphibien und Reptilien aus SO-Iran 1954. Jahrb. Ver. Vaterländ. Naturk. Württemburg, 111(1):90-97.
257. Mertens, R. 1965. Bemerkungen über einige Eidechsen aus Afghanistan. Senckenberg. Biol., 46(1):1-4.
258. Mertens, R. 1969. Die Amphibien und Reptilien West-Pakistans. Stuttgart. Beitr. Naturk., 197:1-96.
259. Mertens, R. 1971. Die Amphibien und Reptilien West-Pakistans. Senckenberg. Biol., 52(1-2):7-15.
260. Mertens, R., and L. Müller. 1928. Liste der Amphibien und Reptilien Europas. Abh. Senckenberg. Naturforsch. Ges., 41(1):1-62.
261. Mertens, R., and L. Müller. 1940. Die Amphibien und Reptilien Europas (Zweite Liste, nach dem Stand vom 1. Januar 1940). Abh. Senckenberg. Naturforsch. Ges., 451:1-56.
262. Minton, S. A. 1962. An annotated key to the amphibians and reptiles of Sind and Las Bela, West Pakistan. Amer. Mus. Novit., 2081:1-60.
263. Minton, S. A. 1966. A contribution to the herpetology of West Pakistan. Bull. Amer. Mus. Nat. Hist., 134(2):27-184.
264. Minton, S. A., and Jer. Anderson. 1965. A new dwarf gecko (*Tropiocolotes*) from Baluchistan. Herpetologica, 21(1):59-61.

265. Minton, S. A., S. C. Anderson, and Jer. Anderson. 1970. Remarks on some geckos from Southwest Asia, with descriptions of three new forms and a key to the genus *Tropiocolotes*. Proc. California Acad. Sci., ser. 4, 37(9):333-362.
266. Müller, L. 1939. über die von den Herren Dr. v. Jordans und Dr. Wolf im Jahre 1938 in Bulgarien gesammelten Amphibien und Reptilien. Mitt. Königl. Nat. Inst., Sofia, 13:1-17.
267. Murray, J. A. 1884. The vertebrate zoology of Sind. London: Richardson, xiv, 424 pages.
268. Murray, J. A. 1892. The zoology of Beloochistan and southern Afghanistan. Bombay: Education Soc., 83 pages.
269. Nader, I. A., and S. Z. Jawdat. 1976. Taxonomic study of the geckos of Iraq (Reptilia: Gekkonidae). Bull. Biol. Res. Cent., Univ. Baghdad, 5:1-41.
270. Pallas, P.-S. 1811. Zoographia Rosso-Asiatica. Volume 3. Petropolis: Caes. Acad. Sci., vii, 428, cxxv pages.
271. Papenfuss, T. J. 1969. Preliminary analysis of the reptiles of arid central west Africa. Wasmann Jour. Biol., 27:249-325.
272. Parker, H. W. 1942. The lizards of British Somaliland. Bull. Mus. Comp. Zool., Harvard Coll., 41:1-101.
273. Pasteur, G. 1960. Redécouverte et validité probable du gekkonidé *Tropiocolotes nattereri* Steind. Comp. Rend. Soc. Sci. Nat. Phys. Maroc., 26(8):143-145.
274. Pasteur, G., and J. Bons. 1960. Catalogue des reptiles actuels du Maroc. Trav. Inst. Sci. Cherif. Rabat, ser. Zool., 21:1-132.
275. Peters, W. C. H. 1869. Über neue Saurier. Monatsber. Knigl. Akad. Wiss. Berlin, 1869:786-790.
276. Peters, W. C. H. 1880. Über die von Hrn. Gerhard Rohlfs und Dr. A. Stecker auf der Reise nach der Oase Kufra gesammelten Amphibien. Monatsber. Knigl. Akad. Wiss. Berlin, 1880:305-309.
277. Peters, G. 1973. Urania Tierreich. Band 4. Reptilia. Leipzig, etc.: Urania Verl., 508 pages.
278. Pope, C. H. 1935. The reptiles of China. Nat. Hist. Centr. Asia, 10: lii, 1-604.
279. Procter, J. B. 1923. Further lizards and snakes from Persia and Mesopotamia. Jour. Bombay Nat. Hist. Soc., 28(1):251-253.
280. Reed, C. A., and H. Marx. 1959. A herpetological collection from northeastern Iraq. Trans. Kansas Acad. Sci., 62(1):91-122.
281. Rössler, H. 1983. Fang, Haltung und Eiblage des Kammzehengeckos *Crossobamon eversmanni* (Wiegmann, 1834). Aquar. Terrar., 31(1):33-36.
282. Scerback [Szczerbak], N. N. 1981. *Cyrtodactylus russowii* (Str., 1887) – Transkaspischer Bogenfingergecko, p. 75-83. In: W. Böhme (ed.), Handbuch der Reptilien und Amphibien Europas. Band 1. Echsen. Wiesbaden: Akad. Verlag.
283. Schlegel, H. 1858. Handleiding tot de Beoefening der Dierkunde. Breda: H. G. Nys for Koninkl. Milit. Akad., vol. 2, xx, 628, (2) pages.
284. Schmidt, K. P. 1939. Reptiles and amphibians from southwestern Asia. Publ. Field Mus. Nat. Hist., ser. Zool., 24(7):49-92.
285. Schmidt, K. P. 1955. Amphibians and reptiles from Iran. Vid. Medd. Dan. Naturhist. Foren. Kobenhaven, 117(3):193-207.
286. Schmidtler, J. J., and J. F. Schmidtler. 1972. Zwerggeckos aus dem Zagros-Gebirge (Iran). Salamandra, 8(2):59-66.
287. Shcherbak [Szczerbak], N. N. 1982. Grundzüge einer herpetogeographischen Gliederung der Palarktis. Vertebr. Hungarica, 21:227-239.
288. Shockley, C. H. 1949. Herpetological notes for Ras Jiunri, Baluchistan. Herpetologica, 5(6):121-123.
289. Seufer, H. 1979. Der Kaspische Geradfinger-Gecko (*Alsophylax pipiens*) Pallas. Herpetofauna, 1:10-15.
290. Smith, M. A. 1933. Remarks on some Old World geckos. Rec. Indian Mus., 35:9-19.
291. Smith, M. A. 1935. Fauna of British India, including Ceylon and Burma. Reptilia and Amphibia. Vol. II. Sauria. London: Taylor and Francis, xiii, 440 pages.
292. Spix, J. B. 1825. Animalia nova sive species novae lacertarum, quas in itinere per Brasiliam annis 1817-1820. Monachii: F. S. Hübschmann, (1), 26 pages.
293. Steindachner, F. 1868. Reptilien, p. 1-98. In: Reise der Österreichischen Fregatte Novara. Wien: Kaiserl.-Königl. Hof-Staatsdruck.
294. Steindachner, F. 1870. Herpetologische Notizen II. Reptilien gesammelt während einer Reise in Senegambien. Sitzungsber. Akad. Wiss. Wien, 62(1):326-335.
295. Steindachner, F. 1900 (1901). Expedition S. M. Schiff "Pola" in das Rothe Meer. 17. Bericht über

die herpetologischen Aufsammlungen. Denkschr. Acad. Wiss. Math.-Naturwiss. Kl. 69(1):325-335.
296. Stepánek, O. 1934. Sur le *Gymnodactylus danilewskii* en Bulgarie. Sbor. Zool. Odd. Národ. Mus. Praze, 1934(1):31.
297. Stepánek, O. 1937. *Gymnodactylus kotschyi* Steindachner und sein Rassenkreis. Arch. Natur. Leipzig, n. f., 6:258-280.
298. Stoliczka, F. 1872. Notes on the reptilian and amphibian fauna of Kachh. Proc. Asiatic Soc. Bengal, 5:71-85.
299. Strauch, A. 1863. Characteristik zweier neuen Eidechsen aus Persien. Bull. Acad. Sci. St.-Petersburg, 4:393-398.
300. Strauch, A. 1887. Bemerkungen über die Geckoniden-Sammlung im zoologischen Museum der kaiserlichen Akademie der Wissenschaften zu St. Petersburg. Mém. Acad. Impér. Sci. St.-Petersburg, ser. 7, 35(2):(2),1-72.
301. Szczerbak, N. N. 1981. Grundzüge der herpetogeographischen Gliederung des palarktischen Gebietes, p. 42. In: O. Gy. Dely (ed.), I Conferencia Herpetologica Respublicarum Socialisticarum, Budapest.
302. Tuck, R. G. 1971*a*. Rediscovery and redescription of the Khuzistan dwarf Gecko, *Microgecko helenae* Nikolsky (Sauria: Gekkonidae). Proc. Biol. Soc. Washington, 83(42):477-482.
303. Tuck, R. G. 1971*b*. Amphibians and reptiles from Iran in the United States National Museum collection. Bull. Maryland Herpetol. Soc., 7(3):48-86.
304. Tuck, R. G. 1973. Additional notes on Iranian reptiles in the United States National Museum collection. Bull. Maryland Herpetol. Soc., 9(1):13-14.
305. Underwood, G. 1954. On the classification and evolution of geckos. Proc. Zool. Soc. London, 124(3):469-492.
306. Vinciguerra, D. 1931. Spedizione scientifica all'Oasi di Cufra - Rettili. Ann. Mus. Civ. Stor. Nat. Genova, 55:248-258.
307. Wermuth, H. 1965. Liste der rezenten Amphibien und Reptilien Gekkonidae, Pygopodidae, Xantusiidae. Das Tierreich, 80:xxii, 1-246.
308. Werner, F. 1917. Reptilien aus Persien (Provinz Fars) gesammelt von Herrn. Prof. Andreas. Verh. Zool.-Bot. Ges. Wien, 67:191-220.
309. Werner, F. 1936 (1937). Reptilien und Gliedertiere aus Persien, p. 193-204. In: Festschrift zum 60. Geburtstage von Prof. Dr. Embrik Strand, vol. 2. Riga: Lizdevnieciba "Latvija."
310. Werner, F. 1938. Reptilien aus Iran und Belutschistan. Zool. Anz., 121(9-10):265-271.
311. Werner, F. 1939. Die Amphibien und Reptilien von Syrien. Abh. Berl. Mus. Nat.-u. Heimat. (Naturk. Vorgesch.) Magdeburg, 7(1):211-223.
312. Werner, Y. L. 1956. Chromosome numbers of some male geckos (Reptilia: Gekkonidae). Bull. Res. Counc. Israel, 5B(3-4):319.
313. Werner, Y. L. 1969. Eye size in geckos of various ecological types (Reptilia: Gekkonidae and Sphaerodactylidae). Israel Jour. Zool., 18:91-316.
314. Wettstein, O. 1937. Vierzehn neue Reptilienrassen von den südlichen Ägäischen Inseln. Zool. Anz., 118(3-4):79-90.
315. Wettstein, O. 1951. Ergebnisse der Österreichischen Iran-Expedition 1949/50, Amphibien und Reptilien. Sitzungsber. Österr. Akad. Wiss. Math.-Naturw. Kl., Abt. 1, 160(5):427-448.
316. Wettstein, O. 1953. Dreizehn neue Reptilienrasse von den Ägäischen Inseln. Anz. Österr. Akad. Wiss, Math.-Naturwiss., 89:251-256.
317. Wettstein, O. 1960. Lacertilia aus Afghanistan. Contribution à l'étude de la faune d'Afghanistan, 3. Zool. Anz., 165(1-2):58-63.
318. Wettstein-Westersheimb, O. 1960. Drei seltene Echsen aus Südwest Asien. Zool. Anz., 165(5-6):190-193.
319. Wiegmann, A. F. A. 1834. Herpetologia Mexicana, seu descriptio amphibiorum Novae Hispaniae. Pars prima, saurorum species. Berlin: C. G. Lüderitz, vi, 54 pages.
320. Witte, G. F. de. 1973. Description d'un Gekkonidae nouveau de l'Iran (Reptilia Sauria). Bull. Inst. Roy. Sci. Nat. Belg., Biol., 49(1):1-6.
321. Witte, G. F. de. 1980. Note relative à *Rhinogekko misonnei* de Witte et *Agamura femoralis* M. Smith (Reptilia Sauria). Bull. Inst. Roy. Sci. Nat. Belg., Biol., 52(8):1-3.
322. Welch, K. R. G. 1983. Herpetology of Europe and southwest Asia: A checklist and bibliography of the orders Amphisbaenia, Sauria and Serpentes. Malabar, Florida: Krieger, viii, 135 pages.
323. Obst, F. U. 1984. Nahodka turetskogo polupalogo gekkona *Hemidactylus turcicus* (Linnaeus, 1758) v Turkmenii [A record of the gecko *Hemidactylus turcicus* (Linnaeus, 1758) in Turkmenia],

p. 142-143. In: L. J. Borkin (ed.), Ecologiya i faunistika amphibij i reptilij SSSR [Ecology and Faunistics of Amphibians and Reptiles of the USSR and Adjacent Countries]. Leningrad: Zool. In-ta AN SSSR [Proc. Zool. Inst. Acad. Sci. USSR] (in Russian).
324. Gruber, U. 1981. Notes on the herpetofauna of Kashmir and Ladakh. British Jour. Herpetol., 6(5):145-150.
325. Leviton, A. E., and S. C. Anderson. 1984. Description of a new species of *Cyrtodactylus* from Afghanistan with remarks on the status of *C. longipes* and *C. fedtschenkoi*. Jour. Herpetol., 18(3):270-276.
326. Reznik, E. P. 1985. K ekologii grebnepalogo gekkona (*Crossobamon eversmanni*) v srednem techenii r. Murgab [On ecology of (Eversmann's) fringe-toed geckos (*Crossobamon eversmanni*) in the middle part of the Murghab River], p. 175-176. In: Voprosi gerpetologii: Avtoref. dokl. VI-j Vsesoyuz. gerpetol. konf. [Problems of Herpetology. Sixth All-Union Herpetol. Conf. Abstracts]. Leningrad: Nauka Press (in Russian).
327. Rzepakovskij, V. T. 1985. K izucheniyu ecologii turkestanskogo golopalogo gekkona v Tadjikistane [On the ecology of the Turkestan thin-toed gecko in Tadjikistan], p. 176-177. In: Voprosi gerpetologii: Avtoref. dokl. VI-j Vsesoyuz. gerpetol. konf. [Problems of Herpetology. Sixth All-Union Herpetol. Conf. Abstracts]. Leningrad: Nauka Press (in Russian).
328. Smirnov, S. I., V. F. Shkunov, and E. Yu. Kudakina. 1985. Gekkony (Gekkonidae) Severnogo Prikaspiya [Geckos (Gekkonidae) of the northern Pre-Caspian region], p. 195-196. In: Voprosi gerpetologii: Avtoref. dokl. VI-j Vsesoyuz. gerpetol. konf. [Problems of Herpetology. Sixth All-Union Herpetol. Conf. Abstracts]. Leningrad: Nauka Press (in Russian).

SUPPLEMENTARY LITERATURE CITED

Anderson, S. C. 1993. A note on the syntopic occurrence of three species of *Teratoscincus* in Eastern Iran. Dactylus, 1(4):8-10.
Anderson, S. C. Lizards of Iran. Soc. Study Amphib. Reptiles, Contr. Herpetol., in preparation.
Autumn, K., and B. Han. 1989. Mimicry of scorpions by juvenile lizards, *Teratoscincus roborowskii* (Gekkonidae). Chinese Herpetol. Res., 2(2):60-64.
Bauer, A. M. 1986. [Review of] The Gekkonid Fauna of the U.S.S.R. and Adjacent Countries, by N. N. Szczerbak and M. L. Golubev. Copeia, 1987(2):525-527.
Bauer, A., and R. Günther. 1991. An annotated type catalogue of the geckos (Reptilia: Gekkonidae) in the Zoological Museum, Berlin. Mitt. Zool. Mus. Berlin, 67(2):279-310.
Biswas, S., and D. P. Sanyal. 1977. Fauna of Rajastan, India. Part Reptilia. Rec. Zool. Surv. India, 73(1-3):247-269.
Böhme, W. 1985. Zur Nomenklatur der Palrctischen Bogenfingergeckos, Gattung *Tenuidactylus* Scerbak & Golubew, 1984 (Reptilia: Gekkonidae). Bonn. Zool. Beitr., 36(1-2):95-98.
Borkin, L. Ya., Kh. Munkhbayar, N. L. Orlov, D. V. Semenov, and Kh. Terbish. 1990. Rasprostranenie reptilij Mongolii [Distribution of reptiles in Mongolia], p. 22-138. Reptiles of Mountain and Arid Territories: Systematics and Distribution. In: Proc. Zool. Inst. Acad. Sci. U.S.S.R., vol. 207 (in Russian).
Bozhanskij, A. T., and V. M. Makeev. 1992. Ecological observations on *Cyrtopodion russowi*. Abstracts First Asian Herpetol. Meeting, 15-20 July 1992, Huangshan, Anhui, China:24.
Brushko, Z. K. 1985. Ekologiia serogo golopalogo gekkona (*Tenuidactylus russowi* v Iliiskoj doline, yuzhnij Kazakhstan [The ecology of the gray thin-toed gecko (*Tenuidactylus russowi*) in the Illy Valley, southern Kazakhstan]. Zool. Jour., 64(5):715-721 (in Russian).
Cherlin, V. A. 1988. K termobiologii serogo gekkona, polosatoj yashchurki i stepnoj agami v vostochnikh Karakumakh [On the thermobiology of the gray thin-toed gecko, sand striped racerunner, and steppe agama in the eastern Karakums]. Izv. AN Turkm. S.S.R. [Proc. Acad. Sci. Turkmenian S.S.R.], ser. Biol. Nauk, 1988(5):36-43 (in Russian).
Cherlin, V. A. 1991. Metodi termobiologicheskogo issledovaniya reptilij [The methods of thermobiological investigations of reptiles], p. 70-97. In: Ecologiya zhivotnih Uzbekistana [Ecology of Uzbek animals]. Tashkent (in Russian).
Cherlin, V. A., Yu. Tsellarius, and A. V. Gromov. 1983. K termobiologii scinkovogo gekkona (*Teratoscincus scincus*) v Karakumakh [On the thermobiology of the common skink gecko (*Teratoscincus scincus*) in the Karakums]. Ecologiya [Ecology], 1983(3):84-87 (in Russian).
Cloudsley-Thompson, J. L. 1987. Bionomics of the rock gecko *Cyrtodactylus scaber* in Qatar. Herpetol. Jour., 1(4):156-157.

Das, I. 1992. *Cyrtodactylus madarensis* Sharma (1980), a junior synonym of *Eublepharis macularius* Blyth (1854). Asiatic Herpetol. Res., 4:55-56.

Das, I. 1996. [Review of] Handbook: Indian Lizards, by B. K. Tikader and R. C. Sharma, 1992. Herpetol. Rev., 27(1):44-46.

Frolov, V. E. 1987. O razmnozhenii scinkovogo gekkona v usloviyakh nevoli [On the breeding of the skink gecko under captive conditions]. Vestn. Zool. [Herald Zool.], 1987(2):86-87 (in Russian).

Golubev, M. L. 1990. Pantsirnij gekkonchik (*Alsophylax loricatus* Str.) - novij vid gekkona dlya fauni Kara-Kalpakskoj SSR [The armored pygmy gecko *Alsophylax loricatus* Str. - a new gecko species for the fauna of Karakalpakia]. Vestn. Zool. [Herald Zool.], 1990(6):67 (in Russian).

Golubev, M. L., and A. B. Streltsov. 1989. K rasprostraneniiu dvukh vidov gekkonov (Reptilia: Gekkonidae) vdol srednego techeniia Amudaryi [On the distribution of two gecko lizard species (Reptilia: Gekkonidae) along the middle part of the Amudarya River]. Vestn. Zool. [Herald Zool.], 1989(2):78-79 (in Russian).

Golubev, M. L., M. S. Khan, and S. C. Anderson. 1995. On the systematics of some Palearctic geckos. Abstracts Second Asian Herpetol. Meeting, 6-10 Sept., 1995, Ashgabat, Turkmenistan:23-24.

Grismer, L. L. 1988. Phylogeny, taxonomy, classification, and biogeography of eublepharid geckos, p. 369-469. In: Estes, R. and G. Pregill (eds.), Phylogenetic Relationships of the Lizard Families. Stanford Univ. Press, Stanford.

Khan, M. S. 1988. A new cyrtodactylid gecko from northwestern Punjab, Pakistan. Jour. Herpetol., 22(2):241-243.

Khan, M. S. 1989. Rediscovery and redescription of the highland ground gecko, *Tenuidactylus montiumsalsorum* (Annandale, 1913). Herpetologica, 45(1):46-54.

Khan, M. S. 1991. A new *Tenuidactylus* gecko from the Sulaiman Range, Punjab, Pakistan. Jour. Herpetol., 25(2):199-204.

Khan, M. S. 1992. Validity of the mountain gecko *Gymnodactylus walli* Ingoldby, 1922. Herpetol. Jour., 2:106-109.

Khan, M. S. 1993. A new angular-toed gecko from Pakistan, with remarks of the taxonomy and a key to the species belonging to genus *Cyrtodactylus* (Reptilia: Sauria: Gekkonidae). Pakistan Jour. Zool., 25(1):67-73.

Khan, M. S. 1993. A new sandstone gecko from Fort Munro, Dera Ghazi Khan District, Punjab, Pakistan. Pakistan Jour. Zool., 25:217-221.

Khan, M. S. 1994. Validity and redescription of *Tenuidactylus yarkandensis* (J. Anderson, 1872). Pakistan Jour. Zool., 26(2):139-143.

Khan, M. S., and K. J. Baig. 1992. A new *Tenuidactylus* gecko from Northeastern Gilgit Agency, North Pakistan. Pakistan Jour. Zool., 24(4):273-277.

Khan, M. S., and R. Tasim. 1990. A new gecko of the genus *Tenuidactylus* from northeastern Punjab, Pakistan, and southwestern Azad Kashmir. Herpetologica, 42(2):142-148.

Kluge, A. G. 1991. Checklist of gekkonid lizards. Smithsonian Herpetol. Informat. Serv., 85:1-35.

Kluge, A. G. 1993. Gekkonid lizard taxonomy. International Gecko Soc., San Diego, California, 254 pages.

Leptien, R., and H. J. Zilder. 1991. Ein seltener Kralengecko, *Bunopus spatalurus hajarensis* Arnold, 1980. Sauria, 13(4):23-25.

Leviton, A. E., S. C. Anderson, K. Adler, and S. A. Minton. 1992. Handbook to Middle East amphibians and reptiles. Soc. Study Amphib. Reptiles, Contr. Herpetol., 8:1-252.

Leviton, A. E., M. L. Golubev, and S. C. Anderson. Herpetofauna of Tarim basin. 1. On geographical variation of *Cyrtopodion elongatus* (Blanf.), with description of a new species. In preparation.

Martens, H., and D. Kock. 1991. Erstnachweise für drei Gecko-Gattungen in Syrien (Reptilia: Sauria: Gekkonidae). Senckenberg. Biol., 71(1-3):15-21.

Mishagina, Zh. V. 1992. Troficheskie svyazi scinkovogo gekkona v vostochnikh Kara-Kumakh [Trophic connections of the common skink lizard, *Teratoscincus scincus* (Gekkonidae) in East Karakum (Repetek)]. Byull. MOIP [Bull. Moscow Soc. Nat.], Otd. Biol., 97(3):34-42 (in Russian).

Moravec, J., and D. Modry. 1994. On the occurrence of *Cyrtopodion heterocercus marginensis* and *Pseudocerastes persicus fieldi* in Syria. Zool. Middle East, 10:53-55.

Moravec, J., and M. Cerny. 1994. Second finding of the Iranian gecko *Tropiocolotes latifi*. Chasopis Narodniho Muzea, Rada prirodovedna [Jour. Nat. Mus., Dept. Sci.], 163(1-4):88.

Schätti, B. 1989. Amphibien und Reptilien aus der Arabischen Republik Jemen und Djibuti. Rev. Suisse Zool., 96(4):905-937.

Schneider, B. 1990. Verbreitung, Unterartgliederung, Ökologie und Shuppen-morphologie der Gekkoniden-Gattung *Tropiocolotes* aus dem Bereich der Sahara. Mitt. Pollichia, 77:409-419.

Selcer, K. W., and R. A. Bloom. 1984. *Cyrtodactylus scaber* (Gekkonidae): a new gecko to the fauna of the United States. Southwest. Natur., 29(4):499-500.

Semenov, D. V. 1989. Recenziya na knigu N. N. Szczerbaka, M. L. Golubeva "Gekkoni fauni SSSR i sopredelnikh stran." Opredelitel [Review of N. N. Szczerbak and M. L. Golubev's book "Gecko Fauna of the U.S.S.R. and Contiguous Countries. The Guide"]. Zool. Zhur. [Zool. Jour.], 68(5):155-157 (in Russian).

Semenov, D. V., and L. Ya. Borkin. 1992. On the ecology of Przewalsky's gecko (*Teratoscincus przewalskii*) in the Transaltai Gobi, Mongolia. Asiatic Herpetol. Res., 4:99-112.

Sivan, N., and Y. L. Werner. 1992. Survey of the reptiles of the Golan Plateau and Mt. Hermon, Israel. Israel Jour. Zool., 37:193-211.

Szczerbak, N. N. 1986. Review of Gekkonidae in fauna of the USSR and neighbouring countries, p. 705-710. In: Rocek, Z. (Ed.). Studies in herpetology. Charles Univ., Prague.

Szczerbak, N. N. 1988. K nomenclature palearkyicheskikh tonkopalikh gekkonov (*Tenuidactylus*, Gekkonidae, Reptilia [On the nomenclature of Palearctic thin-toed geckos (*Tenuidactylus*, Gekkonidae, Reptilia)]. Vestn. Zool. [Herald Zool.], 1988(4):84-85 (in Russian).

Szczerbak, N. N. 1989. Further study of systematics of Palaearctic rock geckoes. First World Congr. Herpetol., Canterbury, 11-19 Sept., 1989. Abstracts, 1 page (unpaginated).

Szczerbak (as Scerbak), N. N. 1991. Eine neue Gecko-Art aus Pakistan: *Alsophylax* (*Altiphylax*) *boemei* sp. nov. Salamandra, 27(1-2):53-57.

Szczerbak, N. N. 1991. Concerning the checklist of the reptile fauna of the U.S.S.R. Abstracts Sixth Ord. Gen. Meeting, Soc. Europ. Herpetol., 19-23 Aug., 1991. Budapest, Hungary:85.

Werner, Y. L. 1991. Notable herpetofaunal records from Transjordan. Zool. Middle East, 5:37-41.

Zhao, E.-M., and K. Adler. 1993. Herpetology of China. Soc. Study Amphib. Reptiles, Contr. Herpetol., 10:1-522.

Zhao, E.-M., and S.-Q. Li. 1987. A new lizard of *Tenuidactylus* and a new Tibetan snake record of *Amphiesma*. Acta Herpetol. Sinica, Chengdu, new ser., 6(1):48-51 (in Chinese).

{p. 229} INDEX TO GENERA AND SPECIES

(Editors' Note: Page numbers given below refer to the original Russian edition. In this English edition, the original pagination is cited in boldface within braces [e.g., {p. 168}])

Agamidae 5
Agamura 3, 12, 17, 18, 22, 23, 207, 208
  agamuroides 196
  cruralis 12, 208, 210
  femoralis 12, 18, 23, 208, 210, 211, 213, 225
  gastropholis 18, 23, 208, 210, 211, 212
  misonnei 18, 23, 208, 210, 211, 213, 214
  persica 12, 18, 23, 208, 209, 210, 212
  persica cruralis 211, 212
Alsophylax 3, 7, 8, 9, 10, 11, 12, 17, 18, 19, 21, 23, 55, 56, 87, 88, 89, 91, 97, 116, 123, 201, 216, 217, 219, 220
  crassicauda 98
  kashkarovi 72, 216
  laevis 8, 18, 19, 20, 55, 69, 72, 73, 78, 80, 82, 120, 123, 216, 217, 218, 220, 228
  laevis tadjikiensis 78
  loricatus 7, 8, 18, 19, 65, 69, 70, 73, 75, 2.7, 219, 220
  loricatus loricatus 55, 67, 69
  loricatus szczerbaki 3, 55, 67, 68, 69, 217, 228
  microtis 56, 62, 72, 123
  persicus 115, 117, 121, 130
  pipiens 8, 18, 19, 55, 56, 59, 61, 63, 65, 66, 72, 78, 121, 123, 216, 224, 228
  przewalski 7, 18, 19, 55, 61, 62, 64, 66, 217
  spinicauda 3, 8, 11, 175
  tadjikiensis 3, 18, 19, 23, 55, 59, 73, 78, 83, 217, 228
  tibetanus 3, 7, 11, 198
  tokobajevi 3, 8, 18, 19, 23, 55, 61, 81, 84, 85, 200, 217, 228
  tuberculatus 89
Altiphylax 3, 8, 11, 18, 19, 81
Ardeosauridae 16
Ascalabotes 7
  pipiens 56
Asiocolotes 3, 19, 20, 21, 101, 120, 121
Athene noctua 140
Boiga trigonatum 146
Bunopus 3, 7, 8, 9, 10, 11, 14, 17, 20, 24, 88, 100, 216, 220, 221, 222
  abudhabi 8, 89, 92, 100
  aspratilis 8, 10, 126, 127
  biporus 89
  blanfordi 7, 8, 89, 92
  crassicauda 8, 20, 89, 91, 94, 98, 99
  persicus 117
  spatalurus 8, 20, 89, 94, 95, 96, 97
  spatalurus hajarensis 95, 97
  spatalurus spatalurus 95, 97
  tuberculatus 3, 7, 8, 20, 88, 89, 90, 91, 93, 94, 228

Carinatogecko 3, 10, 17, 18, 21, 24, 126, 127, 217
  aspratilis 18, 21, 47, 127, 128
  heteropholis 18, 21, 47, 127, 129, 130
Coluber karelini 53, 77, 126, 140, 152, 174
  rhodorhachis 53, 126, 140
Crossobamon 3, 7, 17, 18, 23, 24, 46, 47, 48, 49, 50, 88
  eversmanni 7, 18, 42, 47, 49, 224, 225, 228
  eversmanni eversmanni 49
  eversmanni lumsdenii 49, 50, 57, 58
  lumsdenii 51
  maynardi 50, 51
  orientalis 18, 47, 54, 58
Cyrtodactylus 3, 10, 11, 12, 18, 21, 22, 23, 130, 196, 198, 220
  agamuroides 196
  amictopholis 179, 222
  basoglui 190, 192, 217
  caspius 132, 143
  elongatus 194
  fedtschenkoi 143, 147, 157, 225
  heterocercus 182
  himalayanus 222
  kachhensis 185
  kirmanensis 206
  kotschyi 160, 221
  kotschyi danilewskii 163
  longipes 153, 225
  macularius 5, 26
  montiumsalsorum 192
  persicus 121
  pulchellus 10, 11
  russowi 167, 224
  sagittifer 180
  scaber 190
  stoliczkai 205
  voraginosus 153, 157
  watsoni 188
  yarkandensis 205
Cyrtopodion 12, 20, 22, 23, 130, 185, 207
Diplodactylinae 5, 17
Echis carinatus 77, 140, 152
Elaphe dione 86
Eremias 220
  intermedia 77
  velox 140
Eryx miliaris 53
Eublepharidae 5, 17, 23, 25
Eublepharis 3, 5, 23, 25, 26, 221, 222
  afghanicus 6
  angramainyu 6, 25, 26, 28, 29
  fasciolatus 6, 26
  gracilis 6
  hardwickii 5, 25

macularius 6, 25, 26, 27, 28, 29, 30, 228
macularius fasciolatus 6
macularius fuscus 6
macularius montanus 6
macularius smithi 6
turcmenicus 6, 25, 26, 28, 30, 31, 32, 228
Eumeces schneideri blithanus 126
Gekkonidae 5, 17, 23, 33
Gekkoninae 5, 17, 19
Gymnodactylus 3, 7, 10, 11, 12, 100, 130, 217, 220, 221
  agamuroides 12, 196, 210, 211, 213
  amictopholis 179
  atropunctatus 56
  brevipes 185, 197
  caspius 11, 130, 131, 132
  caspius insularis 3, 137, 216
  chitralensis 201
  colchicus 218
  danilewskii 163, 218, 224
  elongatus 194
  eversmanni 7, 46, 47
  fedtschenkoi 147, 155
  gastropholis 12, 211, 213
  geckoides 10
  heterocercus 183, 184
  heterocercus mardinensis 184
  ingoldbyi 188
  kachhensis 185
  kachhensis kachhensis 185
  kachhensis watsoni 188
  kirmanensis 205
  kotschyi 11, 159, 160, 183, 223, 224
  kotschyi danilewskii 163, 220
  longipes 11, 153, 155, 219
  longipes longipes 156
  longipes microlepis 3, 156
  microlepis 153
  microtis 7, 62
  mintoni 3, 201
  montiumsalsorum 192
  persicus 208
  petrensis 185
  pipiens 56
  russowi 11, 167, 171, 217, 220
  russowi copalensis 171
  russowi zarudnyi 172
  sagittifer 180
  scaber 147, 190, 217
  spinicauda 175
  steydneri 8, 106
  stoliczkai 205
  tenius 223
  tibetanus 198
  turcmenicus 3, 141, 220
  walli 205
  watsoni 188
  zarudnyi 167, 172
Hemidactylus 13, 23
  turcicus 13, 225
Hemiechinus megalotus 126
Iguanidae 5
Lacerta pipiens 7, 55, 56
Lycodon striatus 174, 179
Lytorhynchus ridgewayi 94, 216
Malpolon monspessulanus 140
Mediodactylus 3, 11, 12, 21, 22, 130, 159, 160, 207, 218
Mesodactylus 185
Microgecko 3, 9, 19, 20, 21, 101, 110, 221, 222, 224
  helenae 9, 110, 111, 117
Milvus korshun 140

Ochotona rufescens rufescens 126
Ophisaurus apodus 140
Pristurus karterii 110
Psammophis lineolatum 140, 174, 195
Ptenodactylus 7, 46
Ptyodactylus 221
Pygopodidae 17
Rhinogekko 12, 207
  misonnei 12, 213, 225
Saurodactylus 18
Sphaerodactylidae 5
Sphaerodactylinae 5, 17, 18
Stenodactylus 7, 17, 18, 20, 23, 88, 100, 221
  dunstervillei 54
  lumsdenii 7, 47, 50
  maynardii 7, 50
  orientalis 7, 54
  petersii 106
  scaber 185, 190
  scincus 6, 35, 37
  sthenodactylus 228
Tenuidactylus 3, 11, 12, 14, 17, 18, 20, 21, 22, 23, 87, 130, 131, 132, 196, 220
  agamuroides 12, 22, 23, 185, 196, 197
  amictopholis 21, 22, 91, 160, 164, 179
  caspius 11, 22, 75, 84, 93, 132, 133, 134, 136, 137, 228
  caspius caspius 137
  caspius insularis 137
  chitralensis 21, 22, 198, 200, 201, 202
  dattensis 196
  elongatus 12, 22, 23, 45, 185, 193, 194, 228
  fasciolatus 21, 196, 200, 201
  fedtschenkoi 22, 132, 143, 146, 147, 155, 228
  heterocercus 21, 22, 160, 182, 183
  heterocercus heterocercus 183, 184
  heterocercus mardinensis 183, 184
  himalayanus 21, 196, 200, 201
  kachhensis 12, 22, 185, 186, 187, 190
  kirmanensis 21, 22, 198, 200, 205, 206
  kotschyi 21, 22, 160
  kotshyi danilewskii 160, 161, 162, 163, 164, 165, 166, 228
  lawderanus 21, 196, 198, 200, 201
  longipes 22, 132, 153, 154, 155, 156, 158, 228
  longipes longipes 143, 155, 156
  longipes microlepis 143, 155, 156, 159
  longipes voraginosus 155, 156, 157
  microlepis 155
  mintoni 22, 198, 200, 201, 203
  montiumsalsorum 12, 22, 185, 187, 192, 193
  russowi 21, 22, 42, 84, 160, 167, 168, 170, 171, 174
  russowi russowi 171
  russowi zarudnyi 171, 172, 228
  sagittifer 21, 22, 28, 160, 180, 181
  scaber 12, 22, 185, 189, 190, 191, 192, 193, 194
  spinicauda 22, 32, 160, 170, 175, 176, 178, 228
  stoliczkai 21, 22, 198, 200, 204
  tibetanus 21, 22, 198, 200
  turcmenicus 12, 22, 132, 141, 142, 143, 144, 145, 228
  watsoni 12, 22, 185, 187, 188, 189
Teratoscincus 3, 6, 14, 17, 18, 21, 23, 33, 35, 218
  bedriagai 7, 18, 21, 34, 35, 47
  keyzerlingii 6, 33, 35, 38
  microlepis 7, 18, 21, 35, 43, 44, 47
  przewalskii 7, 18, 21, 35, 42, 44, 45, 228
  scincus 18, 21, 34, 35, 39, 228
  scincus keyzerlingii 6, 33, 37, 38

scincus scincus 37, 219
scincus rustamowi 3, 7, 37, 38, 220
roborowskii 7, 35, 37
zarudnyi 6, 35, 38
Trachydactylus 8
jolensis 8, 94
spatalurus 94
Trapelus 221
Tropiocolotes 3, 8, 9, 10, 17, 19, 20, 21, 23, 100, 101, 117, 217, 221, 222, 223
depressus 9, 19, 20, 21, 113, 120, 121, 124, 125
helenae 9, 20, 21, 111, 112, 113, 117, 118, 120
helenae fasciatus 9
heteropholis 9, 10, 130
latifi 9, 20, 21, 111, 113, 114, 119, 120
levitoni 3, 19, 20, 21, 113, 120, 121, 122, 123
nattereri 8, 9, 106, 224
occidentalis 102, 104
persicus 9, 20, 21, 111, 114, 115, 117, 120
persicus bakhtiari 115, 116, 117
persicus euphorbiacola 115, 116, 117, 118
persicus persicus 115, 116, 117
scorteccl 9, 20, 21, 96, 101, 108, 109, 120
somalicus 102, 104
steudneri 9, 19, 20, 21, 96, 101, 106, 107, 108
tripolitanus 8, 9, 19, 20, 21, 96, 100, 101, 102, 103, 105, 120, 228
tripolitanus algericus 104
tripolitanus apoklomax 9, 105
tripolitanus occidentalis 105
Uroplatidae 17
Vipera lebetina 126, 140
Vulpes vulpes 140
Xantusiidae 17

PUBLICATIONS OF THE
SOCIETY FOR THE STUDY OF AMPHIBIANS AND REPTILES

SOCIETY PUBLICATIONS may be purchased from: Dr. Robert D. Aldridge, Publications Secretary, Department of Biology, Saint Louis University, Saint Louis, Missouri 63103, USA.

Telephone: area code 314, 977–3916 or 977–1710.
Fax: area code 314, 977–3658.
E-mail: ssar@sluvca.slu.edu

Prices are effective through December 1997. Make checks payable to "SSAR." Overseas customers must make payment in USA funds using a draft drawn on American banks (include an additional amount to cover bank conversion charges) or by International Money Order. All persons may charge to MasterCard or VISA (provide account number and expiration date); items marked "out-of-print" are no longer available.

Shipping and Handling Costs

Shipments inside the USA: Shipping costs are in addition to the price of publications. Add an amount for shipping of the first item ($3.00 for a book costing $10.00 or more or $2.00 if the item costs less than $10.00) plus an amount for any additional items ($1.00 each for books costing over $10.00 and $0.50 for each item costing less than $10.00).

Shipments outside the USA: Determine the cost for shipments inside USA (above) and then add 5% of the total cost of the order.

Large prints (marked *): For shipments inside the USA, add $3.00 for any quantity; outside the USA, instead add $7.00 for any quantity.

CONTRIBUTIONS TO HERPETOLOGY

Book-length monographs, comprising taxonomic revisions, results of symposia, and other major works. Prepublication discount to Society members.

Vol. 1. *Reproductive Biology and Diseases of Captive Reptiles*, by James B. Murphy and Joseph T. Collins (eds.). 1980. Results of a Society-sponsored symposium, including papers by 37 leading specialists. 287 p., illus. Paperbound. Out-of-print.

Vol. 2. *The Turtles of Venezuela*, by Peter C. H. Pritchard and Pedro Trebbau. 1984. An exhaustive natural history covering half of the turtle species of South America. 414 p., 48 color plates (25 watercolor portraits by Giorgio Voltolina and 165 photographs of turtles and habitats) measuring $8\frac{1}{2} \times 11$ inches, keys, 16 maps. Regular edition, clothbound $45.00; patron's edition, two leatherbound volumes in cloth-covered box, signed and numbered by authors and artist $300.00. (*Also*: set of 25 color prints of turtle portraits on heavy paper stock, in protective wrapper $30.00.)

Vol. 3. *Introduction to the Herpetofauna of Costa Rica / Introducción a la Herpetofauna de Costa Rica*, by Jay M. Savage and Jaime Villa R. 1986. Bilingual edition in English and Spanish, with distribution checklist, bibliographies, and extensive illustrated keys. 220 p., map. Clothbound $30.00.

Vol. 4. *Studies on Chinese Salamanders*, by Ermi Zhao, Qixiong Hu, Yaoming Jiang, and Yuhua Yang. 1988. Evolutionary review of all Chinese species with keys, diagnostic figures, and distribution maps. 80 p., 7 plates (including 10 color photographs of salamanders and habitats). Clothbound $12.00.

Vol. 5. *Contributions to the History of Herpetology*, by Kraig Adler, John S. Applegarth, and Ronald Altig. 1989. Biographies of 152 prominent herpetologists (with portraits and signatures), index to 2500 authors in taxonomic herpetology, and academic lineages of 1450 herpetologists. International coverage. 202 p., 148 photographs, 1 color plate. Clothbound $20.00.

Vol. 6. *Snakes of the* Agkistrodon *Complex: A Monographic Review*, by Howard K. Gloyd and Roger Conant. 1990. Comprehensive treatment of 33 taxa of pitvipers included in four genera: *Agkistrodon*, *Calloselasma*, *Deinagkistrodon*, and *Hypnale*. Also includes nine supplementary chapters by leading specialists. 620 p., 33 color plates (247 photographs of snakes and habitats), 20 uncolored plates, 60 text figures, checklist and keys, 6 charts, 28 maps. Clothbound $75.00. (*Also*: separate set of the 247 color photographs of snakes and habitats [on 32 plates], in protective wrapper. $30.00; limited-edition print of the book's frontispiece illustrating snakes of all four genera, from watercolor by David M. Dennis. Signed individually by Roger Conant and the artist $25.00.)

Vol. 7. *The Snakes of Iran*, by Mahmoud Latifi. 1991. Review of the 60 species of Iranian snakes, covering general biology, venoms, and snake bite. Appendix and supplemental bibliography by Alan E. Leviton and George R. Zug. 167 p., 22 color plates of snakes (66 figures), 2 color relief maps, 44 species range maps. Clothbound $22.00.

Vol. 8. *Handbook to Middle East Amphibians and Reptiles*, by Alan E. Leviton, Steven C. Anderson, Kraig Adler, and Sherman A. Minton. 1992. Annotated checklist, illustrated key, and identification manual covering 148 species and subspecies found in region from Turkish border south through the Arabian Peninsula (including Bahrain, Qatar, and United Arab Emirates) and the Arabian (Persian) Gulf. Chapters on venomous snakes and snakebite treatment plus extensive bibliography. 264 p., 32 color plates (220 photographs), maps, text figures. Clothbound $30.00.

Vol. 9. *Herpetology: Current Research on the Biology of Amphibians and Reptiles*. 1992. Proceedings of the First World Congress of Herpetology (1989), with a foreword by H.R.H. Prince Philip, Duke of Edinburgh. Includes the plenary lectures, a summary of the congress, and a list of delegates with their current addresses. 225 p., 28 photographs. Clothbound $28.00.

Vol. 10. *Herpetology of China*, by Ermi Zhao and Kraig Adler. 1993. Comprehensive review of Chinese amphibians and reptiles, including Hong Kong and Taiwan. 522 p., 48 color plates (371 photographs illustrating all 164 genera and half of the 661 species), portraits, text figures, maps. Clothbound $60.00.

Vol. 11. *Captive Management and Conservation of Amphibians and Reptiles*, by James B. Murphy, Kraig Adler, and Joseph T. Collins (eds.). 1994. Results of a Society-sponsored symposium, including chapters by 70 leading specialists. Foreword by Gerald Durrell. 408 p., 35 photographs, 1 color plate. Clothbound $58.00.

Vol. 12. *Contributions to West Indian Herpetology*, by Robert Powell and Robert W. Henderson (eds.). 1996. Results of a Society-sponsored symposium, including research chapters by 59 authors and a checklist of species with complete citations. Foreword by Thomas W. Schoener. 457 p., 28 photographs, 70 color photographs, index. Clothbound $60.00.

Vol. 13. *Gecko Fauna of the USSR and Contiguous Regions*, by Nikolai N. Szczerbak and Michael L. Golubev. 1996. Covers the systematics, natural history, and conservation of the gecko fauna of the former Soviet Union and related species in surrounding regions from Mongolia through Pakistan, the Middle East, and northern Africa. 245 p., 24 color and numerous uncolored photographs, spot maps, bibliography, index. Clothbound $48.00.

FACSIMILE REPRINTS IN HERPETOLOGY

Exact reprints of classic and important books and papers. Most titles have extensive new introductions by leading authorities. Prepublication discount to Society members.

ANDERSON, J. 1896. *Contribution to the Herpetology of Arabia*. Introduction and new checklist of Arabian amphibians and reptiles by Alan E. Leviton and Michele L. Aldrich. 160 p., illus. (one plate in color), map. Clothbound $25.00.

BELL, T. 1842–1843. *Herpetology of the "Beagle."* Part 5 of Charles Darwin's classic, "Zoology of the Voyage of H.M.S. Beagle," containing descriptions of amphibians and reptiles collected on the expedition. Introduction by Roberto Donoso-Barros. 100 p., 20 plates (measuring 8 1/2 × 11 inches), map. Paperbound $13.00.

BOGERT, C. M., and R. MARTÍN DEL CAMPO. 1956. *The Gila Monster and Its Allies*. The standard work on lizards of the family Helodermatidae. New preface by Charles M. Bogert and retrospective essay by Daniel D. Beck. 262 p., color plate, 62 photographs, 35 text figures, index. $38.00.

BOJANUS, L. H. 1819–1821. *Anatome Testudinis Europaeae*. Introduction by Alfred Sherwood Romer. Out-of-print.

BOULENGER, G. A. 1877–1920. *Contributions to American Herpetology*. A collection of papers (from various journals) covering North, Central, and South American species, with an introduction by James C. Battersby. Complete in 18 parts totalling 880 p., numerous illustrations, index. Paperbound. Complete set: 18 parts plus index and two tables of contents (for binding in two volumes), in parts as issued $55.00.

BULLETIN OF THE ANTIVENIN INSTITUTE OF AMERICA. Volumes 1–5, 1927–1932. Introduction by Sherman A. Minton. Out-of-print.

CAMP, C. L. 1923. *Classification of the Lizards*. New preface by Charles L. Camp and an introduction by Garth Underwood. Out-of-print.

CHANG, M. L. Y. 1936. *Amphibiens Urodèles de la Chine*. With a new checklist by Arden H. Brame. Out-of-print.

COPE, E. D. 1864. *Papers on the Higher Classification of Frogs*. Reprinted from Proceedings of the Academy of Natural Sciences of Philadelphia and Natural History Review. 32 p. Paperbound $3.00.

COPE, E. D. 1871. *Catalogue of Batrachia and Reptilia Obtained by McNiel in Nicaragua; Catalogue of Reptilia and Batrachia Obtained by Maynard in Florida*. 8 p. Paperbound $1.00.

COPE, E. D. 1892. *The Osteology of the Lacertilia*. An important contribution to lizard anatomy, reprinted from Proceedings of the American Philosophical Society. 44 p., 6 plates. Paperbound $4.00.

COWLES, R. B., and C. M. BOGERT. 1944. *A Preliminary Study of the Thermal Requirements of Desert Reptiles*. The foundation of thermoregulation biology, with extensive review of recent studies by F. Harvey Pough. Reprinted from Bulletin of American Museum of Natural History. 52 p., 11 plates. Paperbound $5.00.

DUNN, E. R. 1926. *Salamanders of the Family Plethodontidae*. Introductions by David B. Wake and Arden H. Brame. Out-of-print.

ESCHSCHOLTZ, F. 1829–1833. *Zoologischer Atlas* (herpetological sections). Descriptions of new reptiles and amphibians from California and the Pacific. Introduction by Kraig Adler. 32 p., 4 plates (measuring 8 1/2 × 11 inches). Paperbound $3.00.

ESPADA, M. JIMÉNEZ DE LA. 1875. *Vertebrados del Viaje al Pacifico: Batracios*. Major taxonomic work on South American frogs. Introduction by Jay M. Savage. 208 p., 6 plates, maps. Clothbound $20.00.

FAUVEL, A.-A. 1879. *Alligators in China*. Original description of *Alligator sinensis*, including classical and natural history. 42 p., 3 plates. Paperbound $5.00.

FITZINGER, L. J. 1843. *Systema Reptilium*. An important nomenclatural landmark for herpetology, including Amphibia as well as reptiles; world-wide in scope. Introduction by Robert Mertens. 128 p., index. Paperbound $15.00.

GLOYD, H. K. 1940. *The Rattlesnakes, Genera* Sistrurus *and* Crotalus. Introduction and new checklist by Hobart M. Smith and Herbert M. Harris. Out-of-print.

GRAY, J. E. 1825. *A Synopsis of the Genera of Reptiles and Amphibia*. Reprinted from Annals of Philosophy. 32 p. Paperbound $3.00.

GRAY, J. E. 1831–1844. *Zoological Miscellany*. A privately printed journal, devoted mostly to descriptions of amphibians, reptiles, and birds from throughout the world. Introduction by Arnold G. Kluge. 86 p., 4 plates. Paperbound $6.00.

GRAY, J. E., and A. GÜNTHER. 1845–1875. *Lizards of Australia and New Zealand*. The reptile section from "Voyage of H.M.S. Erebus and Terror," together with Gray's 1867 related book on Australian lizards. Introduction by Glenn M. Shea. 82 p., 20 plates (measuring 8 1/2 × 11 inches). Clothbound $20.00. (*Also*: set of the 20 plates in protective wrapper $12.00.)

GÜNTHER, A. 1885–1902. *Biologia Centrali-Americana. Reptilia and Batrachia*. The standard work on Middle American herpetology with 76 full-page plates measuring 8 1/2 × 11 inches (12 in color). Introductions by Hobart M. Smith, A. E. Gunther, and Kraig Adler. 575 p., photographs, maps. Clothbound $50.00. (*Also*: separate set of the 12 color plates, in protective wrapper $18.00.)

HOLBROOK, J. E. 1842. *North American Herpetology*. Five volumes bound in one. The classic work by the father of North American herpetology. Exact facsimile of the definitive second edition, including all 147 plates, measuring 8 1/2 × 11 inches (20 reproduced in full color). Introduction and checklists by Richard and Patricia Worthington and by Kraig Adler. 1032 p. Leatherbound patron's edition, out-of-print; regular edition, clothbound $60.00.

JUNIOR SOCIETY OF NATURAL SCIENCES (CINCINNATI, OHIO). 1930–1932. Herpetological papers from the society's Proceedings, with articles by Weller, Walker, Dury, and others. 56 p. Paperbound $3.00.

KIRTLAND, J. P. 1838. *Zoology of Ohio* (herpetological portion). 8 p. Paperbound $1.00.

LeCONTE, J. E. 1824–1828. *Three Papers on Amphibians*, from the Annals of the Lyceum of Natural History, New York. 16 p. Paperbound $2.00.

LINNAEUS, C. 1766–1771. *Systema Naturae* (ed. 12) and *Mantissa Plantarum* (herpetological portions from both). Introduction by Kraig Adler. Out-of-print.

LOVERIDGE, A. 1946. *Reptiles [and Amphibians] of the Pacific*. The standard review of the herpetofauna of the Pacific region including Australia and extending from Indonesia to Hawaii and the Galápagos Islands. 271 p., 7 plates, 1 double-page map, index. Out-of-print.

McILHENNY, E. A. 1935. *The Alligator's Life History*. The most complete natural history of the American alligator. Introduction by Archie Carr and a review of recent literature by Jeffrey W. Lang. 125 p., 18 photographs and a portrait. Clothbound $20.00.

McLAIN, R. B. 1899. *Contributions to North American Herpetology* (three parts). 28 p., index. Paperbound $2.00.

ORBIGNY, A. D' [and G. BIBRON]. 1847. *Voyage dans l'Amérique Méridionale*. This extract comprises the complete section on reptiles and amphibians from this voyage to South America. 14 p., 9 plates measuring 8½ × 11 inches. $3.00.

PETERS, W. 1838–1883. *The Herpetological Contributions of Wilhelm C. H. Peters (1815–1883)*. A collection of 174 titles, world-wide in scope, and including the herpetological volume in Peters' series, "Reise nach Mossambique." Biography, annotated bibliography, and synopsis of species by Aaron M. Bauer, Rainer Günther, and Meghan Klipfel. 714 pages, 114 plates, 9 photographs, maps, index. Clothbound $75.00.

RAFINESQUE, C. S. 1820. *Annals of Nature* (herpetological and ichthyological sections), 4 p. Paperbound $1.00.

RAFINESQUE, C. S. 1822. *On Two New Salamanders of Kentucky*. 2 p. Paperbound $1.00.

RAFINESQUE, C. S. 1832–1833. *Five Herpetological Papers from the Atlantic Journal*. 4 p. Paperbound $1.00.

SOWERBY, J. DeC., E. LEAR, and J. E. GRAY. 1872. *Tortoises, Terrapins, and Turtles Drawn From Life*. The finest atlas of turtle illustrations ever produced. Introduction by Ernest E. Williams. 26 p., 61 full-page plates (measuring 8½ × 11 inches). Clothbound $25.00.

SPIX, J. B. VON, and J. G. WAGLER. 1824–1825. *Herpetology of Brazil*. The most comprehensive and important early survey of Brazilian herpetology. Introduction by P. E. Vanzolini. 400 p., 98 plates, one in color (each measuring 8½ × 11 inches), map. Clothbound $36.00.

STEJNEGER, L. 1907. Herpetology of Japan and Adjacent Territory. Introduction by Masafumi Matsui. Also covers Taiwan, Korea, and adjacent China and Siberia. 684 pages, 35 plates, 409 text figures, keys, index. Clothbound $58.00.

TROSCHEL, F. H. 1850 [1852]. Cophosaurus texanus, *neue Eidechsengattung aus Texas*. 8 p. Paperbound $1.00.

TSCHUDI, J. J. VON. 1838. *Classification der Batrachier*. A major work in systematic herpetology, with introduction by Robert Mertens. 118 p., 6 plates. Paperbound $18.00.

TSCHUDI, J. J. VON. 1845. *Reptilium Conspectus*. New reptiles and amphibians from Peru. 24 p. Paperbound $2.00.

VANDENBURGH, J. 1895–1896. *Herpetology of Lower California*. Herpetology of Baja California, Mexico (collected papers). 101 p., 11 plates, index. Paperbound $8.00.

WAITE, E. R. 1929. *The Reptiles and Amphibians of South Australia*. Introduction by Michael J. Tyler and Mark Hutchinson. 282 p., color plate, portrait, 192 text figures including numerous photographs. Clothbound $35.00.

WIEGMANN, A. F. A. 1834. *Herpetologia Mexicana*. Introduction by Edward H. Taylor. Out-of-print.

WILCOX, E. V. 1891. *Notes on Ohio Batrachians*. 3 p. Paperbound $1.00.

WILLISTON, S. W. 1925. *Osteology of the Reptiles*. With introduction by Claude W. Hibbard. Out-of-print.

WRIGHT, A. H., and A. A. WRIGHT. 1962. *Handbook of Snakes of the United States and Canada, Volume 3, Bibliography*. Cross-indexed bibliography to Volumes 1 and 2. 187 p. Clothbound $18.00.

HERPETOLOGICAL REVIEW
AND H.I.S.S. PUBLICATIONS

The Society's official newsletter, international in coverage. In addition to news notes and feature articles, regular departments include regional societies, techniques, husbandry, life history, geographic distribution, and book reviews. Issued quarterly as part of Society membership or separately by subscription. All numbers are paperbound as issued and measure 8½ × 11 inches. In 1973, publications of the Herpetological Information Search Systems (*News-Journal* and *Titles and Reviews*) were substituted for *Herpetological Review*; content and format are the same.

Volume 1 (1967–1969), numbers 1–9, $5.00 per number.
Volumes 2–27 (1970–1996), four numbers in each volume (except volumes 3–4, with 6 numbers each), $5.00 per number.
The following numbers are out-of-print and *no longer available*: Volume 1(number 7), 3(2), 4(1), 5(1, 2), 6(1, 2, 4), 7(3, 4), and 10(2).
Cumulative Index for Volumes 1–7 (1967–1976), 60 pages, $5.00.
Cumulative Index for Volumes 1–17 (1967–1986), 90 pages, $8.00.
H.I.S.S. Publications: News-Journal, volume 1, numbers 1–6, and Titles and Reviews, volume 1, numbers 1–2 (all of 1973–1974), complete set, $10.00.
Index to Geographic Distribution Records for Volumes 1–17 (1967–1986), including H.I.S.S. publications, 44 pages, $6.00.

CATALOGUE OF
AMERICAN AMPHIBIANS AND REPTILES

Loose-leaf accounts of taxa (measuring 8½ × 11 inches) prepared by specialists, including synonymy, definition, description, distribution map, and comprehensive list of literature for each taxon. Covers amphibians and reptiles of the entire Western Hemisphere. Issued by subscription. Individual accounts are not sold separately.

CATALOGUE ACCOUNTS:
  Complete set: Numbers 1–640, $340.00.
  Partial sets: Numbers 1–190, $60.00.
         Numbers 191–410, $70.00.
         Numbers 411–640, $210.00.
INDEX TO ACCOUNTS 1–400: Cross-referenced, 64 pages, $6.00; accounts 401–600: Cross-referenced, 32 pages, $6.00.
IMPRINTED POST BINDER: $35.00. (*Note*: one binder holds about 200 accounts.)
SYSTEMATIC TABS: Ten printed tabs for binder, such as "Class Amphibia," "Order Caudata," etc., $6.00 per set.

JOURNAL OF HERPETOLOGY

The Society's official scientific journal, international in scope. Issued quarterly as part of Society membership. All numbers are paperbound as issued, measuring 7 × 10 inches.

Volume 1 (1968), numbers 1–4 combined.
Volumes 2–5 (1968–1971), numbers 1–2 and 3–4 combined, $8.00 per double number.
Volume 6 (1972), numbers 1, 2, and double number 3–4.
Volumes 7–30 (1973–1996), four numbers in each volume, $8.00 per single number.
The following volumes and numbers are out-of-print and are *no longer available*: Volume 1, 2, 3, 4, 5(3, 4), 6, 7(1), 8(1), 9(1, 2, 4), 10(1, 4), 11(4), and 12(1, 2).
Cumulative Index for Volumes 1–10 (1968–1976), 72 pages, $8.00.

SOCIETY PUBLICATIONS

HERPETOLOGICAL CIRCULARS

Miscellaneous publications of general interest to the herpetological community. All numbers are paperbound, as issued. Prepublication discount to Society members.

No. 1. *A Guide to Preservation Techniques for Amphibians and Reptiles* by George R. Pisani. 1973. 22 p., illus. $3.00.

No. 2. *Guía de Técnicas de Preservación de Anfibios y Reptiles* por George R. Pisani y Jaime Villa. 1974. 28 p., illus. $3.00.

No. 3. *Collections of Preserved Amphibians and Reptiles in the United States* compiled by David B. Wake (chair) and the Committee on Resources in Herpetology. 1975. 22 p. Out-of-print.

No. 4. *A Brief Outline of Suggested Treatments for Diseases of Captive Reptiles* by James Murphy. 1975. 13 p. Out-of-print.

No. 5. *Endangered and Threatened Amphibians and Reptiles in the United States* compiled by Ray E. Ashton, Jr. (chair) and 1973–74 SSAR Regional Herpetological Societies Liaison Committee. 1976. 65 p. Out-of-print.

No. 6. *Longevity of Reptiles and Amphibians in North American Collections* by J. Kevin Bowler. 1977. 32 p. $3.00. (See also number 21.)

No. 7. *Standard Common and Current Scientific Names for North American Amphibians and Reptiles* (1st ed.) by Joseph T. Collins, James E. Huheey, James L. Knight, and Hobart M. Smith. 1978. 36 p. $3.00. (See also numbers 12 and 19.)

No. 8. *A Brief History of Herpetology in North America Before 1900* by Kraig Adler. 1979. 40 p., 24 photographs, 1 map. $3.00.

No. 9. *A Review of Marking Techniques for Amphibians and Reptiles* by John W. Ferner. 1979. 42 p., illus. $3.00.

No. 10. *Vernacular Names of South American Turtles* by Russell A. Mittermeier, Federico Medem and Anders G. J. Rhodin. 1980. 44 p. $3.00.

No. 11. *Recent Instances of Albinism in North American Amphibians and Reptiles* by Stanley Dyrkacz. 1981. 36 p. $3.00.

No. 12. *Standard Common and Current Scientific Names for North American Amphibians and Reptiles* (2nd ed.) by Joseph T. Collins, Roger Conant, James E. Huheey, James L. Knight, Eric M. Rundquist, and Hobart M. Smith. 1982. 32 p. $3.00. (See also numbers 7 and 19.)

No. 13. *Silver Anniversary Membership Directory*, including addresses of all SSAR members, addresses and publications of the herpetological societies of the world, and a brief history of the Society. 1983. 56 p., 4 photographs. $3.00.

No. 14. *Checklist of the Turtles of the World with English Common Names* by John Iverson. 1985. 14 p. $3.00.

No. 15. *Cannibalism in Reptiles: A World-Wide Review* by Joseph C. Mitchell. 1986. 37 p. $4.00.

No. 16. *Herpetological Collecting and Collections Management* by John E. Simmons. 1987. 72 p., 6 photographs. $6.00.

No. 17. *An Annotated List and Guide to the Amphibians and Reptiles of Monteverde, Costa Rica* by Marc P. Hayes, J. Alan Pounds, and Walter W. Timmerman. 1989. 70 p., 32 figures. $5.00.

No. 18. *Type Catalogues of Herpetological Collections: An Annotated List of Lists* by Charles R. Crumly. 1990. 50 p. $5.00.

No. 19. *Standard Common and Current Scientific Names for North American Amphibians and Reptiles* (3rd ed.) compiled by Joseph T. Collins (coordinator for SSAR Common and Scientific Names List). 1990. 45 p. Out-of-print. (See also numbers 7 and 12.)

No. 20. *Age Determination in Turtles* by George R. Zug. 1991. 32 p., 6 figures. $5.00.

No. 21. *Longevity of Reptiles and Amphibians in North American Collections* (2nd ed.) by Andrew T. Snider and J. Kevin Bowler. 1992. 44 p. $5.00. (See also number 6.)

No. 22. *Biology, Status, and Management of the Timber Rattlesnake* (Crotalus horridus): *A Guide for Conservation* by William S. Brown. 1993. 84 p., 16 color photographs. $12.00.

No. 23. *Scientific and Common Names for the Amphibians and Reptiles of Mexico in English and Spanish / Nombres Científicos y Comunes en Ingles y Español de los Anfibios y los Reptiles de México* by Ernest A. Liner. Spanish translation by José L. Camarillo R. 1994. 118 p. $12.00.

No. 24. *Citations for the Original Descriptions of North American Amphibians and Reptiles*, by Ellin Beltz. 1995. 48 p. $7.00.

PUBLICATIONS OF THE OHIO HERPETOLOGICAL SOCIETY

OHS was the predecessor to the Society for the Study of Amphibians and Reptiles. All publications international in scope. Paperbound as issued.

Volume 1 numbers 1–4, plus Special Publications 1–2 (all 1958), facsimile reprint, out-of-print.

Volume 2 (1959–1960), four numbers, $2.00 per number; numbers 3 and 4 out-of-print.

Volume 3 (1961–1962), four numbers, $1.00 per number; numbers 1 and 3 out-of-print.

Volume 4 (1963–1964), four numbers; double number 1–2, $4.00, numbers 3 and 4 $2.00 each.

Volume 5 (1965–1966), four numbers, $2.00 per number.

Special Publications 3–4 (1961–1962), $2.00 per number; number 3 out-of-print.

OTHER MATERIALS AVAILABLE FROM THE SOCIETY

The following color prints and brochures may be purchased from the Society. (*Extra postage required; see "Shipping and Handling Costs.")

*SILVER ANNIVERSARY COMMEMORATIVE PRINT. Full-color print ($11^{1}/_{2} \times 15^{1}/_{4}$ inches) of a Gila Monster (*Heloderma suspectum*) on natural background, from a watercolor by David M. Dennis. Issued as part of Society's 25th Anniversary in 1982. Edition limited to 1000. $6.00 each or $5.00 in quantities of 10 or more.

*WORLD CONGRESS COMMEMORATIVE PRINT. Full-color print ($11^{1}/_{2} \times 15$ inches) of an Eastern Box Turtle (*Terrapene carolina*) in a natural setting, from a watercolor by David M. Dennis. Issued as part of SSAR's salute to the First World Congress of Herpetology, held at Canterbury, United Kingdom, in 1989. Edition limited to 1500. $6.00 each or $5.00 in quantities of 10 or more.

*WATER SNAKE PRINT. Full-color print ($9^{1}/_{2} \times 12$ inches) of endangered water snakes (*Nerodia erythrogaster neglecta* and *N. sipedon insularum*) described by Roger Conant. Edition limited to 450 copies, each individually signed and numbered by Conant and the artist, David M. Dennis. $25.00.

GUIDELINES FOR THE USE OF LIVE AMPHIBIANS AND REPTILES IN FIELD RESEARCH, by George R. Pisani, Stephen D. Busack, Herbert C. Dessauer, and Victor H. Hutchison, representing a joint committee of ASIH, HL, and SSAR. 1987. Brochure covers animal care, regulations, collecting, restraint and handling, marking, housing and maintenance in field, and final disposition of specimens. 16 p. $4.00 ($3.00 each in quantities of five or more copies).

PRODUCTION SPECIFICATIONS

WORD PROCESSING AND FORMATTING OF TEXT: California Academy of Sciences, San Francisco, California, USA (Alan E. Leviton). Type was set in Adobe Systems Times Roman face using the Ventura Publisher program, version 4.2, on an ALR VEISA 486 computer and camera-ready copy printed on a LEXMARK Optra R+ laser printer.

PRINTING OF TEXT AND BINDING: Thomson-Shore, Inc., Dexter, Michigan, USA (Ned Thomson, Lana Paton, Merry Sumner). The text is printed on 60-pound Joy White Offset paper, a recycled stock. The book is covered in Roxite C cloth (vellum finish) with Multicolor Antique endpapers.

COLOR PLATES: Four Colour Imports, Ltd., Louisville, Kentucky, USA (Petrice Dahl, Katherine Peterson), agents for Everbest Printing Co., Ltd., Hong Kong. The color plates were laser-scanned from color transparencies by All Systems Color of Miamisburg, Ohio (Thomas Kiser, Mindy Stelzer) and are printed on 128-gsm Korean Chon Ju gloss art paper.

PAPER: All paper in this book is acid- and groundwood-free. It meets the guidelines for permanence and durability of the Committee on Publication Guidelines for Book Longevity of the Council on Library Resources.

DATE OF PUBLICATION: 31 December 1996.

PLACE OF PUBLICATION: Ithaca, New York, USA.

NUMBER OF COPIES: 1500.